Interfaces in Cementitious Composites

Recent RILEM Publications

High Performance Fiber Reinforced Cement Composites

Edited by H.W. Reinhardt and A.E. Naaman

Advanced composites and the fundamental understanding of their behaviour is a rapidly expanding area of civil engineering. These materials can be designed to have outstanding combinations of strength (five-to ten-times that of conventional concrete) and energy absorption capacity (up to 1000-times that of plain concrete). Exciting engineering applications are therefore being developed to take advantage of these properties for blast resistant, earthquake resistant and offshore structures, and also for building components such as cladding, pipes, tiles and roofing.

This book provides a compendium of the most recent research advances and reviews presented at an international workshop held under the auspices of RILEM and the American Concrete Institute in Mainz, Germany in June 1991. It includes over 40 contributions from the leading international specialists and researchers from USA, Europe, Japan and elsewhere and is an essential reference for engineers and researchers who need to be at the forefront of developments in cement composites.

Among the materials discussed are composites containing steel, polypropylene, polyacryonitrile, carbon, glass and cellulose fibres. SIFCON is extensively covered. The other main topics considered are: technologies of production; composite optimization; mechanical fracture and interface properties; modelling; structural applications and implications.

RILEM Proceedings 15, Published 1992, 584 pages, ISBN 0 419 39270 4

Fibre Reinforced Cement and Concrete

Edited by R.N. Swamy

Fibre reinforced cement-based materials are being increasingly used throughout the world, for applications which range from simple shelter structures to advanced construction engineering. This book provides a comprehensive updating on recent research and development into these materials from over 100 leading specialists from all parts of the world.

Because of concern about long-term performance of fibre composites, special emphasis is given to ageing and durability. Information on newer, high performance fibres and reinforcing elements as well as the results of studies of composites made using the more established steel and glass fibres, are presented in the book. Natural fibres and wood/cellulose fibres are readily available in developing countries, and these, together with ferrocement, form important themes.

Fabrication techniques, engineering properties, dynamic and structural behaviour, fracture behaviour, modelling, and applications of fibre reinforced cement and concrete and ferrocement are all covered.

The book forms the Proceedings of the Fourth RILEM International Symposium on the subject, previous ones being held in London in 1975, and Sheffield in 1978 and 1986. The book will be valuable for all those involved with research into the applications of new construction materials who need to keep up to date with the latest advances in this field.

RILEM Proceedings 17, Published 1992, 1376 pages, ISBN 0 419 18130 X

Interfaces in Cementitious Composites

Proceedings of the International Conference held by RILEM (The International Union of Testing and Research Laboratories for Materials and Construction) at the Université Paul Sabatier, Toulouse, and organised by RILEM Technical Committee 108 and the Laboratory for Materials and Durability of Constructions (LMDC), INSA-UPS, Toulouse, France

Toulouse
October 21-23, 1992

EDITED BY

J.C. Maso

LMDC, INSA-UPS, Toulouse, France

CRC Press
Taylor & Francis Group
Boca Raton London New York

CRC Press is an imprint of the
Taylor & Francis Group, an **informa** business
A TAYLOR & FRANCIS BOOK

CRC Press
Taylor & Francis Group
6000 Broken Sound Parkway NW, Suite 300
Boca Raton, FL 33487-2742

First issued in paperback 2019

ISBN-13: 978-0-419-18230-6 (hbk)
ISBN-13: 978-0-367-86343-2 (pbk)

A catalogue record for this book is available from the British Library.

Library of Congress Cataloging-in-Publication data available.

Publisher's Note
This book has been produced from camera ready copy provided by the individual contributors in order to make the book available for the Conference.

Publisher's Note
The publisher has gone to great lengths to ensure the quality of this reprint but points out that some imperfections in the original may be apparent

Visit the Taylor & Francis Web site at
http://www.taylorandfrancis.com

and the CRC Press Web site at
http://www.crcpress.com

Contents

Preface xiii

PART 1.1 INTERFACIAL MICROSTRUCTURE AND PROPERTIES: CHARACTERIZATION OF INTERFACIAL MICROSTRUCTURE 1
(Microstructure et propriétés de l'auréole de transition: Characterisation de la microstructure de l'auréole)

1 **Formation and development of interfacial zones between aggregates and Portland cement pastes in cement-based materials** 3
S. CHATTERJI and A. D. JENSEN
Danish Technological Institute, Taastrup, Denmark

2 **SEM investigations of the contact zones between rock surfaces and cement paste** 13
S. DIAMOND
School of Civil Engineering, Purdue University, West Lafayette, Indiana, USA
S. MINDESS and LIE QU
Department of Civil Engineering, University of BC, Vancouver, Canada
M. G. ALEXANDER
University of the Witwatersrand, Johannesburg, South Africa

3 **Fibre-cement paste transition zone: characterization and effect on mechanical properties** 23
H. SAVASTANO Jr
Air Force Academy, Pirassununga, Brazil
V. AGOPYAN
Escola Politécnica, University of São Paulo (EPUSP), Brazil
F. de A. S. DANTAS
Instituto de Pesquisas Tecnológicas (IPT), São Paulo, Brazil

4 **Effects of sodium chloride on the hydration products in the interfacial zone between cement paste and alkali-reactive aggregate** 33
M. SAITO
Kanazawa Institute of Technology, Nonoichi, Ishikawa, Japan
M. KAWAMURA
Department of Civil Engineering, Kanazawa University, Ishikawa, Japan

5 **Influence de l'origine des cendres volante employees comme ajouts sur l'auréole de transition des bétons hydrauliques qui en renferment** 43
(Influence of the origin of fly ashes used as partial replacement of cement in concretes on the transition zone)
B. HUSSON and A. CARLES-GIBERGUES
LMDC, INSA-UPS, Toulouse, France

6 **Effects of pozzolanic and non-reactive microfillers on the transition zone in high strength concretes** 53
A. GOLDMAN and A. BENTUR
Technion - IIT, Haifa, Israel

PART 1.2 INTERFACIAL MICROSTRUCTURE AND PROPERTIES: DEVELOPMENT AND NATURE OF INTERFACIAL MICROSTRUCTURE 63
(Microstructure et propriétés de l'auréole de transition: Formation de l'auréole)

7 **Influence of mineral admixtures on the transition zone in concrete** 65
U. NILSEN, P. SANDBERG and K. FOLLIARD
Department of Civil Engineering, University of California, Berkeley, USA

8 **Studies of the interfacial bond between cement pastes and expanded shale** 71
QI YUN and CHU QINGLI
Chongqing Institute of Architecture and Engineering, Chongqing, China

9 **Etude des interfaces ciment-fibres organiques** 79
(Interfaces between synthetic fibers and cement paste)
V. HOUGET, M. CHABANNET and J. PERA
Laboratoire des Matériaux Minéraux, INSA de Lyon, Villeurbanne, France

10 **Adhesion of bonded mortar to polymer-cement paste coated mortar substrates** 89
S. N. PAREEK, Y.OHAMA and K. DEMURA
College of Engineering, Nihon University, Koriyama, Japan

11 **The interface and adhesive strength between reactive resin and concrete** 99
S.-W. CHEN
Chongqing Institute of Architecture and Engineering, Chongqing, P.R. China; at present TU Vienna, Austria
W. PERBIX and K. P. GROSSKURTH
Institute for Building Materials, Constructions and Fire Resistance in Civil Engineering, Technical University of Braunschweig, Germany

12 **Computer modelling of the interfacial zone in concrete** 107
D. P. BENTZ, J. GARBOCZI and P. E. STUTZMAN
National Institute of Standards and Technology, Gaithersburg, Maryland, USA

PART 1.3 INTERFACIAL MICROSTRUCTURE AND PROPERTIES: MECHANICAL PROPERTIES - TECHNICAL METHODS 117
(Microstructure et propriétés de l'auréole de transition: Characterisation de la microstructure de l'auréole: Propriétés mecaniques - methodes d'étude)

13 **A study of properties of the paste-aggregate interface** 119
K. MITSUI
Takenaka Corporation, Tokyo, Japan
ZONGJIN LI, D. A. LANGE and S. P. SHAH
Northwestern University, Evanston, Illinois, USA

14 **The influence of rock and cement types on the fracture properties of the interfacial zone** 129
M. G. ALEXANDER
University of the Witwatersrand, Johannesburg, South Africa
S. MINDESS and LIE QU
University of British Columbia, Vancouver, Canada

15 **Fracture of mortar-aggregate interfaces in concrete composites** 139
O. BUYUKOZTURK and K. M. LEE
Department of Civil Engineering, Massachusetts Institute of Technology, Cambridge, Massachusetts, USA

16 **The interface zone around expanded shale grain in hardened cement paste** 149
U. SCHNEIDER and S. CHEN
Technische Universität Wien, Vienna, Austria

PART 1.4 INTERFACIAL MICROSTRUCTURE AND PROPERTIES: MECHANICAL MODELLING 157
(Microstructure et propriétés de l'auréole de transition: Modélisation)

17 **Numerical simulation of altering mechanical properties of concrete due to material behaviour of the interfaces** 159
H. SCHORN
University of Bochum, Germany

18 **Mechanical modelization of interfaces in cementitious fiber composites** 169
P. HAMELIN and R. M. COURTADE
Materials and Mechanical Engineering Laboratory, Université Claude Bernard - IUTI, Lyon, France

PART 2.1 RELATIONSHIPS BETWEEN MICROSTRUCTURE OF THE INTERFACIAL ZONE AND PROPERTIES OF THE COMPOSITE: MECHANICAL PROPERTIES 177
(Relations entre la microstructure de l'aureole et les propriétés du composite: Les propriétés mecaniques)

19 **Effect of aggregate type on the mechanical behaviour of concrete** 179
H. KAWAKAMI
Fukui University, Fukui, Japan

20 **Some mechanical effects of interface debonding in plain concrete** 187
P. STROEVEN
Delft University of Technology, Delft, The Netherlands

21 **Modèle expérimental d'étude mecanique et physico-chimique de l'interface acier-béton** 197
(Experimental model for mechanical, physical and chemical study of steel-concrete interface)
R. FRANCOISE, M. ARNAUD, G. ARLIGUIE and A. AYED
Laboratoire Matériaux et Durabilité des Constructions, INSA-UPS, Toulouse, France

22 **Mechanical properties and microstructure of polyester-cementitious composites** 207
F. SANDROLINI, M. ROSAS ROSAS, V. BONORA and A. SACCANI
Dipartimento di Chimica Applicata e Scienza dei Materiali, University of Bologna, Bologna, Italy

23 **Etude des interfaces dans des composites ciment-fibres de polyester** 217
(Study of the interfaces of polyester fibre reinforced cement composites)
A. JELIDI, M. CHABANNET, J. AMBROISE and J. PERA
Laboratoire des Matériaux Minéraux, INSA de Lyon, Villeurbanne, France

24 **Modélisation du comportement mecanique de composites ciment-fibres de verre en liaison avec la nature des interfaces** 227
(Mechanical model for the tensile behaviour of glass fiber reinforced modified cementitious composites)
J. PERA, J. DEJEAN and J. AMBROISE
Laboratoire des Matériaux Minéraux, INSA de Lyon, Villeurbanne, France

25 **Numerical study of the influence of interfacial properties on the mechanical behaviour of cement-based composites** 237
E. SCHLANGEN and J. G. M. van MIER
Stevin Laboratory, Delft University of Technology, Delft, The Netherlands

26 **Une méthode intégrale de frontière tenant compte du comportement mécanique de la liaison hétérogénéité-matrice** 247
(A boundary integral method taking into account the mechanical behaviour of the matrix-inhomogenity interface)
L. ULMET and S. CAPERAA
Laboratoire Génie Civil, University Limoges, Egletons, France

PART 2.2 RELATIONSHIPS BETWEEN MICROSTRUCTURE OF THE INTERFACIAL ZONE AND PROPERTIES OF THE COMPOSITE: TRANSPORT PROPERTIES 257
(Relations entre la microstructure de l'aureole et les propriétés du composite: les propriétés de transfert)

27 **Interfacial zone percolation in cement-aggregate composites** 259
K. A. SNYDER, D. P. BENTZ and E. J. GARBOCZI
National Institute of Standards and Technology, Gaithersburg, Maryland, USA
D. N. WINSLOW
Purdue University, West Lafayette, Indiana, USA

28 **Diffusivité des ions chlore dans la zone de transition entre pâte ciment et roche granitique** 269
(Diffusivity of chloride ions in the transition zone between cement paste and granite)
D. BRETTON
CRIB, Sherbrooke University, Canada
J.-P. OLLIVIER
LMDC, INSA-UPS, Toulouse, France
G. BALLIVY
CRIB, Sherbrooke University, Canada

29 **Influence of aggregate concentration on the diffusion of CO_2 and O_2** 279
Y. F. HOUST
Federal Institute of Technology (EPFL), Lausanne, Switzerland
H. SADOUKI and F. H. WITTMANN
Federal Institute of Technology (ETHZ), Zurich, Switzerland

30 Interface influence upon thermal conductivity of lightweight concretes 289
V. S. GRYSLOV
R.I. Laboratory for Testing Building Materials, Cherepovets, Russia

PART 2.3 RELATIONSHIPS BETWEEN MICROSTRUCTURE OF THE INTERFACIAL ZONE AND PROPERTIES OF THE COMPOSITE: ENVIRONMENTAL CONDITIONS 295
(Relations entre la microstructure de l'auréole et les propriétés du composite: Influence de l'environnement)

31 Influence of the interfacial zone between cement paste and aggregate on the corrosive attack on concrete 297
K. P. GROSSKURTH, W. MALORNY and W. RUTHMANN
Institut für Baustoffe, Massivbau und Brandschutz, Technical University of Braunschweig, Germany

Abstracts of French papers in English 307

Author index 309
Subject index 310

Scientific Council

J. C. Maso (President), Laboratoire Matériaux et Durabilité des Constructions, INSA-UPS, Toulouse, France
J. P. Ollivier (Secretary), Laboratoire Matériaux et Durabilité des Constructions, INSA-UPS, Toulouse, France
M. G. Alexander, University of Witwatersrand, South Africa
A. Bentur, Building Research Station, Israel Institute of Technology, Haifa, Israel
F. Massazza, Italcementi, Laboratorio Chimico Centrale, Bergamo, Italy
S. Mindness, University of British Columbia, Vancouver, B.C., Canada
P. J. M. Monteiro, University of California, Berkeley, USA
I. Odler, Institute of Nonmetallic Materials, Technical University Clausthal, Germany
P. L. Pratt, Imperial College of Science, Technology and Medicine, London, UK
S. P. Shah, Northwestern University, Evanston, Illinois, USA

Organizing Committee

M. Arnaud, LMDC, France
A. Fenouil, LMDC, France
M. Fickelson, RILEM
J. C. Maso, LMDC, France
J. P. Ollivier, LMDC, France
H. Paillares, ADERMIP, France
M. Vie, LMDC, France

Conference Secretariat

LMDC, INSA-UPS, Complexe Scientifique de Rangeuil,
31077 Toulouse Cedex, France

Preface

The first significant studies on interfaces between cement paste and associated materials were made during the 1950s. Against a difficult economic background, with the second world war and its train of destruction still recent, Electricity of France's engineers in charge of building dams had to solve the problem of making concrete of adequate performance but at the lowest cost, with very varied rocks sometimes of very poor quality found in the vicinity of the construction sites.

These studies resulted in the discovery of the transition zone between non-porous aggregates and cement paste. The transition zone results, essentially, on the one hand from a wall effect which produces, initially, a porosity gradient and therefore a gradient of W/C ratio and, on the other hand, from differences in the mobility of ions. This zone is richer than the rest of the cement paste in hydrated compounds formed with the most mobile ions: ettringite and portlandite with Portland cements for example. The crystals are generally larger and better formed, and at all ages the porosity remains greater in the transition zone than in the bulk.

As a result, the transition zone is the weakest part of the composite under mechanical action and the first microcracks start there before propagation from aggregate to aggregate. It is also a preferential route for penetration of aggressive compounds. So, the transition zone is of great importance for the durability of concrete and for its association with steel in reinforced concrete.

For porous aggregates, suction by dry aggregates or diffusion through saturated aggregates modify the process of development of the transition zone that no longer has these unfavourable characteristics.

The addition of submicron additives and of superplaticizers modifies the wall effect drastically reducing the porosity gradients. The transition zone no longer has the defects it has in traditional concretes. We can observe this in high strength concrete.

If these general mechanisms at the interface between cement pastes and associated materials are globally now well known and well accepted, there are several items missing in the train of knowledge, with the microstructures at the interface, their formation and their influence on the properties of the composite.

Since 1969, researchers have had the opportunity to present their results and to exchange their points of view at several occasions in international meetings partially or totally concerned with the subject. The RILEM International Conference in October 1992 in Toulouse is in keeping with this train of meetings. It has a double objective: to review on one hand the state of the art on microstructures at the interfaces, their characteristics, their influence on properties of concrete and reinforced concrete and, on the other hand the latest scientific results on these subjects.

Professor J C Maso
Toulouse

Préface

C'est dans le courant des années 50 que les premières études significatives concernant les liaisons entre la pâte de ciment et les matériaux associés ont été entreprises. Dans un contexte économiquement difficile - la 2ème guerre mondiale et son cortège de destructions étaient encore tout récents - les ingénieurs d'Electricité de France chargés de la construction des barrages se trouvaient confrontés au problème de la réalisation de bétons de performances suffisantes mais les moins coûteux possibles, avec des roches très differentes et de qualité parfois très médiocre au voisinage immédiat des sites d'implantation. Il leur fallait dès lors maîtriser l'influence de la nature minéralogique des granulats sur les propriétés des bétons. Ils associèrent à leur démarche les minéralogistes et géotechniciens de l'Université de Toulouse.

Les études entreprises débouchèrent sur la mise en évidence de l'existence d'une auréole de transition autour des granulats non-poreux. Les recherches initiées à Toulouse puis développées dans le monde entier permirent de montrer que celle-ci résulte essentiellement d'une part d'un effet de paroi qui induit des gradients de E/C et, d'autre part, des différences de mobilité des ions issus des grains de ciment anhydre. Elle est plus riche que la masse de la pâte en constituants hydratés formés à partir des ions les plus mobiles: ettringite et portlandite avec les ciments Portland par exemple. Les cristaux y sont généralement plus gros et mieux formés. Enfin, et à tous les âges, sa porosité, à quelques microns de l'interface est toujours plus élevée que dans le reste de la matrice du béton.

Ces caractéristiques font de l'auréole de transition la partie la moins résistante à des actions mécaniques et les premières fissurations y prennent naissance, avant de se propager par pontage de granulat en granulat. Elles en font vraisemblablement aussi une voie de pénétration préférentielle des agents chimiquement agressifs. L'auréole de transition joue donc un rôle doublement prépondérant dans la durabilité du matériau.

Dans le cas de granulats poreux à porosité ouverte, les phénomènes de succion et de diffusion empêchent la formation d'une auréole présentant les caractères défavorables décrits précédemment. La zone interfaciale n'est alors plus la partie la plus fragile du composite.

L'addition d'éléments solides submicroniques et de fluidifiants réducteurs d'eau vient modifier l'effet de paroi en réduisant de façon drastique les gradients de porosité. L'auréole de transition ne présente alors plus les caractères défaborables décrits plus haut. C'est ce que l'on observe avec les bétons de hautes performances.

Si ces mécanismes généraux aux interfaces entre les pâtes de ciment et les matériaux qui lui sont associés sont globalement connus désormais et bien admis, des lacunes subsistent dans la chaîne des connaissances tant en ce qui concerne les microstructures aux interfaces et leur formation que leur influence sur les propriétés du composite.

Depuis 1969 les chercheurs ont présenté leurs résultats et échangés leurs points de vue à plusieurs reprises dans des réunions partiellement ou totalement consacrées à ce thème. La Conférence Internationale d'octobre 1992 à Toulouse, organisée à l'initiative de la RILEM, s'inscrit dans cette suite de réunions d'échange. Elle a pour double objectif de faire l'état de l'ensemble de nos connaissances sur les microstructures aux interfaces, leurs caractéristiques, leur influence sur les propriétés des bétons, armés ou non, et de présenter les derniers développements scientifiques sur ce sujet.

Professeur J C Maso
Toulouse

PART 1.1

INTERFACIAL MICROSTRUCTURE AND PROPERTIES: CHARACTERIZATION OF INTERFACIAL MICROSTRUCTURE

(Microstructure et propriétés de l'auréole de transition: Characterisation de la microstructure de l'auréole)

1 FORMATION AND DEVELOPMENT OF INTERFACIAL ZONES BETWEEN AGGREGATES AND PORTLAND CEMENT PASTES IN CEMENT-BASED MATERIALS

S. CHATTERJI and A. D. JENSEN
Danish Technological Institute, Taastrup, Denmark

Abstract
In connection to our study of durability of cement based materials, the formation and development of the interfacial zones between different types of aggregates and cement paste has been investigated. This investigation has been carried out from the green, freshly mixed stage through to the hardened stage of the materials. A freshly mixed cement based material is a particulate suspension which is very far from its equilibrium state and possesses very non-linear and irreversible characteristics. One of these properties is that a particulate suspension has a tendency to differentiate into two phases: one richer in the liquid phase and the other richer in the solid phase than the original mix composition. This separation is facilitated by low frequency, low amplitude movements of the paste and especially those between the aggregate particles and the paste. The liquid rich phase collects round the aggregate particles. If the aggregate particles are non-absorbent then it forms rather weak, porous zones round the aggregates. If the aggregate particles are absorbent then these interfacial layers will be less porous. In most cases these interfacial zones will be weaker than either the paste or the aggregate particles. This initially formed interfacial zones are further widened by various processes like the autogenous shrinkage of the cement paste, the differential thermal shrinkage between the aggregate particles and the cement paste etc. During subsequent period of cement hydration and hardening the above formed interfacial zones get partly filled up by large and well-formed crystals like $Ca(OH)_2$, ettringite, etc. However these zones still remain more porous than the cement paste. These interfacial zones form easy passages for the penetration of degrading agents e.g. water, CO_2, etc. An improved technique for the visualization of these interfacial zones will be described. Evidences will be provided for the flow of the degrading agents through these interfacial zones.
Key words: Interfacial zone, Cement paste, Aggregate, Durability, Visualization, Technique.

Interfaces in Cementitious Composites. Edited by J. C. Maso. © RILEM.
Published by E & FN Spon, 2 - 6 Boundary Row, London SE1 8HN. ISBN 0 419 18230 6.

1 Introduction

All the durability problems of cement-based materials may be considered to be the consequence of materials and energy exchange between the hardened masses and their environments. In the case of a perfectly formed structure, without any crack and differential porosity at different places, the attacking agent or process must progress evenly from the surface inwards. Unless the agent or the process is highly energetic e.g. strong acid etc. its progress will be slow and uniform. If however there are flaws in the structure then the agent or the process may penetrate deep inside the structure selectively through those flaws making extensive damage to the structure. We have studied the flaw making processes for about 20 years and time to time published our results on various aspects (Chatterji and Jeffery, 1971, Wheeler and Chatterji, 1972, Christensen et al 1981). In this report we summarise our present position on interfacial zones only.

2 Instability of green mix and the formation of initial interfacial zones

2.1 Instability of green mixes

Densities of concrete making materials vary over a wide limits. As a result of the differences in densities and the chemical reactivities of the anhydrous cement phases a green mix is far from its equilibrium state and develop different types of instabilities. A green mix will always show its chemical instability by reacting with water forming a set mass. The manifestation of mechanical instability will depend on the rate of chemical reaction and setting. A common type of mechanical instability of particulate suspension like concrete is its tendency to separate into two phases: one richer in the fluid phase and the other richer in the solid phase than the original mix composition.

2.2 Bleeding of green mixes

A green mix, with a water/cement ratio of about 0.4, when left standing gradually accumulates a layer of water at its free surfaces i.e.bleeds. If the bleed water collects slowly and uniformly over the whole surface then it is called the normal bleeding. If water surfaces at few points with sufficient velocity to carry cement and fine sand particles then it is called channel bleeding. Channel bleeding may not contribute to the formation of aggregate-paste interfaces.

First systematic study of bleeding was carried out by Powers (1939). Steinour (1945) followed up this work and derived the fundamental equation governing the rate of bleeding. A solid particle submerged in a freshly made green mix is subjected to a downward driving force due to its submerged weight and a viscous drag opposing the downward movement. When these two are in balance then the particle settles at a constant speed. Steinour showed that the correct submerged weight to be used is ρ_c-ρ_m, where ρ_m is the density of the matrix. For a fully dispersed paste Steinour derived the following equation:

$$U = (K_1/\sigma^2).(g.(\rho_c.\rho_f)/\eta).e^3/(1\text{-}e) \quad (1)$$
$$= (K/\sigma^2).e^3/(1e) \quad (1')$$

where U is the bleeding rate, σ is the specific surface of the solid particles, ρ_c is the density of the particle, ρ_f is the density of the liquid, g is the gravitational constant,η is the viscosity of the fluid, e is the volume of water per unit volume of the mix, and K_1 and K are constants.

However for cement pastes Steinour had to modify the above equation into:

$$U = (K_2/\sigma^2).((1/(1\text{-}e_i)).(g.(\rho_c\text{-}\rho_f))/\eta).(e\text{-}e_i)^3/(1\text{-}e) \quad (2)$$

where K_2 is another constant and the other symbols have their usual meanings. The empirical constant e_i has to be introduced to conform the calculated results to the observed values as well as to make the bleeding rate vs time curve pass through the origin. The origin of e_i remained a matter of speculation.

Wheeler and Chatterji (1972) and Chatterji (1988) showed that the equations 1 and 2 need modification. Chatterji modified the equation 1 to the following:

$$U = (K_3/\sigma^2).((e^3/(1\text{-}e)).\{1+ (dp/dz).(K_4/((1\text{-}e).e)\} \quad (3)$$

where K_3 and K_4 are two new constants. It can be seen that first part of the equation 3 has the same form as that of the equation 1'. The part containing the term dp/dz reduces the rate of bleeding.

In a concrete sample the submerged weight of an aggregate particle, ρ_a-ρ_m where ρ_a is the density of the aggregate particle, is lower than that of the cement particles and as a result it will settle at a slower rate than the surrounding cement particles. The void between the aggregate particle and the surrounding paste will be filled up by the bleed water. Notice this collection of bleed water under the aggregate particles does not depend on the aggregate to aggregate bridge formation. If however aggregate to aggregate bridge formation occurs then collection of bleed water will be augmented. This is the first process in the formation of weak interfacial zones.

2.3 Formation of water-rich layer due to low amplitude, low frequency vibration

Even those low water/cement ratio cement pastes which do not show any bleeding may differentiate in to two phases when subjected to low frequency, low amplitude vibrations. This is best seen by slow tapping at the sides of a low water/cement paste. Before tapping the top surface will have a dry appearance and after the tapping the top surface will have a shining appearance indicating the formation of a water-rich layer. A similar formation of water-rich layer occurs when any non-absorbent surface is slowly moved on the top surface of a cement paste. This phenomenon seems to have not been studied systematically though it is very common phenomenon in particulate pastes.

2.4 Effects of compaction and standing

During compaction of a freshly made concrete will be subjected to some form of mechanical vibration which will always produce a relative movement between aggregates and the paste. This relative movement will encourage the formation of a water-rich layer round individual aggregate particles; the larger the aggregate particle more water-rich area will form. This will, at least partly, explain some of the recently reported results on the interfacial zones using large, flat surfaced artificial, aggregates. Incidentally the formation of this water-rich layer at the start of a viscosity measurement will also vitiate the subsequent results.

After compaction the green mix is generally left to harden. In the mean time bleeding may cause water to collect underneath aggregate particles. As a result of the above phenomena water-rich layers will always form round aggregate particles in field concretes. If aggregate is non-absorbent then these layers are thicker and more porous than in the case of absorbent aggregate.

In any case the cement-rich paste will be separated from aggregate particles by these weaker and porous interfacial layers i.e on the both sides of these weak layers are strong materials. When these strong materials change volume they do so on themselves thereby changing the thickness of these interfacial layers.

3 Altering or widening of the interfacial zones

Any process which causes a differential volume change between the aggregates and the surrounding cement paste will cause an alteration in the interfacial zones between them. Many of the causes which alter or widen the already formed interfacial zones will also cause alteration or micro or macro-cracking of cement paste surrounding the aggregate particles.

3.1 The effect of pore solution chemistry

In an actual concrete structure, its temperature remains high for 72 hours or longer. In the mean time the pore solution, inside the structure, is changing its nature. Ultimately the pore solution becomes a solution of alkali hydroxides only. The ultimate alkali hydroxide concentration will depend of the soluble alkali salts present in the concrete and the amount of free water left after cement hydration and loss due to evaporation. The alkali hydroxide concentration can quite easily go above 1 Normal.

An highly alkaline pore and bleed solution may react with aggregates present in the concrete. If the pore solution is stronger than about 0.7 Normal alkali hydroxide solution and the temperature of the hardening concrete is above about 40°C then it may attack the surfaces of calcium carbonate aggregate forming calcium hydroxide according to the following equation (Partington, 1961).

$$CaCO_3 + 2\,NaOH = Ca(OH)_2 + Na_2CO_3 \qquad (4)$$

This newly formed calcium hydroxide will of course be precipitated near the aggregate giving an illusion that calcium hydroxide has been formed topo-tactically or has been preferentially nucleated on the carbonate aggregate.

Another affect of highly alkaline pore solution will be its attack on silicious aggregate surfaces (Chatterji and Jeffery, 1971).

3.2 The effect of the autogenous shrinkage

During hydration a cement-water system undergoes an absolute volume decrease. Most of the autogenous shrinkage occurs after concrete has hardened. As only the cement paste is shrinking and the aggregate-paste interfacial zones are weak, the paste, especially near bigger particles, tends to get detached from the aggregates. This process also tend to produce micro-cracks.

3.3 Mismatch of the thermal expansion coefficients between aggregate particles and cement paste

During the setting and the initial stages of hardening the concrete temperature goes up and all the ingredients of concrete expand according to their thermal expansion coefficients. In this heating up period all the constituents keep some form of mechanical contiguity. However during cooling period the mechanical contiguity is lost and the individual components start to shrink according to their own thermal expansion coefficients. If a concrete is assumed to be made with 16 mm maximum size granite aggregates of thermal expansion coefficient of $6x10^{-6}/^{o}C$ and that the Portland cement paste has a thermal expansion coefficient of $15x10^{-6}/^{o}C$ (a mid value between the reported extremes). If concrete is cooling down from 60°C to 20°C then it can be calculated that in cooled concrete a strain of 0.005 mm will occur across 16 mm aggregates. A higher strain will develop if larger sized aggregates were used. Up to a critical level this strain could be accommodated by altering the width of the interfacial zone. If however the strain is larger, it could not be accommodated and micro-cracks will form. This widening process will not be operative with small sized aggregates and in the laboratory made concrete samples because of their small temperature rise during setting and hardening. The importance of size is best emphasised in the case of reinforcing iron rods. The difference of the thermal expansion coefficient between the rebars and set cement paste is not very high; however the size of rebars is effectively infinitely big. One would therefore expect a substantial widening at their interfaces. This is evident from the following: "In comparison with the hydraulic resistance (against water permeation) offered by the concrete itself, the resistance along the bars is negligible." (US Bureau of Reclamation, 1975).

3.4 First drying of concrete

When the form-work is removed from a hardened concrete structure it gradually looses a part of its free water which in its turn causes a volume diminution of the structure. Most of this volume diminution occurs in the cement paste. This differential strain, during the first drying of a concrete structure, will cause widening of the interfacial zone as well as micro-cracking.

Above points are best seen if we consider the Pickett's equation relating the drying shrinkage of a cement paste to that of a mortar of concrete (Pickett, 1956):

$$S_c = S_p (1-V_a)^{\underline{n}} \qquad (5)$$

where S_c is the shrinkage of concrete, S_p is the shrinkage of the cement paste, V_a is the fractional volume occupancy of the aggregate and $\underline{n}$ is constant value of which varies between 1.2 to 1.7 (L'Hermite, 1960). The equation 5 indicates that a volume of the paste, in excess of that replaced by the aggregate, is not contributing to the observed shrinkage. The cement paste content of most usual concrete is about 30%. From the equation 5 it can be calculated that in a normal concrete a fractional paste content between 0.236 and 0.129, depending up on the value of $\underline{n}$, effectively contributes to the drying shrinkage of the concrete. In other words between 21 and 57% of the paste content i.e. on the average about 40%, of a concrete has zero effective shrinkage. Perhaps more realistically the whole of the paste content has been altered such that it has an effective drying shrinkage of about 60% of the unaltered cement paste. The above change could also be interpreted in the following manner. Due to the presence of weak interfacial zones between the bulk cement paste and the aggregate particles the volume of the paste has been effectively divided into a large number cells. The weak interfacial zones will hinder a complete transfer of shrinkage of paste in one cell to the next thereby reducing the observable shrinkage. This interpretation is consistent with the recent mercury porosimetric results of Winslow and Liu (1990) that the presence aggregate particles makes a cement paste more porous. (See also Chatterji, 1991).

4 Healing and filling up of interfacial zones during extended curing of a concrete structure

4.1 The autogenous healing

During curing cement paste hydrates and expands to a limited extend thereby decreasing the width of the interfacial zones. Furthermore a part of the cement hydration products are deposited in the interfacial zones. This filling up of the interfacial zones led to some strengthening or healing of the interfacial zones.

5 Techniques of visualization of the interfacial zones

5.1 General

The most direct and least controversial method for the identification and location of interfacial zones is the petrographic examination of the concrete sample in question. In addition to the petrographic method, a method based on the electrical resistance measurements has also been proposed recently (Ping, Beaudoin, Brousseau, 1991).

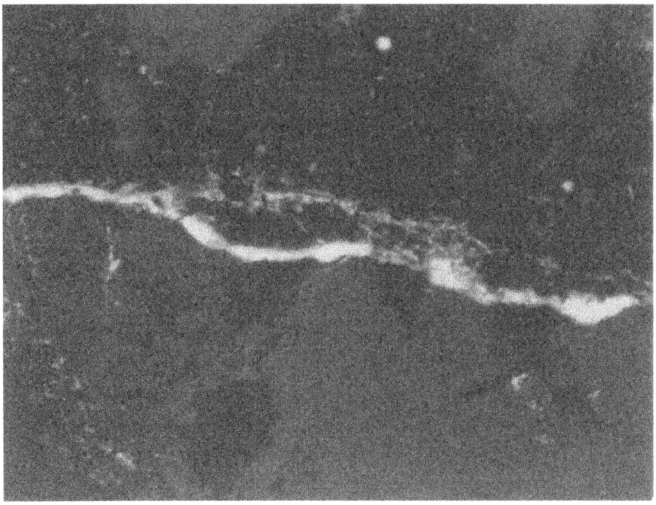

Fig. 1. A photo-micrograph of a concrete section taken in the transmission fluorescent mode. The bright wavy line is an empty zone between a coarse aggregate and the cement paste. Magnification 60X.

5.2 Petrographic technique and its improvement

In the petrographic technique for the examination of concrete, the sample is first embedded in a resin. This impregnation preserves the spatial relationship, including the porosities, within the sample unaltered through the process of making a thin section. This spatial relationship among the different constituents is then observed microscopically with or without the use of polarised light and other accessories. However the detection of porosities could be improved by using a resin containing a fluorescent dye and viewing the thin section in the fluorescent mode. In that case the degree of porosity could be detected from the intensity distribution of the fluorescent light (Fig. 1) and even be quantified using proper standards.

If however one is only interested in the study of the formation and development of the interfacial zones or fine micro-cracks then one could use a very low viscosity resin which does not penetrate into the matrix but only the voids and cracks. For this purpose we have used a Japanese cyano-acrylic acid resin, Aronalpha (Toagosei Chemical Industry, Japan.) and dissolved a fluorescent dye in it. This resin has a very low viscosity and has the property of hardening very quickly (within about 30 sec.) when it comes in contact of water. We first soaked the concrete sample in alcohol for about 24 hours to remove excess water and to eliminate further crack formation, then it was vacuum impregnated with this resin. Under this condition the resin penetrates even very small voids and microcracks, but does not penetrate the matrix or in the cement paste. After grinding off the excess resin and a few μm of concrete, the ground surface could be examined with incident ultra-violet light. The interfacial zones and fine micro-

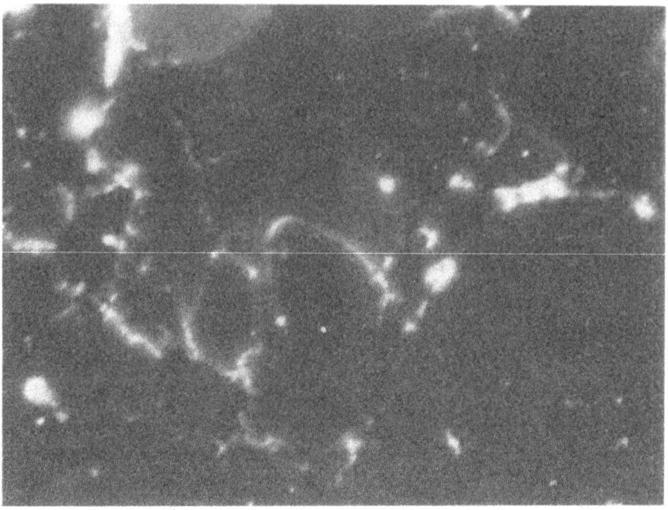

Fig. 2. A photo-micrograph of a concrete sample taken in the incident light fluorescent mode. The bright lines are due to penetration of the resign in the interfacial zones between aggregates and the cement paste. Magnification 50X.

cracks show up as bright areas or lines against a dark background. Fig. 2 shows a micro-photograph of such a sample. If an aggregate is transparent like a clear quartz sand then one can see the empty interfacial zone underneath the sand grain; though this is difficult to show in two-dimensional micrograph like Fig. 2.

In a well cured concrete or a concrete which has been subsequently exposed to an outside source of water e.g. rain or ground water etc. the interfacial zones will tend to be filled up with well formed crystalline materials like calcium hydroxide crystals. These crystals can easily be localised if a concrete thin section is examined under polarizing microscope. Fig. 3 shows such formation of calcium hydroxide crystals adjacent to sand grains; a similar formation can also be seen around large aggregate particles.

6 Easy flow of degrading agents through these porous interfacial zones

Compared to the high resistance offered by the bulk of the cement paste these porous interfacial zones offer considerably lower resistance to any flow of aggregative materials. Attention has already been drawn to the observation of high permeability of water through reinforcing iron bar concrete interfaces (US Bureau of Reclamation, 1975). Another example would be the observation of Uchikawa et al (1989) that in the diffusion of alkali hydroxide through mortar and concrete most of alkali ions diffuse through the porous interfacial zones between aggregates and cement paste.

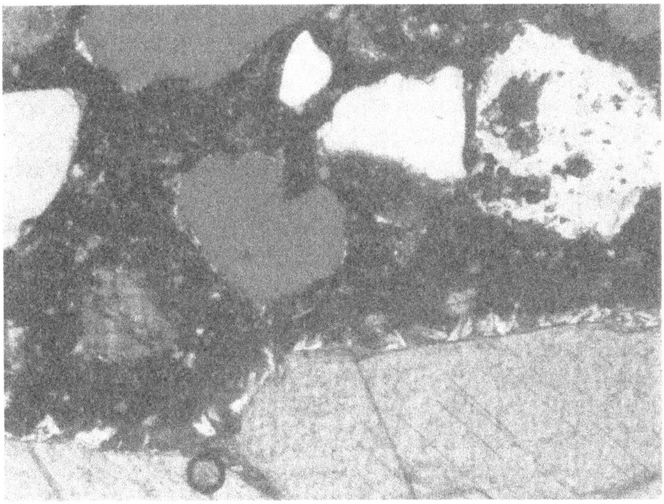

Fig. 3. A photo-micrograph showing the formation of large crystals of $Ca(OH)_2$ as bright patches in an interfacial zone between a coarse aggregate and the cement paste. The formation of $Ca(OH)_2$ can also be seen round sand grains as white patches. Viewed under cross-polaroid. Magnification 60X.

7 Acknowledgements

The authors are grateful to Dr. S. Tanikawa of Toagosei Chemical Industry Co., Ltd., Nagoya, Japan for the donation of the samples of Aron-alpha.

8 References

Chatterji, S. and Jeffery, J.W. The nature of bond between different types of aggregates and Portland cement. Indian Concrete J. Vol. 45, 346-349, 1971.
Chatterji, S. On the properties of freshly made Portland cement paste. Part 2. Cement and Concrete Research. Vol. 18, 615-620, 1988.
Chatterji, S. A discussion of the paper "The pore structure of paste in concrete" by Winslow and Liu. Cement and Concrete Research. Vol. 21, 196-197, 1991.
Christensen, P., Chatterji, S., Thaulow, N., Jensen, A.D. and Gudmundsson, H. Microscopic studies of aggregates in Danish concrete. Proc. 3rd International Conference Cement Microscopy. Houston, 1981. pp. 374-385.
L'Hermite, R. Volume changes of concrete. Proc 4th Intern. Symp. Chemistry of Cements. Washington D.C., 1960. pp. 659-94.
Partington, J.R. "General and Inorganic Chemistry." Macmillan and Co. Ltd. London. 1961. p. 302.

Pickett, G. Effect of aggregate on shrinkage of concrete and hypothesis concerning shrinkage. J. Amer. Concrete Institute. Vol. 52, 581-90, 1956.
Ping, X., Beaudoin, J.J., Brousseau, R. Flat aggregate-Portland cement paste interfaces, I. Electrical conductivity Models. Cement and Concrete Research. Vol. 21, 515-522, 1991.
Powers, T.C. The bleeding of Portland cement paste, mortar, concrete. Bulletin 2. Portland Cement Association. Chicago. 1939.
Steinour, H.H. Further studies of the bleeding of Portland cement paste. Bulletin 4. Portland Cement Association. Chicago. 1945.
Uchikawa, H., Uchida, S., Hanehara, S. Relationship between Structure and Penetrability of Na Ions in Hardened Blended Cement Paste, Mortar and Concrete. Proc. 8th Intern. Conf. Alkali-Aggregate Reaction. Kyoto.1989. 121
US Bureau of Reclamation. Concrete Manual. 1975.
Wheeler, J., Chatterji, S. Settling of particles in fresh cement pastes. J. American Ceramics Soc. Vol. 55, 461-464, 1972.
Winslow and Liu. The pore structure of paste in concrete. Cement and Concrete Research. Vol. 20, 227-235. 1990.

2 SEM INVESTIGATIONS OF THE CONTACT ZONES BETWEEN ROCK SURFACES AND CEMENT PASTE

S. DIAMOND
School of Civil Engineering, Purdue University, West Lafayette, Indiana, USA
S. MINDESS and LIE QU
Department of Civil Engineering, University of British Columbia, Vancouver, B.C., Canada
M. G. Alexander
University of the Witwatersrand, Johannesburg, South Africa

Abstract
We have examined contact zones and traces of the fracture surfaces produced in specimens in which cement paste has been cast against previously fractured rock specimens and fractured again after appropriate curing. Fracturing of both the original rock and the composite specimens was by ISRM Method 1, using chevron-notched cylinders. Specimens were prepared with both dolomite and andesite rock, and cement paste with and without silica fume. Slices were cut normal to the original interface, and polished and coated for SEM investigation, using both secondary and backscatter electron modes. Failure cracks in some cases ran at the original interface; in others, presumably as the result of better bonding, the cracks left the interface at the notch, travelled 1 to 2 cm into the paste, and then turned and ran parallel to the interface. Silica fume bearing paste and andesite rock tended to favour the second mode. The fracture cracks, whether within the paste or at the interface, were found to show significant branching. Examination of the paste region near the original interface did not provide any particular indication of a transitional zone different in character from bulk paste away from the rock interface.
<u>Keywords</u>: cement/rock interface; SEM; fracture surfaces; transition zone; dolomite; andesite.

1 Introduction

In a recent study, Diamond and Mindess (1992) presented stereoscopic scanning electron micrographs of failure surfaces of rock and of cement paste specimens. Subsequently, Mindess and Diamond (1992) published similar stereoscopic scanning electron micrographs for fracture surfaces of composite specimens of cement paste cast against previously fractured rock specimens. In each case the fracture specimens studied were chevron-notched cylinders prepared and tested in flexure according to Method 1 of the International Society of Rock Mechanics (ISRM). Loading was in a very stiff machine, initially to the point of

Interfaces in Cementitious Composites. Edited by J. C. Maso. © RILEM.
Published by E & FN Spon, 2 - 6 Boundary Row, London SE1 8HN. ISBN 0 419 18230 6.

first deviation from linearity. The specimen was then unloaded and reloaded in the same manner repeatedly until failure occurred.

In these studies, rock specimens were prepared from dolomite and from andesite rock used commercially for concrete aggregate. Cement paste specimens were prepared with and without silica fume, using superplasticizer and a water:cement ratio of 0.30. After demoulding they were aged under water for approximately 3 weeks before testing. Pastes of both types were also used for composite specimens cast against previously fractured rock specimens. A more complete description of the specimen preparation and the test method has been provided by Diamond and Mindess (1992).

Figure 1 shows the overall appearance of the fracture surface of an andesite rock test specimen after this kind of fracture. The crack in this case starts at the intersection of the pre-cut chevron notches at the left of the figure and travels across the specimen from left to right.

Fig. 1 Appearance of the fracture surface of an andesite rock specimen illustrating the overall geometry.

In the present paper we report the results of further observations of the nature of the fracture surfaces and of the contact surfaces between rock and cement paste in rock - cement paste composite specimens of the type described by Mindess and Diamond (1992). These observations were made by scanning electron microscope (SEM) studies of polished slices cut normal to the fracture surface, i.e. parallel to the axis of the composite specimens. Slicing was done along a line midway between the traces of the chevron notches shown in Figure

l. The slices were mounted on glass slides, and polished using the normal polishing technique for geological specimens at the University of British Columbia. They were then carbon coated and examined in a SEM equipped with both backscatter and secondary electron detectors.

2 Observations

The specimens examined were all of composite character and involved four types: plain cement paste against andesite rock; silica fume - bearing cement paste against andesite rock, plain cement paste against dolomite rock, and silica fume - bearing cement paste against dolomite rock. Generally speaking, fracture occurred either in the form of a parting of the rock and the paste at the general level of the interface, or else the failure crack left the notch at an appreciable angle, penetrated into the paste for a considerable distance, and then turned and ran more or less parallel to the original interface. The type of fracture surface produced depended on both the rock type and whether the cement paste contained silica fume or not.

Failure with andesite rock tended to be within the paste, indicative of stronger bonding to this rock. However, with andesite - plain cement paste specimens there was considerable variation, some specimens parting at the interface. Figure 2(a) shows a mounted slice in which this mode of failure occurred. In contrast, the appearance of a slice from a replicate specimen in which failure was entirely through the paste is shown in Figure 2(b). All of the andesite - silica fume bearing cement specimens tested also failed entirely through the paste, similar to the specimen shown in Figure 2(b).

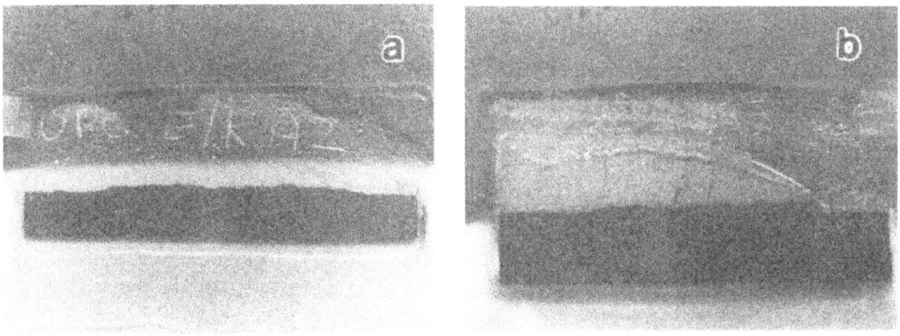

Fig. 2. Macroscopic appearances of specimen slices of (a) plain cement paste against andesite, with fracture at the interface and only the rock showing; and (b) plain cement paste against andesite, with failure through the paste.

The dolomite - plain cement paste specimens all showed parting at the interface, as did some dolomite - silica fume bearing specimens. The other dolomite - silica fume bearing paste specimens failed within the paste.

Figure 3 shows a secondary mode SEM photo taken at low magnification of the notch area of an andesite - plain paste specimen that had failed within the paste. The lower part of the figure, located by "A", is the rock. The approximately horizontal line above it indicated by the arrows is the boundary between the rock and the cement paste, which was the original rock fracture surface against which the paste was cast. Above this is the cement paste, designated "B", that remained below the fracture path and in contact with the rock. The upper boundary of the cement paste is the trace of the failure crack produced on loading the composite specimen.

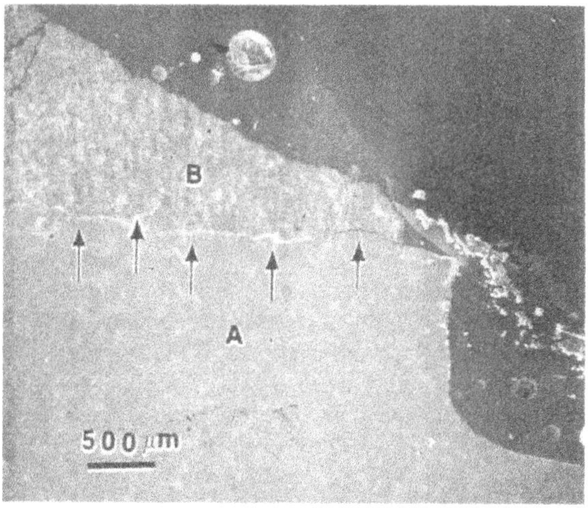

Fig. 3. Low magnification secondary electron micrograph of root of the fracture crack in andesite-plain cement paste specimen. The interface (indicated by the arrows) separates andesite rock ("A") and cement paste ("B"). The upper boundary of the paste is the trace of the failure crack.

The trace of the original notch sawn into the rock is visible at the extreme right of the figure. This represents the origin of the fracture path, i.e. the point where the chevron notches intersect. There is some debris resting on the original point of fracture. The branched crack running downward through the cement paste from the fracture crack surface was produced by the fracture process, and is not an artifact of specimen preparation.

A closer look at a portion of the cement paste - rock boundary of Figure 3 is shown at much higher magnification as a backscatter - SEM micrograph of Figure 4. Here the internal structure of the rock can be seen clearly. The

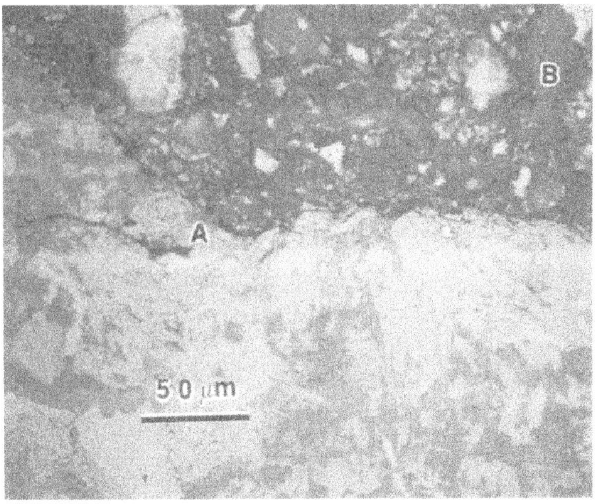

Fig. 4. Higher magnification backscatter SEM of the cement paste-rock boundary region of Figure 3. Bond crack branches at "A"; shrinkage cracks are prominent around "B" and elsewhere in paste.

darker component is a feldspar, presumably a member of the series between albite and anorthite, and from the range of its energy dispersing x-ray analyses likely andesine. The brighter component appears to be a melanocratic mineral, likely either augite or hornblende.

Even though the failure crack in this and similar specimens ran entirely through the cement paste at some distance from the original paste-rock boundary, the boundary is definitely cracked. A branch of this bond crack occurs where the boundary abruptly changes direction at "A", and runs through the rock. Additionally, a number of narrow cracks, probably due to shrinkage, are found in the rightmost part of the paste itself; the largest is at "B".

Figures 5 and 6 are analogs of Figures 3 and 4, for a dolomite - silica fume bearing cement paste composite. In Figure 5 the bond crack (between the rock below and the cement paste above) is again evident. There is also a crack at "A" running downward from the fracture surface and past a large air void. This again is the trace of a branch of the main fracture crack that defines the upper surface of the cement paste.

Figure 6 shows the main bond crack very clearly. This time the bond crack branches into the paste rather than into the rock. The dolomite rock appears chemically homogeneous but shows some local pits and imperfections. There are some very short crack segments in the rock near the rock - paste boundary.

Contrary to what might be concluded from the above, the cement paste - rock interface does not always show a bond crack. Sometimes there is a crack parallel to the actual bond interface but well within the paste. We found this especially true with silica fume bearing cement paste cast against andesite rock.

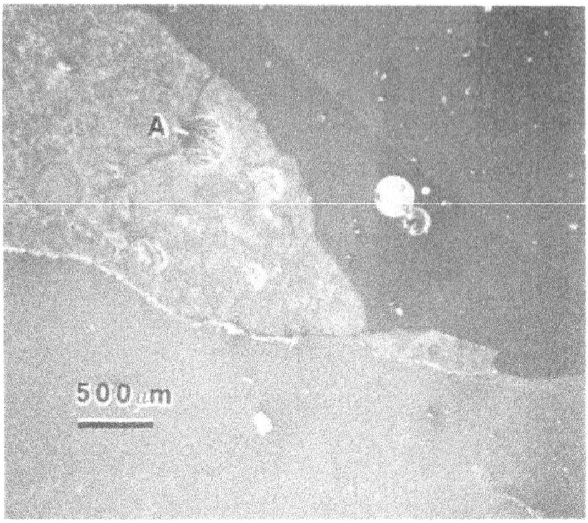

Fig. 5. Low magnification view of slice of silica fume bearing paste against dolomite rock, showing failure through the paste. "A" marks location of branch crack that travelled downward from the failure surface.

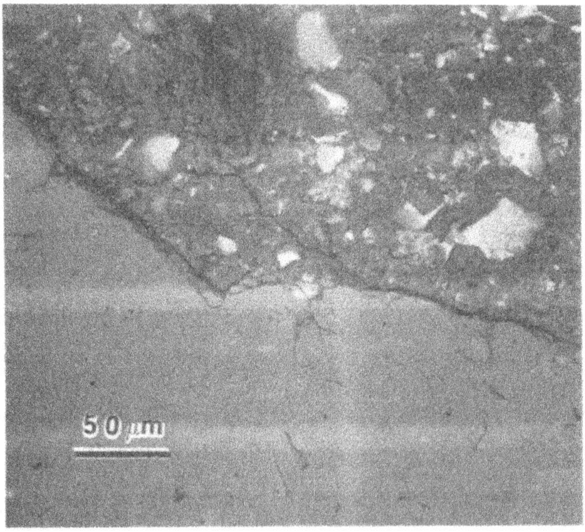

Fig. 6. Higher magnification view of an area of Figure 5, showing bond cracking, and a branch of the bond crack through the paste.

Another region of interest showing the interfacial zone of the silica fume bearing cement paste against andesite rock is pictured in Figure 7. Here again the actual interface is not cracked, but shrinkage cracks occur in the nearby cement paste. Also of interest is the observation that the silica fume bearing cement paste within the pictured zone (i.e. within 20 μm or so from the actual interface) seems to provide a uniformly dense paste structure right up to the contact with the rock. This seems to be true even where individual rock crystals, such as at "A" and "B" project into the paste and would seem to create local sites around which uniform paste packing might be difficult.

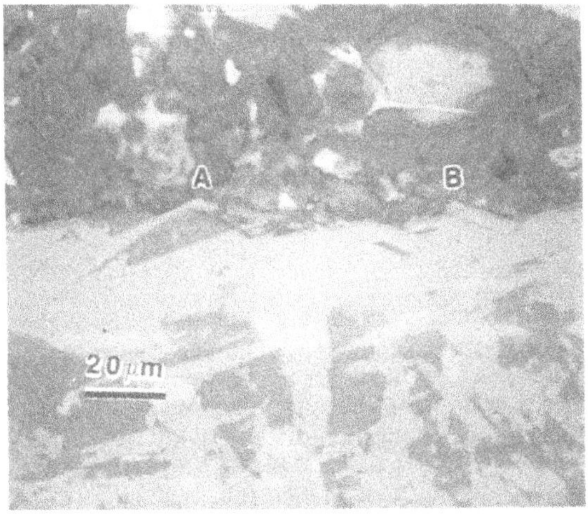

Fig. 7 Silica fume bearing cement paste against andesite rock, showing dense character of the silica fume bearing cement paste close to the uncracked interface, even around projecting feldspar crystals at "A" and "B".

All of the above illustrations represent specimens whose fracture surface ran within the paste rather than at the interface. As indicated earlier, some specimens, especially dolomite-plain cement paste composites, failed at the interface.

Figure 8(a) shows the general appearance of the dolomite rock side of one such specimen. The oscillatory character of the trace of the failure surface is apparent at this low magnification. At the higher magnification of Figure 8(b) some additional roughness of the surface can be seen, as well as the existence of a local crack segment produced within the rock.

The geometry of the trace of the failure surface of such a specimen from the paste side is naturally similar, as can be seen in Figure 9(a). Higher magnification views are provided in Figures 9(b) and (c). Figure 9(c) shows the trace of an obvious branch crack running upward from the main bond failure crack that constitutes the lower surface of the cement paste.

This sequence of three figures provides some insight concerning possible differences in the paste structure between regions near and regions well away from the former rock-paste interface. Polishing here also was not successful in producing a flat surface. The elevated grains are either larger unhydrated cement grains or larger cement grains that have hydrated in a compact mode in situ. The lower level of elevation marks finer textured and much more porouse bulk or "matrix" cement paste. It appears from these micrographs that the visible features of the internal structure of the paste near the interface were not markedly different from those of the bulk paste far away from the interface. In particular, the frequency of coarse cement grains shows little observable difference. This is surprisint in that one commonly advanced explanation for the development of a microstructurally different aureole de transition has to do with the postulated difficulty of packing the larger cement particles near the actual interface.

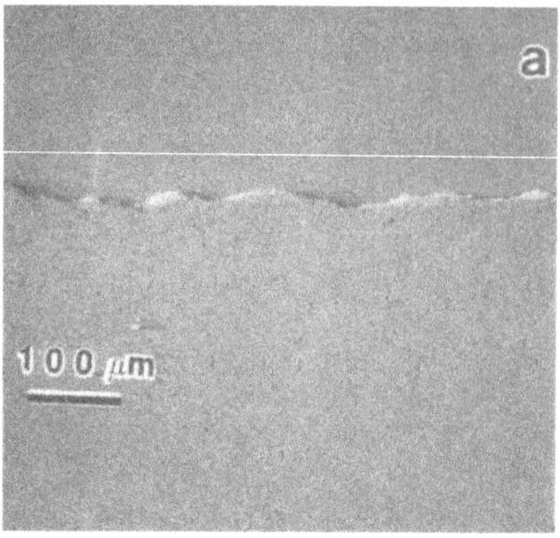

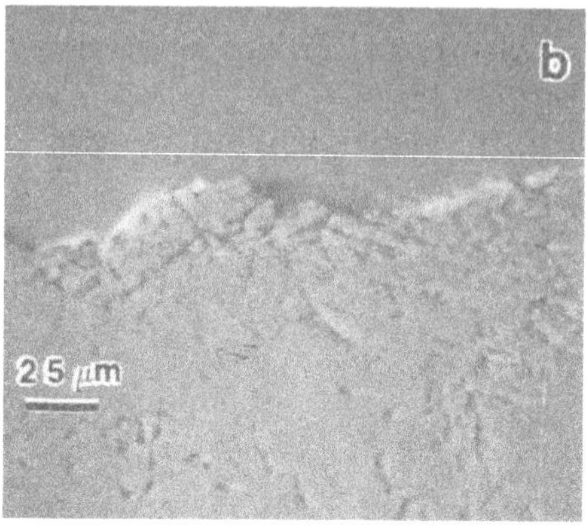

Fig. 8 Dolomite side of a cement paste-dolomite speciment that failed at the original interface : (a) at low magnification, illustration the oscillatory character of the fracture surface ; and (b) higher magnification showing an only modest increase in visible roughness of the fracture surface.

3 Discussion

The observation described here can be summarized as follows :

1) For cement paste cast against previously fracture rock surfaces, the failure crack starts at the defined notch and either runs at the original interface, or else departs from it and enters the paste at an appreciable angle. If the latter occurs, the crack penetrates the paste 1 to 2 cm, then turns and runs parallel to the interface. Presumably the mode depends on a balance between adhesive (bond) strength and the strength of the paste.

2) Silica fume bearing pastes tend to genrate a sufficiently strong bond to favour the second mode, especially when cast against the andesite rock.

3) In this "within paste" type of failure, the crack passing through the paste tends to produce branches that travel through the bonded paste layer toward the original interface ; these may turn and run parallel to and near the interface for some distance.

4) In the parting failure mode, the trace of the failed interface appears to be oscillatory at low magnification, but shows some roughness at higher magnification. Crack branching may occur on either the paste or the rock side of the failure crack.

5) In both failure modes, crack braching out of the failure plane is extensively documented, and obviously constitutes an important energy absorbing mechanism.

6) Some insight can be obtained by examining the paste region near the original paste-rock interface. There seems to be little visual indication of a transition zone different from the bulk paste well removed from the original interface in thes specimens.

It is also pertinent to point out that the observational technique itself seems to provide an appropriate way of relating the trace of the failue crack to the trace of the pre-existing paste-rock interface. However, better polishing techniques appears to be necessary to relate the crack more adequately to local features of the internal paste morphology. Such techniques are under development.

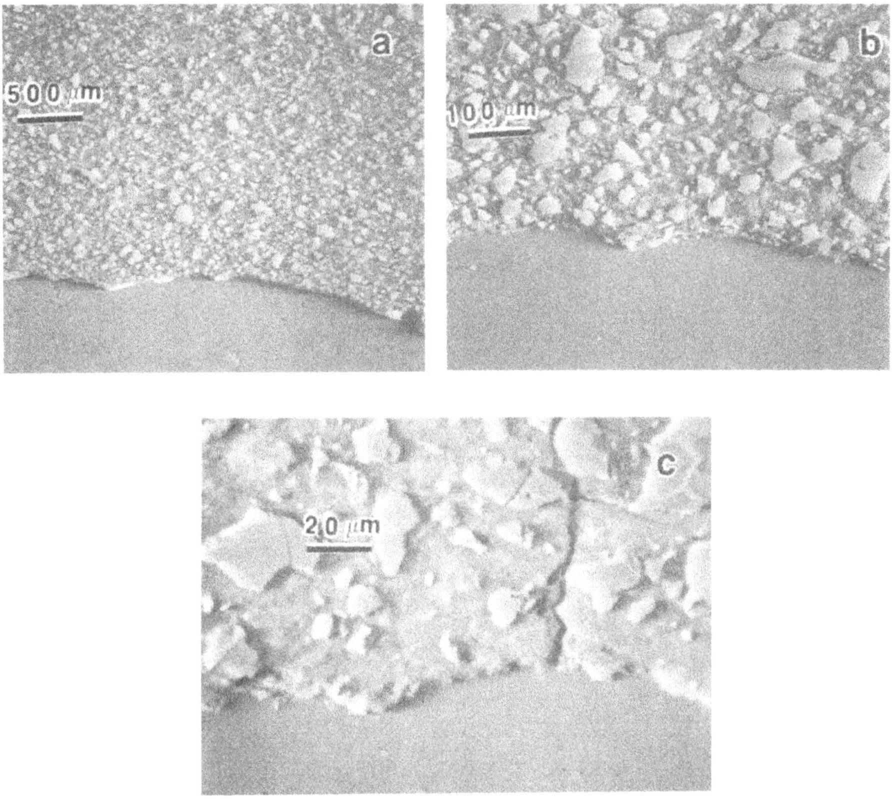

Fig. 9 (a) Cement paste side of a cement paste-dolomite specimen that failed at the original interface, at low magnification. Note the apparent uniformity of the paste near and away from the original interface. (b) Higher magnification view. (c) Very high magnification view. Note the branch crack travelling upward into the paste from the failure crack that constituted the original interface.

4 Acknoledgments

This paper represents a joint contribution of the N.S.F. Center fo Advanced Cement Based Materials at Purdue University and of the Canadian Centers of Excellence in High Performance Concrete at the University of British Columbia. The work was carried out at the University of British Columbia. We thank Ms. Yvonne Dumas for preparing the polished specimens, and Ms. Mary Mager for technical assistance in scanning electron microscopy.

5 References

Diamond, S. and Mindess, S. (1992) SEM investigations of fracture surfaces using stereo pairs. I. Fracture surfaces of rock and cement paste. **Cem. Conc. Res., 22, 67-68.**

Mindess, S. and Diamond, S., SEM of fracture surfaces using stereo pairs : II. Fracture surfaces of rock-cement paste composite specimens. **Cem. Concr. Res., 22** (in press).

3 FIBRE-CEMENT PASTE TRANSITION ZONE: CHARACTERIZATION AND EFFECT ON THE MECHANICAL PROPERTIES

H. SAVASTANO Jr
Air Force Academy, Pirassununga, Brazil
V. AGOPYAN
Escola Politécnica, University of Sao Paulo (EPUSP), Brazil
F. de A. S. DANTAS
Instituto de Pesquisas Tecnologicas (IPT), Sao Paulo, Brazil

Abstract
The interface of short monofilament fibres randomly dispersed in a paste of ordinary portland cement (without carbonate addition) is analyzed. The water-cement ratio of the matrix varies from 0.30 to 0.46 and the age of the specimens varies from 7 to 180 days. The fibres are coir, malva, sisal, chrysotile asbestos and polypropylene. The analysis is done by backscattered electron image (BSEI) and energy dispersive spectroscopy (EDS), as well as by mechanical tests. The interface of vegetable fibres is thicker and more porous than that of the other fibres.
Keywords: Transition Zone, Vegetable Fibres, Asbestos Fibres, Polypropylene Fibres, Composites, Fibrecement.

1 Introduction

Several studies have been done on fibre-matrix transition zone of composites. Bentur et al. (1985) shows steel fibre reinforced cement where there is a duplex film close to the fibre with a thickness from 1 to 2 micrometers. The first layer consists of a continuous film on the fibre surface of portlandite crystals with their C-axis perpendicular to this surface. The second layer consists of a film of hydrated calcium silicate, but this film is not continuous and is very difficult to recognize with a microscope. On the duplex layer film they found a portlandite zone with a thickness of 10 micrometers. Covering this zone there is a porous transition zone with a thickness from 10 to 20 micrometers, which becomes more compacted as it farthens from the fibre.

More recently Bentur et al. (1989) searching the fibrillated polypropylene fibre reinforced concrete found a homogeneous and dense region of a thickness varying from 10 to 40 micrometers with moderate amount of portlandite crystals which helps the better bonding between the two phases.

Interfaces in Cementitious Composites. Edited by J. C. Maso. © RILEM.
Published by E & FN Spon, 2 - 6 Boundary Row, London SE1 8HN. ISBN 0 419 18230 6.

Akers and Garret (1983) have not found the increase of porosity and portlandite crystals close to the fibres for chrysotile asbestos cement. They concluded that this phenomenon may happen because they cast their specimens by pressure; it can also be due to the fact that the asbestos fibres are defibrillated so their surface area has increased and the wall effect has been reduced.

This study aims to characterize the transition zone of composites with vegetable fibres in comparison with those of asbestos and polypropylene fibres.

2 Experimental work

The matrix of the specimens is an ordinary portland cement (without carbonate addition) paste. The selected fibres for this study are: chrysotile asbestos type 4Z (QAMA classification), polypropylene (filament), coir (Cocos **nucifera** Linn.), malva (Urena **lobata** Linn.) and sisal (Agave **sisalana** Perrine). All the fibres have not been treated.

The water absorption of the vegetable fibres is higher than 100%, moreover, more than 60% of absorption occurs at the first 15 minutes of immersion. This characteristic affects directly the behaviour of the fibres immersed in the fresh matrix. The durability problem of these vegetable fibres is discussed elsewhere (Agopyan, 1988 and John et al., 1990).

Fibre lengths vary from 15 to 30 mm, with the exception of asbestos fibres which are 5 mm long. The fibres are randomly dispersed in the cement paste matrix. The fibre volume for the composites with asbestos and polypropylene fibres is 1% and for the vegetable fibres composites is 4%; these volumes are enough to change the properties of the matrices.

Thirty series of specimens of vegetable fibre reinforced cement paste and six series of the unreinforced matrix were prepared to be tested at the ages of 7, 28, 90 and 180 days. The water-cement ratio was 0.38 and with the age of 28 days specimens with 0.30 and 0.46 were made.

This experimental work adopts the procedures mentioned by Savastano Jr. and Agopyan (1992) and includes the comparison of composites with fibres other than vegetable.

3 Results and comments

3.1 Bending test

The adopted method is the four point bending test, following the RILEM (1984) recommendation for this type of material. Figures 1 and 2 present the specific energy in function to the water-cement ratio and age, respectively.

As it was expected the short asbestos fibres induce a small energy absorption of their composites, the best results are obtained with the polypropylene fibres.

The effect of age of the composites is not noticed, except for those with vegetable fibres which present a reduction of the ductility with the increase of age due to deterioration of the fibres in an alkaline medium.

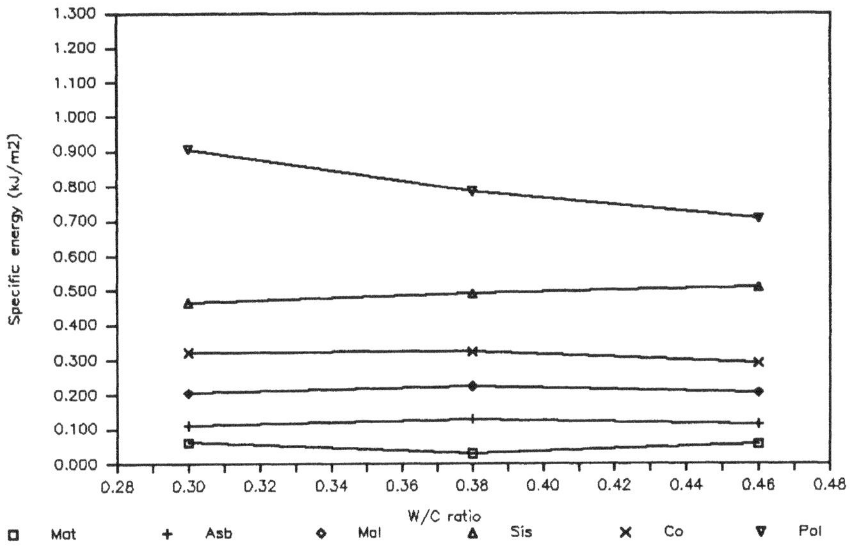

Fig.1. Bending test: effect of water-cement ratio on specific energy at the age of 28 days.

3.2 Direct tensile test

The ASTM C 190 test method for mortars is applied for the composites. In Figure 3, as was expected the tensile strength is reduced with the increase of water-cement ratio. The increase in strength with age is not clearly seen in Figure 4, except for the composites with asbestos.

3.3 Pull-out test

Following Tattersall and Urbanowicz (1974) casting procedure, specimens were prepared with a single filament of 100 mm long fibres (the asbestos fibres were not tested). The sisal fibres have bonding strengths with similar values of those of polypropylene fibres (about 140 kPa) although the absorbed specific energy of sisal fibre composites are lower than those of polypropylene fibres. The coir fibres which have less rough surface than the sisal ones have lower strength (about 80 kPa).

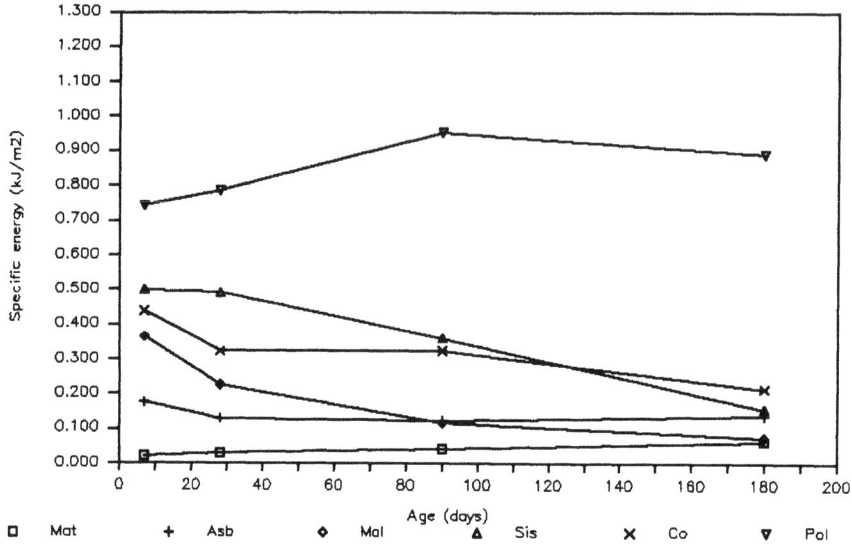

Fig.2. Bending test: effect of the age on specific energy with water-cement ratio of 0.38.

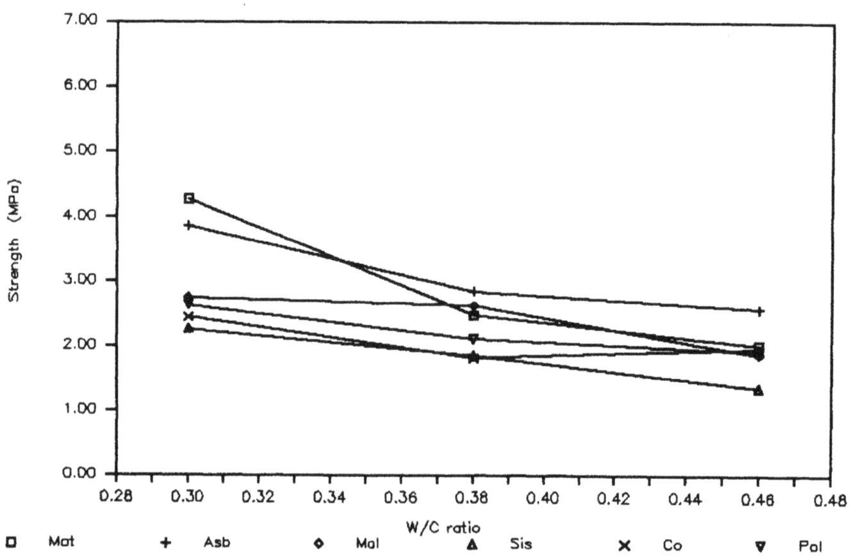

Fig.3. Direct tensile test: effect of water-cement ratio on the strength at the age of 28 days.

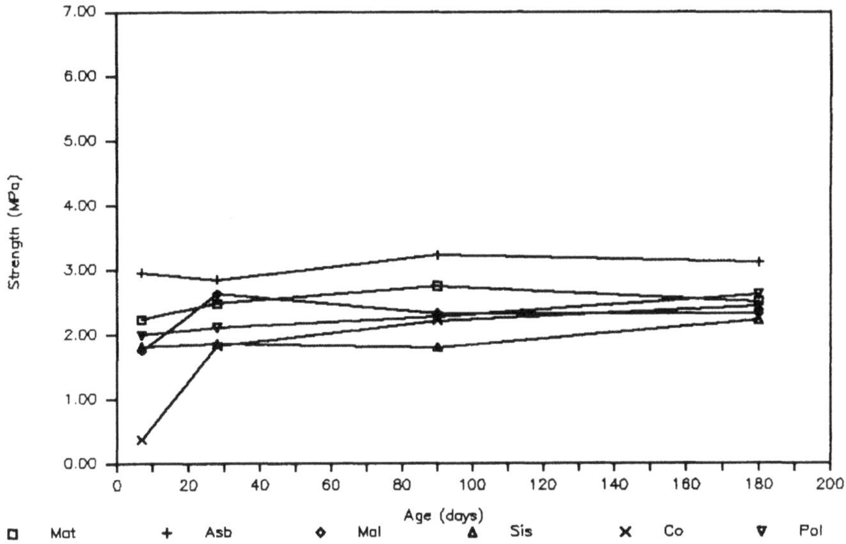

Fig.4. Direct tensile test: effect of the age on the strength with water-cement ratio of 0.38.

3.4 Microstructural analysis

The fibre-matrix transition zone is analyzed through BSEI following Scrivener et al. (1988) procedures done for high strength concretes. The EDS analysis is also done to identify the major chemical elements.

The specimens were prepared following Kjellsen et al. (1991) recommendations; this preparation consists basically of: a) stoppage of cement hydration with acetone; b) impregnation of the specimens with a low viscosity resin in a vacuum chamber; c) polishing the analyzed surface with silicon carbide and later with a diamond paste; d) recovering the surface with a carbon layer.

For asbestos fibres a 5 micrometers portlandite crystal layer can be seen in Figure 5 (the light gray shaded region), with no change of porosity. This contradicts Akers and Garret (1983) study results probably because they casted the specimens with pressure.

The efficiency of polypropylene fibres clearly presented by the mechanical properties (Figures 1 and 2) can also be confirm by microstructural analysis as shown in Figure 6 where a crack has been moved by fibre action.

For the malva composite, at the age of 7 days and with a water-cement ratio of 0.38, the transition zone is highly porous and has portlandite macrocrystals at the thickness of about 50 micrometers (Figure 7). This zone is clearly fissured probably due to lower strength than the matrix itself. Some microcracks cross even the portlandite crystals, and calcium hydroxide plates can also be noticed in a perpendicular direction from the analyzed surface.

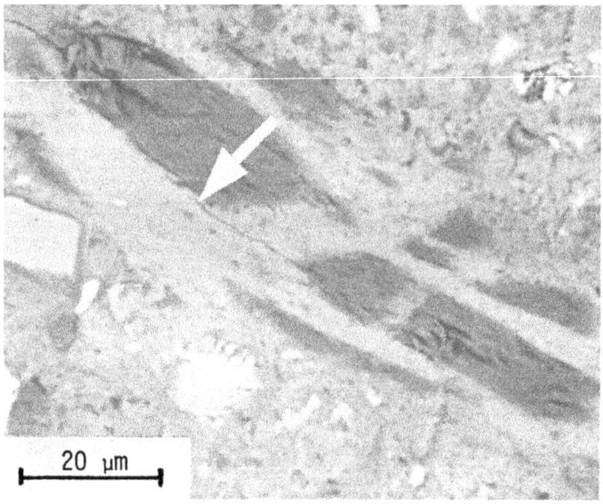

Fig.5. BSEI of composite with asbestos fibre, water-cement ratio of 0.30, age of 28 days (1230x). The arrow indicates the portlandite layer close to the fibre.

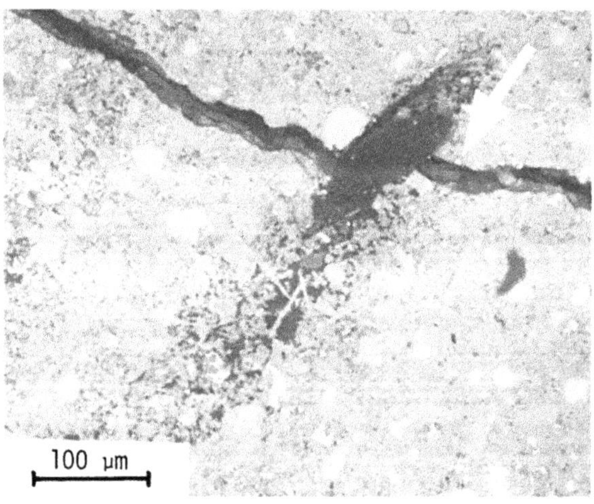

Fig.6. BSEI of composite with polypropylene fibre, water-cement ratio of 0.38, age of 180 days (203x). The arrow indicates a crack being deviated by the fibre.

Moreover the fibres are almost completely debonded from the matrix. This phenomenon is probably due to the fact that there is a large amount of water in the transition zone due to high water absorption of the fibre. When the composite begins to dry, the fibres reduce their volume and separate from the matrix impairing their action as reinforcement.

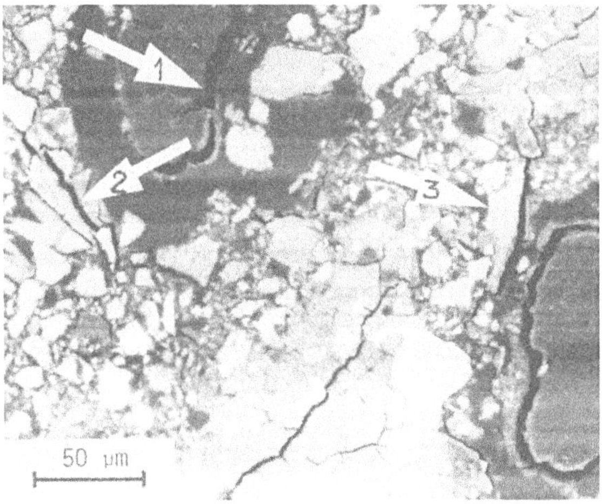

Fig.7. BSEI of composite with malva fibre, water-cement ratio of 0.38, age of 7 days (354x). Arrow 1: fibre debonding; arrow 2: microcracks; arrow 3: portlandite macrocrystal.

The sisal fibre is also debonded from the matrix and their transition zone presents high porosity, even at the age of 180 days. Figure 8 shows a section of a sisal fibre (dark area) with a transition zone of about 200 micrometers in thickness. The EDS analysis of this zone indicates a calcium-silica ratio higher than 4.4 (Table 1)

Table 1. EDS: atomic element composition of two spots in the transition zone of sisal fibre composite (see Figure 8)

Spot	Ca(%)	Si(%)	Mg(%)	S(%)	Al(%)	Fe(%)
1	70.0	11.1	5.3	3.3	6.5	3.8
2	52.5	12.0	17.4	5.0	10.2	2.9

Figures 9 and 10 present coir fibre composites with different water-cement ratios. The high cracking rate of the matrix close to the fibres has not shown reduction with the decrease of water-cement ratio, it seems to be mainly a function of the fibre absorption rate. However, with higher water-cement ratios the portlandite layer seems to increase as can be noticed in Figure 10.

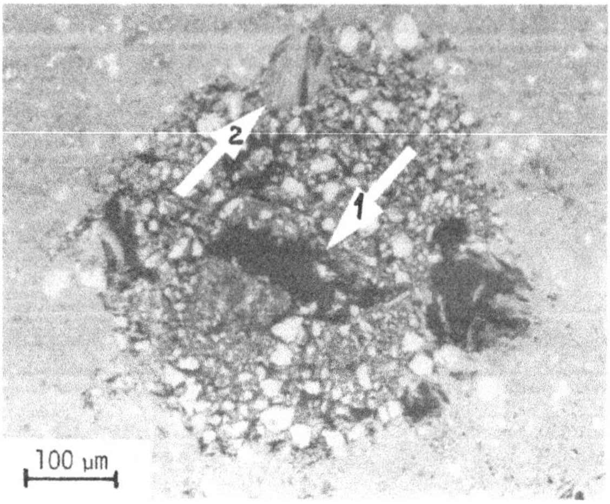

Fig.8. BSEI of composite with sisal fibre, water-cement ratio of 0.38, age of 180 days (146x). Arrow 1: fibre; arrow 2: portlandite macrocrystal.

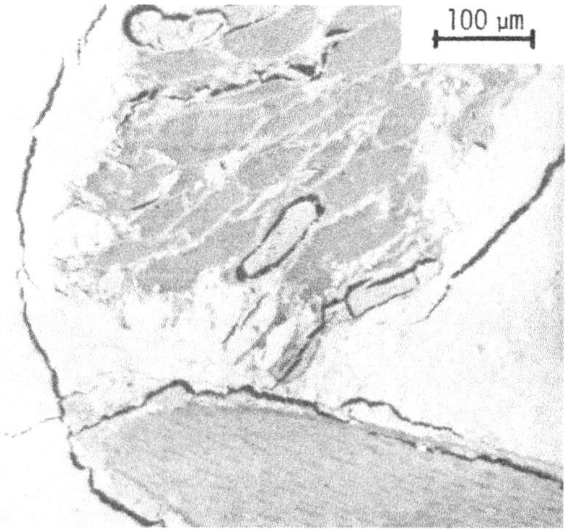

Fig.9. BSEI of composite with coir fibres, water-cement ratio of 0.30, age of 28 days (140x). High cracking rate close to the fibre.

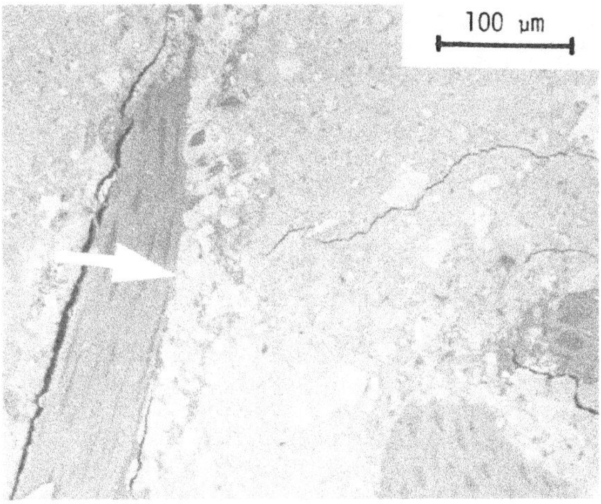

Fig.10. BSEI of composite with coir fibre, water-cement ratio of 0.46, age of 28 days (235x). The arrows indicates the concentration of portlandite crystals in the transition zone.

4 Conclusions

The mechanical test results show better performance of composites with polypropylene and asbestos fibres. These fibres do not absorb water as do the vegetable fibres.

The transition zone of vegetable fibre composites is porous, thick and rich of random portlandite crystals. The volume variation of these fibres must be the reason for their debonding. The high water absorption of these fibres could be the reason for the cracking of the transition zone due to differential shrinkage.

High water-cement ratio induces to an increase of the concentration of portlandite and porosity in this zone, probably due to higher mobility of calcium ions.

For vegetable fibre composites the duplex film of portlandite and hydrated calcium silicate has not been noticed due to inexistence of the wall effect.

The age of the composite has a positive effect as it usually reduces the porosity of the interface, however the microstructure of that transition zone seems to be defined at the age of 7 days.

To improve the mechanical performance of vegetable fibre composites it is recommended to: a) use a casting method with pressure to remove the excessive water close to the fibres; b) avoid long term humid curing; c) reduce the fibre water absorption perhaps by impregnating them with coating agents.

5 References

Agopyan, V. (1988) Vegetable fibre reinforced building materials - developments in Brazil and other Latin-American countries, in **Natural Fibre-Reinforced Cement and Concrete** (Concrete Technology Design 5, ed. R. N. Swamy), Blackie, Glasgow, pp. 208-242.

Akers, S. A. S. and Garret, G. G. (1983) Fibre-matrix interface effects in asbestos-cement composites. **J. Mater. Sci.**, 18, 2200-2208.

Bentur, A.; Diamond, S. and Mindess, S. (1985) The microstructure of the steel fibre-cement interface. **J.Mater. Sci.**, 20, 3610-3620.

Bentur, A.; Mindess, s. and Vondran, G. (1989) Bonding in polypropylene fibre reinforced concretes. **Int. J. Cem. Compos. Lightweight Concr.**, 11, 153-158.

John, V. M.; Agopyan, V. and Derolle, A. (1990) Durability of blast furnace slag based cement mortar reinforced with coir fibres, in **Vegetable Plants and their Fibres as Building Materials** (RILEM Proceedings 7, ed. H. S. Sobral), Chapman and Hall, London, pp. 87-97.

Kjellsen, K. O.; Detwiler, R. J. and Gjorv, O. E. (1991) Backscattered electron image analysis of cement paste specimens: specimen preparation and analytical methods. **Cem. Concr. Res.**, 21, 388-390.

RILEM Technical Committee 49 TFR (1984) Testing methods for fibre reinforced cement based composites. **Matériaux et Constructions**, 17, 441-456.

Savastano Jr., H. and Agopyan, V. (1992) Transition zone of the hardened cement paste and vegetable fibres, in 4th Int. Symp. on Fibre Reinforced Cement and Concretes, Chapman and Hall.

Scrivener, K. L.; Bentur, A. and Pratt, P. L. (1988) Quantitative characterization of the transition zone in high strength concretes. **Advances in Cement Research**, 1.

Tattersal, G. H. and Urbanowicz, C. R. (1974) Bond strength in steel-fibre reinforced concrete. **Mag. Concr. Res.**, 26, 105-113.

Acknowledgments

The authors would like to thank Professor Paulo S. C. Pereira da Silva of the Metallurgical and Materials Engineering Department of EPUSP, for his help with the microstructural analysis. They are also in debt to the IPT where the main research work has been carried out with the help of Dr. Maria A. Cincotto and Mr. Pedro C. Bilesky.

4 EFFECTS OF SODIUM CHLORIDE ON THE HYDRATION PRODUCTS IN THE INTERFACIAL ZONE BETWEEN CEMENT PASTE AND ALKALI-REACTIVE AGGREGATE

M. SAITO
Kanazawa Institute of Technology, Nonoichi, Ishikawa, Japan
M. KAWAMURA
Department of Civil Engineering, Kanazawa University,
Kanazawa, Ishikawa, Japan

Abstract
X-ray diffraction and quantitative thermogravimetry analysis were carried out to evaluate the amount of $Ca(OH)_2$, ettringite and Friedel's salt, and the degree of orientation of $Ca(OH)_2$ crystals in the cement paste-reactive aggregate interfacial zone. The analysis showed that the use of reactive aggregate significantly affected the amount and the degree of orientation of $Ca(OH)_2$ crystals in the interfacial zone. The result that the intrusion of 1N NaCl solution promoted the precipitation of $Ca(OH)_2$ crystals in the zone was also obtained. In most composite specimens, ettringite was found to precipitate in the interfacial zone. The specimens soaking 1N NaCl solution had greater amounts of ettringite in their interfacial zone than those soaking water. Friedel's salt was also formed in the interfacial zone of the composite specimens containing NaCl at the stage of mixing and those without added NaCl soaking 1N NaCl solution for 8 weeks.
Keywords: Alkali-silica reaction, Calcium hydroxide, Ettringite, Friedel's salt, Sodium chloride, X-ray diffraction, Thermogravimetry

1 Introduction

Chemical and microstructural characteristics of the interfacial zone between paste and aggregate have been considered to influence the durability as well as the mechanical properties of concrete. Diamond (1989a) suggested that $Ca(OH)_2$ in the interfacial zone took part in the alkali-silica reaction in concrete. It is also well known that NaCl supplied from the surrounding media promotes the expansion of concretes and mortars containing reactive aggregate (Hobbs, 1988). Recently, Kawamura and Diamond (1991) found that great amounts of chloride-bearing ettringite and Friedel's salt existed within mortar specimens containing a Japanese opal stored in 1N NaCl solution at 38 °C for longer than one year. However, at present, it is ambiguous whether the formation of the chloride-bearing ettringite is related to the expansion of the mortars in the NaCl solution. On the other hand, Monteiro and Mehta (1985) reported that greater amounts of ettringite were formed in the interfacial zone than in bulk cement paste. The chloride-bearing ettringite may be formed in the interfacial zone around reactive aggregate grains in mortars in NaCl solution, although no sulfate ions are supplied.

Interfaces in Cementitious Composites. Edited by J. C. Maso. © RILEM.
Published by E & FN Spon, 2 - 6 Boundary Row, London SE1 8HN. ISBN 0 419 18230 6.

This study aims at elucidating the orientation and the amount of $Ca(OH)_2$, and the formation of ettringite in the cement paste-reactive and -non-reactive aggregate interfacial zone by the use of model prismatic composite specimens into which Na^+ and Cl^- ions are allowed to intrude from their one face.

2 Experimental procedures

2.1 Materials

The cement used was a high-alkali ordinary Portland cement. Two types of pyroxene andesitic and granitic rocks were used as a reactive and a non-reactive aggregate, respectively. The chemical composition of the cement and the rocks is given in Table 1. An X-ray diffraction pattern for the pyroxene andesite showed the existence of cristobalite and tridymite. The reduction in alkalinity (Rc) and the dissolved silica (Sc) of the reactive andesitic aggregate obtained in accordance with ASTM C-289 test were 208 mmol/l and 514 mmol/l, respectively.

Table 1. Chemical composition of cement and rocks used

	Cement	Reactive pyroxene andesite (%)	Granite (%)
Ig. loss	0.5	1.7	0.2
Insol.	0.1	-	-
SiO_2	21.2	60.3	76.4
Al_2O_3	5.0	17.3	12.8
Fe_2O_3	3.1	5.6	1.3
CaO	64.9	6.4	0.0
MgO	1.5	4.3	0.2
SO_3	1.9	0.0	-
Na_2O	0.51	3.82	4.40
K_2O	0.70	0.89	4.60
Total	99.41	100.31	99.90

2.2 Preparation of samples

The examinations of the cement paste-aggregate interfacial zone by the X-ray diffraction and the thermogravimetry analysis were carried out on 72 composite specimens. 40 x 15 x 25 mm composite prisms were made by bonding 40 x 15 x 15 mm rock prisms cut from the rock blocks and the cement paste prisms with the water-cement ratio of 0.35, as shown in Fig. 1(a). The surfaces of rock prisms were polished by hand on a smooth iron plate using water and three carborundum powders of #180, #400, and # 800. The rock prisms had been stored in water for at least 24 hours before the cement paste was cast on them.

Immediately after demoulding, 5 faces of the composite specimens other than the cement paste face of 15 by 40 mm (Fig. 1(a)) were coated with a silicone sealant to prevent them from drying and carbonation, and then stored in the following different conditions at 40 °C; placed on the setup so as to be in contact with water (A) and 1N NaCl solution

(B) for 8 weeks immediately after demoulding, with 1N NaCl solution for 4 weeks after curing in the fog box for 4 weeks (E), and with water (F) and 1N NaCl solution (G) for 8 weeks after curing in the fog box for 8 weeks (Fig. 1(b)). The composite specimens made with the cement paste

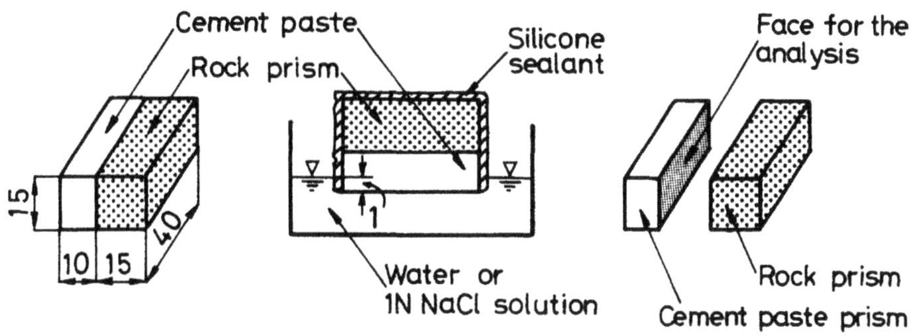

(a) Composite specimen (b) Setup for soaking water or 1N NaCl solution (c) Sample for the analysis

Fig. 1. Composite specimen for the analysis (unit : mm).

Table 2. Storage conditions and notation of composite specimens

Aggregate type		Reactive (R)				Non-reactive (N)			
Time (weeks)		2	4	8	16	2	4	8	16
Storage condition									
On water	(A)	RA2	RA4°	RA8°	-	NA2	NA4°	NA8°	-
On 1N NaCl solution	(B)	RB2	RB4°	RB8°	-	NB2	NB4°	NB8°	-
In a fog box	(C)	RC2	RC4	RC8	RC16	NC2	NC4	NC8	NC16
In a fog box*	(D)	RD2	RD4	RD8	-	ND2	ND4	ND8	-
On 1N NaCl solution for 4 weeks after storing in a fog box for 4 weeks	(E)	-	-	RE8	-	-	-	NE8	-
On water for 8 weeks after storing in a fog box for 8 weeks	(F)	-	-	-	RF16	-	-	-	NF16
On 1N NaCl solution for 8 weeks after storing in a fog box for 8 weeks	(G)	-	-	-	RG16	-	-	-	NG16

* : Composite specimens containing NaCl at the stage of mixing.
° : TG analysis is conducted.

with and without added NaCl as the controls, which are designated as (D) and (C), respectively, were cured in the fog box at 40 °C for 16 weeks. In the composite specimens (D) cured in the fog box, the amount of NaCl added at the stage of mixing was 2.55 % by weight of cement. The notation of composite specimens prepared is given in Table 2.

After curing, the composite specimens were broken by flexural loads to produce cement paste samples for the analyses as shown in Fig. 1(c). Thereafter, it was confirmed with an optical microscope that no remnants of the cement paste on the aggregate surface were found or thinly scattered.

2.3 X-ray diffraction and thermogravimetry analysis procedures

The amount of ettringite and Friedel's salt, and the orientation of $Ca(OH)_2$ crystals in the zone up to about 100 μm away from the interface were evaluated by the X-ray diffraction technique developed by Grandet and Ollivier (1980). Surfaces of the cement paste samples to be analyzed were prepared by the step by step abrasion with a #400 emery paper. Carles-Gibergues et al. (1982) confirmed that the crystals arrangement was not destroyed by the abrasion. The weight of materials removed by abrasion was measured with a balance with 0.05 mg sensitivity. The distance from the interface to the surface to be analyzed was calculated on the basis of the weight of materials removed and the density of bulk cement paste. The preferential orientation of $Ca(OH)_2$ crystals was determined from the ratio of the height of (001) peak (I(001)) to that of (101) peak (I(101)). The orientation index I is defined as Equation (1).

$$I = I(001) / \{0.74 \times I(101)\} \tag{1}$$

The thermogravimetry analysis was conducted to determine the amount of $Ca(OH)_2$ in the interfacial zone at the heating rate of 10 °C per minute using the Shimadzu TGA-50H setup. Cement paste prisms obtained by braking composite specimens were allowed to dry for 24 hours in a vacuum drying chamber at a room temperature. After drying, cement paste powders of about 12 mg corresponding to the portion up to about 10 μm from the interface were carefully collected from broken surfaces by hand using a sharp-edged tool.

3 Results and discussion

3.1 Orientation of $Ca(OH)_2$ crystals in the interfacial zone

The typical variations in the orientation index I of $Ca(OH)_2$ crystals within the cement paste-aggregate interfacial zone are shown in Fig. 2. I value at the immediate vicinity of the interface and the maximum I value in the interfacial zone for all samples are also given in Table 3 and Table 4, respectively. I values in Table 3 and Table 4 are the average of two measurements.

The plots in Fig. 2 show that $Ca(OH)_2$ crystals within about 50 μm from the interface are highly oriented as compared to those in the bulk cement paste. It is found from these results that there are differences in I value in the interfacial zone between the samples made using the reactive and the non-reactive aggregate in all storage conditions.

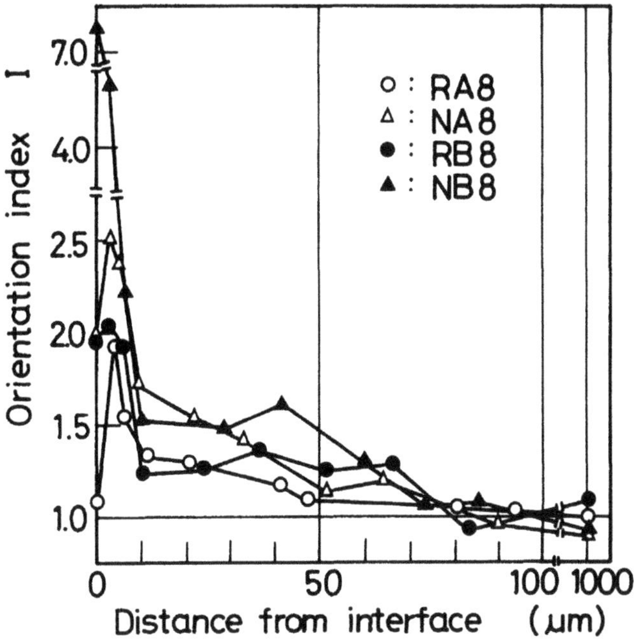

(a) RA8, NA8, RB8, and NB8 series

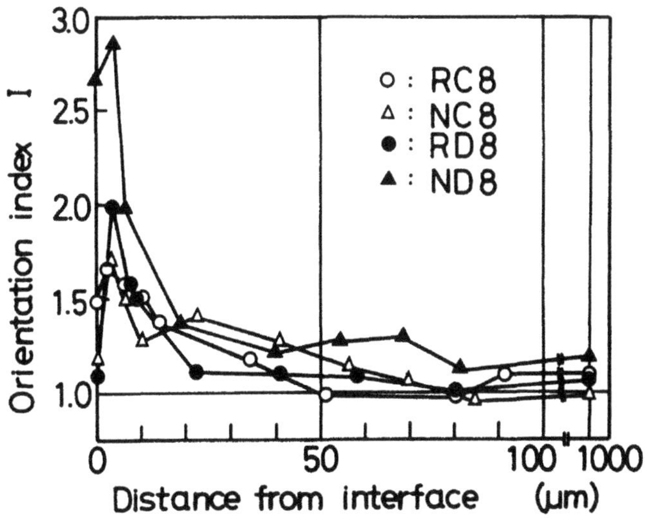

(b) RC8, NC8, RD8, and ND8 series

Fig. 2 Variations in the orientation index I of $Ca(OH)_2$ in the interfacial zone.

I values of $Ca(OH)_2$ on the reactive aggregate blocks are lower than on the non-reactive aggregate ones.

It is apparent from the data given in Table 3 and Table 4 that both the intrusion of NaCl from the outside and the addition of NaCl at the stage of mixing significantly increases the value of I of $Ca(OH)_2$ in the interfacial zone between the cement paste and the non-reactive aggregate at the age of 8 weeks. The greatest increase in I value of $Ca(OH)_2$ due to the supply of NaCl is found in the non-reactive aggregate samples immersed in 1N NaCl solution for 8 weeks. On the other hand, there is a little increase in I value in the samples contaminated with NaCl in all samples with the reactive aggregate. These results indicate that less oriented $Ca(OH)_2$ crystals are formed on the reactive aggregate blocks and that the orientation of $Ca(OH)_2$ crystals on the reactive aggregate is not so greatly influenced by NaCl as on the non-reactive one.

Table 3. Values of the orientation index I of $Ca(OH)_2$ at the immediate vicinity of the interface

Aggregate type		Reactive (R)				Non-reactive (N)			
Storage period (weeks)		2	4	8	16	2	4	8	16
Storage condition	A	1.66	1.11	1.17	-	2.20	1.58	2.12	-
	B	1.53	1.91	1.82	-	1.86	2.24	5.14	-
	C	1.08	1.24	1.40	1.14	1.58	1.41	1.13	1.49
	D	1.32	1.15	1.15	-	1.30	1.70	1.94	-
	E	-	-	1.68	-	-	-	2.05	-
	F	-	-	-	1.34	-	-	-	1.68
	G	-	-	-	1.28	-	-	-	1.67

Table 4. Maximum values of the orientation index I of $Ca(OH)_2$ in the interfacial zone

Aggregate type		Reactive (R)				Non-reactive (N)			
Storage period (weeks)		2	4	8	16	2	4	8	16
Storage condition	A	1.87	1.74	2.09	-	2.33	2.02	2.53	-
	B	2.19	2.14	2.33	-	2.16	2.24	5.14	-
	C	1.91	1.83	1.62	1.47	1.81	1.84	1.60	1.86
	D	1.74	1.33	1.71	-	1.75	1.81	2.05	-
	E	-	-	2.01	-	-	-	2.15	-
	F	-	-	-	1.77	-	-	-	2.06
	G	-	-	-	1.78	-	-	-	1.99

3.2 Amount of $Ca(OH)_2$ in the interfacial zone

The amount of $Ca(OH)_2$ in the interfacial zone up to about 10 μm from the interface and in the bulk cement paste at the portion of 5 mm from the interfacial zone obtained by the TG analysis is given in Table 5. Table 5 shows that the amounts of $Ca(OH)_2$ in the interfacial zone in both the reactive and non-reactive rock-cement paste composite specimen are considerably greater than those in the bulk cement paste, and that the more the period of storage, the greater the amount of $Ca(OH)_2$ in the interfacial zone. The amount of $Ca(OH)_2$ in the cement paste-reactive aggregate interfacial zone is also found to be considerably less than those in the interfacial zone between cement paste and non-reactive aggregate. The latter result indicates that $Ca(OH)_2$ in the region around reactive aggregate grains are consumed by the alkali-silica reaction. This result is an explicit evidence for the proposition by Diamond (1983b) that a thin-layer of $Ca(OH)_2$ deposited around the surfaces of aggregate particles is source of calcium taking part in the alkali-silica reaction.

It is found from the data given in Table 5 that smaller amounts of $Ca(OH)_2$ existed in the bulk cement paste in the specimens on the NaCl solution than in those on water, but that there is only a little difference in $Ca(OH)_2$ content in the interfacial zone between both except in the composite specimens made with the reactive aggregate stored for 8 weeks. The former result shows that greater amounts of $Ca(OH)_2$ were dissolved from the cement paste on the NaCl solution than on water. The latter one is found to indicate that the precipitation of $Ca(OH)_2$ in the interfacial zone was promoted by the intrusion of Na^+ and Cl^- ions, taking account of the fact that amounts of Friedel's salt produced in the interfacial zone were greater than those in the bulk cement paste (Fig. 3).

Table 5. $Ca(OH)_2$ content in the interfacial zone and the bulk cement paste determined by TG analysis

		Interfacial zone (%)				Bulk cement paste* (%)	
		Reactive aggregate (R)		Non-reactive aggregate (N)			
Storage period (weeks)		4	8	4	8	4	8
Storage condition	A	10.6	12.4	13.9	17.1	9.0	7.9
	B	9.6	15.0	14.0	17.6	5.9	6.1

* : Samples taken from the portion of 5 mm from the interface.

3.3 Formation of ettringite and Friedel's salt

The amount of ettringite in the interfacial zone was determined by measuring the diffraction peak height at 9.7 Å. The intensity of the diffraction peak of ettringite at the immediate vicinity of the interface in various specimens is given in Table 6. The formation of ettringite was restricted within the zones up to about 5 μm from the interface. It is found from Table 6 that in the storage condition of 'on water' and 'on 1N NaCl solution', the interfacial zone between cement paste and reactive aggregate shows the larger diffraction peak intensity of ettringite than that between cement paste and non-reactive aggregate. Furthermore, it is also apparent that the peak intensity of ettringite in the sample on 1N NaCl solution (B) is higher than that in the corresponding sample on water (A). In other storage conditions (E, F, G) in which specimens had been cured in the fog box for a relatively long period before they were placed on 1N NaCl solution and water, there is little difference in the intensity of the X-ray diffraction peak of ettringite between the samples made using the two types of aggregates, but the samples on the NaCl solution (G) show considerably greater X-ray diffraction peak intensity than those on water (F).

Friedel's salt was found to be also formed in the interfacial zone in the composite specimens containing NaCl at the stage of mixing (RD2-8 and ND2-8 series) and added NaCl-free ones in contact with 1N NaCl solution for 8 weeks (RB8 and NB8 series). The interfacial zone in other composite specimens did not show any X-ray diffraction peak of Friedel's salt. The variations in the intensity of the X-ray diffraction peak of Friedel's salt at 7.9 Å in the interfacial zone are presented in Fig. 3. The relatively high diffraction peak of Friedel's salt was found within about 10 μm from the interface, compared to that in the bulk cement paste.

Table 6. X-ray diffraction peak intensity (cps) of ettringite at the immediate surface of the interface

Aggregate type		Reactive (R)				Non-reactive (N)			
Storage period (weeks)		2	4	8	16	2	4	8	16
Storage condition	A	14	10	24	-	None	None	None	-
	B	21	30	41	-	11	23	None	-
	C	None	14	43	16	24	34	None	28
	D	21	16	45	-	None	37	24	-
	E	-	-	21	-	-	-	25	-
	F	-	-	-	36	-	-	-	30
	G	-	-	-	52	-	-	-	57

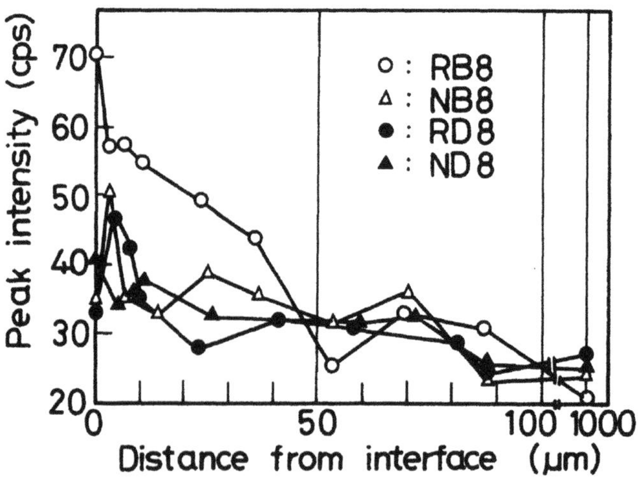

Fig. 3. Variations in the X-ray diffraction peak intensity of Friedel's salt in the interfacial zone.

4 Conclusions

The conclusions drawn from this study are summarized as follows;

1. The degree of orientation of $Ca(OH)_2$ crystals in the interfacial zone between cement paste and the reactive aggregate is significantly less than between cement paste and the non-reactive aggregate. Particularly, the reduction in the degree of orientation of $Ca(OH)_2$ crystals in the interfacial zone in the samples made with the reactive aggregate immersed in 1N NaCl solution for 8 weeks is remarkable.

2. Amounts of $Ca(OH)_2$ in the interfacial zone between the cement paste and the reactive aggregate are considerably less than in the zone between the cement paste and the non-reactive aggregate. The intrusion of 1N NaCl solution appears to promote the precipitation of $Ca(OH)_2$ in the interfacial zone.

3. In the interfacial zone in most composite specimens, ettringite is formed. The formation of ettringite is restricted within the zone up to 5 μm from the interface. The composite specimens on 1N NaCl solution show the higher X-ray diffraction peak of ettringite in the interfacial zone than those on water.

4. Friedel's salt is formed in the interfacial zone in both the composite specimens containing NaCl at the stage of mixing and added NaCl-free ones placed on 1N NaCl solution for 8 weeks.

5 References

Carles-Gibergues, A. Grandet, J. and Ollivier, J.P. (1982) Contact zone between cement paste and aggregate. **Bond in Concrete.** Editor: P. Bartos, Applied Science Publishers Ltd., London, pp. 24-33.

Diamond, S. (1989a) ASR-another look at mechanisms, in **Proc. 8th International Conference on Alkali-Aggregate Reaction**, Kyoto, pp. 83-94.

Diamond, S. (1983b) Alkali reactions in concrete-pore solution effects, in **Proc. 6th International Conference on Alkalis in Concrete,** Copenhagen, pp. 155-166.

Grandet, J. and Ollivier, J.P. (1980) New method for the study of cement-aggregate interfaces, in **Proc. 7th International Congress on the Chemistry of Cement,** Vol. III, Paris, pp. VII.85-VII.89.

Hobbs, D.W. (1988) **Alkali-Silica Reaction in Concrete.** Thomas Telford Ltd., London.

Kawamura, M. and Diamond, S. (1991), in the course of preparation.

Monteiro, P.J.M. and Mehta, P.K. (1985) Ettringite formation on the aggregate-cement interface. **Cement and Concrete Research,** 15, 378-380.

5 INFLUENCE DE L'ORIGINE DES CENDRES VOLANTES EMPLOYEES COMME AJOUTS SUR L'AUREOLE DE TRANSITION DES BETONS HYDRAULIQUES QUI EN RENFERMENT

(Influence of source of fly ash additions on the transition zone of hardened concrete)

B. HUSSON and A. CARLES-GIBERGUES
LMDC, INSA-UPS, Toulouse, France

Résumé *(English abstract, page 307)*
Dans cette publication, nous avons étudié l'influence de trois cendres volantes d'origines différentes sur la structuration de l'auréole de transition : la première, CC, issue d'une chaudière sans désulfuration,les deux autres, CD et LFC, obtenues par désulfuration respectivement par voie semi-humide et par voie sèche.

Ces cendres sont différentes par leurs caractéristiques physico-chimiques : la cendre CD a une finesse beaucoup plus importante, les cendres CD et LFC sont enrichies en soufre respectivement sous forme de sulfite et de sulfate de calcium. Leur comportement rhéologique est également modifié : l'introduction des cendres CD et LFC dans les pâtes accroît la demande en eau du matériau, alors que la cendre CC a un rôle légèrement fluidifiant.

Par contre, ces différences de comportement rhéologique ne modifient pas de façon différentielle l'auréole de transition : dans les trois cas, on observe une diminution de l'indice d'orientation de la portlandite associé à un accroissement de l'épaisseur de l'auréole.
Mots-clés : Sulfite de calcium, Lit Fluidisé Circulant, Comportement rhéologique, Auréole de transition, Indice d'orientation.

1 Introduction

Les réglementations plus sévères concernant la dépollution des fumées des centrales thermiques ont amené des changements profonds dans la technologie des chaudières et, par voie de conséquence, dans la constitution des cendres volantes.

C'est notamment le cas des cendres sulfitiques et des cendres produites par chaudières à lit fluidisé circulant (LFC).

Ces cendres sont nouvelles par leur granularité, par la forme des particules (grains sphériques plus rares ou absents) et par la nature minéralogique de leurs constituants.

Leur comportement rhéologique dans les mortiers et les bétons est modifié en conséquence.Ces modifications rhéologiques ont-elles des répercussions au niveau de la structure de l'auréole de transition ?

Cet article apporte des éléments de réponse qui concernent essentiellement la structuration de l'auréole et ses caractéristiques initiales.

Interfaces in Cementitious Composites. Edited by J. C. Maso. © RILEM.
Published by E & FN Spon, 2 - 6 Boundary Row, London SE1 8HN. ISBN 0 419 18230 6.

2 Origine des cendres volantes

La fixation du soufre des fumées de centrales thermiques nécessite un apport de calcaire ou de chaux (Odler et Zysk) qui peut se faire suivant deux voies indiquées ci-dessous (Gokhale et Burnet).

* La voie sèche : le calcaire est introduit au niveau du foyer, c'est-à-dire dans des zones à température toujours supérieure à 850°C (foyers à Lit Fluidisé Circulant).
* La voie semi-humide ou humide : à leur sortie de la chaudière, donc à température inférieure à 150°C, les gaz traversent un réacteur d'échange où ils sont "arrosés" par une solution alcaline de chaux ou de calcaire (sous forme de spray humide).

Nous avons étudié trois cendres :
- une cendre issue d'une chaudière classique sans désulfuration (Carling), référencée CC,
- une cendre de chaudière à Lit Fluidisé Circulant (Carling), référencée LFC,
- une cendre sulfitique provenant d'une centrale thermique allemande à désulfuration semi-humide, référencée CD.

3 Programme expérimental

Après avoir déterminé les caractéristiques physiques, chimiques et minéralogiques des trois cendres, nous avons étudié le comportement rhéologique de pâtes mixtes de ciment + cendres à l'appareil de VICAT. Le temps de prise de ces pâtes a également été mesuré.

Dans une troisième étape, enfin, nous avons déterminé l'influence de ces matériaux sur la structuration de l'auréole de transition. Pour cela, nous avons mené l'étude sur un modèle mis en oeuvre au sein du Laboratoire : le béton est modélisé sous la forme d'une éprouvette mixte, consistant en de la pâte coulée sur un granulat, plan, unique. Ces essais conduisent à la quantification de deux paramètres : l'orientation de la portlandite (*) et l'épaisseur de l'auréole.

4 Résultats expérimentaux

4.1 Caractéristiques des cendres anhydres

4.11 Caractéristiques physiques

Les valeurs de la masse volumique et de la surface spécifique sont reportées dans le tableau 1.

(*) La notion d'orientation de la portlandite couramment utilisée jusqu'à présent est toutefois à considérer avec prudence. Des études en cours au sein du LMDC laissent à penser que l'importance de la réflexion (001) de la portlandite sur un diffractogramme de rayons X ne serait pas nécessairement due à une orientation préférentielle des cristaux mais à l'existence de cristaux de portlandite mieux formés au contact du granulat. En tout état de cause, une variation de l'indice d'orientation I traduit une modification structurale.

Tableau 1. Caractéristiques physiques des cendres

échantillon	CC	CD	LFC
masse volumique (10^3kg/m^3)	2,237	2,376	2,240
surface spécifique (10^3 m^2/kg)	0,33	1,40*	0,39

Les courbes granulaires des différents échantillons (granulomètre à laser) sont reportées sur la figure 1.

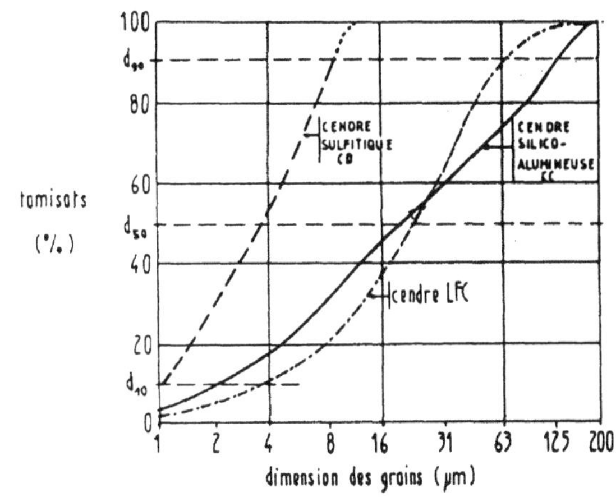

Fig.1. Courbes granulaires des échantillons CC, CD et LFC

4.12 Composition chimique

Les compositions chimiques des trois cendres sont reportées sur le tableau 2.

Tableau 2. Compositions chimiques de CC, CD et LFC

oxyde	SiO_2	Al_2O_3	Fe_2O_3	CaO	MgO	K_2O
CC	54,7	24,2	8,2	1,5	2,5	4,5
CD	13,2	6,9	1,9	49,6	0,9	1,0
LFC	39,1	17,2	7,0	7,9	3,2	3,2

oxyde	Na_2O	MnO	TiO_2	SO_2	SO_3	p.f.
CC	0,4	0,2	0,8	0	1,4	2,1
CD0	0,2	0,1	0,2	16,0	3,2	6,8
LFC	0,3	0,2	-	0	4,5	17,2

* Cette valeur n'est à considérer qu'à titre indicatif car elle a été mesurée par l'appareil de Blaine et sort par valeur supérieure du domaine de validité de la méthode (0,1-1 m2/g)

4.13 Constitution minéralogique

Elle a été déterminée par diffractométrie de rayons X (figure 2).

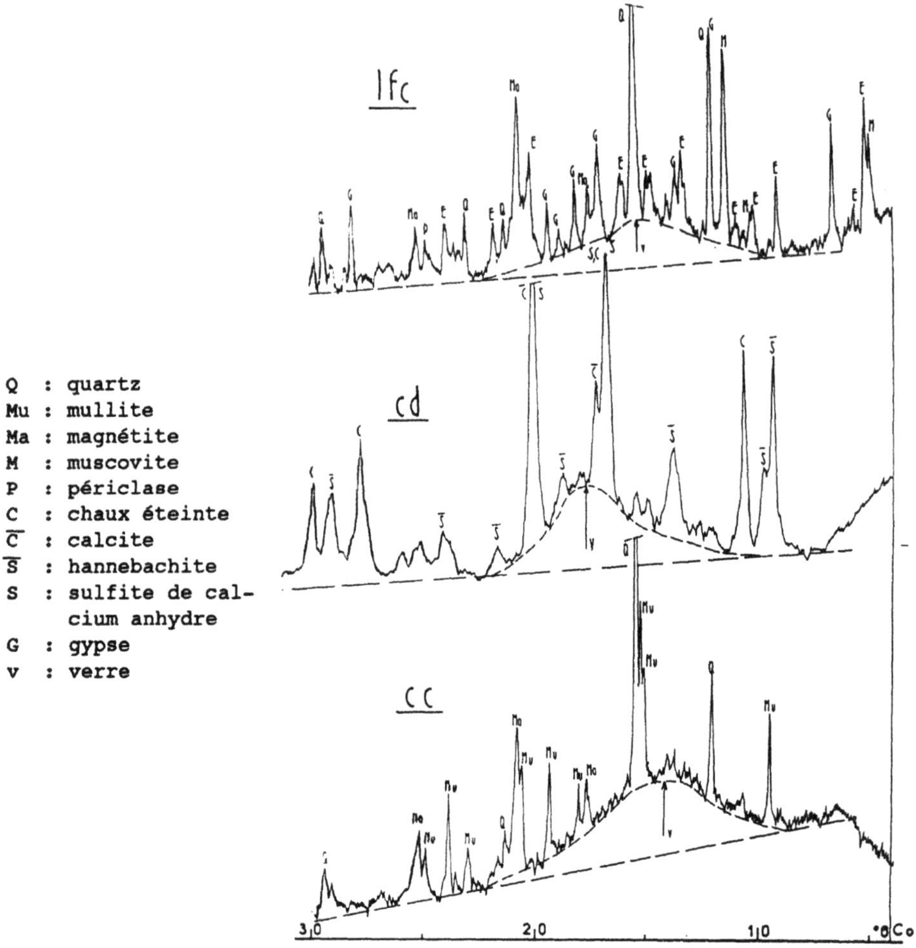

Fig.2. Diffractogrammes de rayons X des cendres CC, CD, LFC

Chaque cendre comporte une phase vitreuse et une phase polycristalline. Les minéraux essentiels sont les suivants :

* pour la cendre CC : quartz, mullite et magnétite.
* pour la cendre CD : chaux éteinte, hannebachite (ou sulfite de calcium hémi-hydraté) et sulfite de calcium anhydre. En plus faibles quantités, on note la présence de calcite et d'anhydrite. Magnétite et hématite sont aussi détectées à des concentrations nettement moindres.
* pour la cendre LFC : quartz, magnétite, muscovite, périclase, ettringite et gypse. La présence de ces deux derniers minéraux est imputable à une hydratation en cours de stockage.

4.2 Modification du comportement rhéologique de pâtes de ciment et d'eau après incorporation de cendres volantes

4.21 Consistance des pâtes
La consistance des pâtes de ciment CPA 55 pur et de pâtes mixtes ciment (70 % en volume) + cendre volante (30 % en vol.) a été mesurée à l'appareil de Vicat. Les valeurs des taux de cisaillement pour un rapport E/S croissant ont été reportées sur la figure 3 pour les différents mélanges.

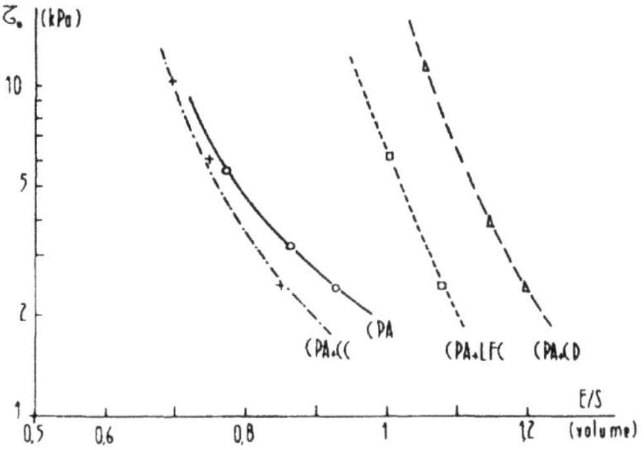

Fig.3. Consistance des pâtes CPA et CPA + CV

L'examen de cette figure permet de noter que l'introduction de cendre CC a une légère activité fluidifiante ; par contre,la substitution d'une partie du ciment soit par la cendre LFC, soit par la cendre CD accroît la demande en eau.

4.22 Durcissement des pâtes
La vitesse de prise de la pâte pure de CPA 55 ou des pâtes ciment (70 % en volume et cendre volante (30 %) a été mesurée à l'aiguille de Vicat. Les quantités d'eau mises en oeuvre pour gâcher ces pâtes sont telles que l'on obtient une concentration volumique en solide à peu près constante et égale à 0,52 ; cela nous conduit pour les cendres CC et LFC à un rapport E/S en volume de 0,93 et 1,0. Seule la cendre CD a nécessité une quantité d'eau plus importante (E/S = 1,21) pour une mise en place correcte, soit une concentration volumique égale à 0,45. Les résultats de cet essai sont reportés figure 4.

L'examen de ces courbes permet de mettre en évidence :

* un comportement similaire de la pâte pure de CPA et de la pâte mixte CPA + LFC,

* le début de prise des pâtes CPA + CC et CPA + CD est retardé respectivement 3 h 20 et 4 h 30 ; par contre, le temps de prise est comparable.

4.3 Influence des cendres volantes sur l'auréole de transition pâtes-granulats
Nous avons modélisé le béton sous la forme d'une éprouvette mixte : pâte coulée sur un granulat.

4.31 Matériaux
a) Pâte pure
Nous avons travaillé sur des pâtes de ciment CPA 55 pur gâchées, au rapport pondéral eau/ciment égal à 0,30.
b) Pâtes mixtes
Ce sont les mêmes pâtes que celles utilisées pour les mesures à la sonde de Vicat.

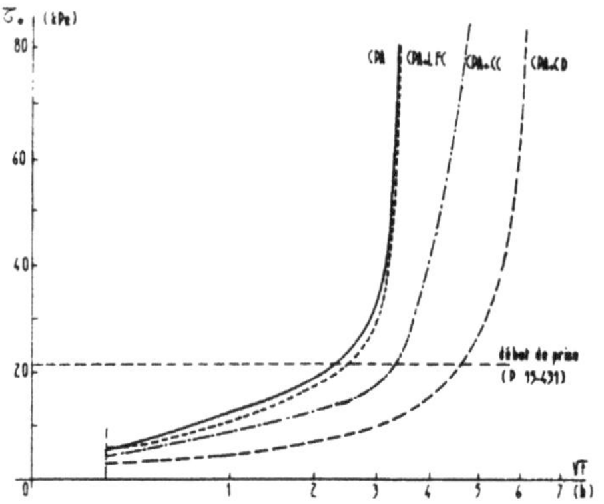

Fig.4. Evolution du seuil de cisaillement des pâtes CPA ou CPA + CV en fonction du temps

c) Granulat
Il est constitué par une carotte de marbre de Carrare dont une face a été rectifiée puis soigneusement polie.

4.32 Etude de l'auréole de transition de la pâte pure de CPA
Notre étude a porté sur les points suivants : morphologie de la surface de rupture, orientation de la portlandite, épaisseur de l'auréole et nature des hydrates présents dans l'auréole à l'exclusion de la phase C-S-H difficilement observable par diffractomètrie de rayons X. Les essais ont été faits à 1, 7 et 28 jours.

Les diffractogrammes de rayons X effectués sur les surfaces polies du granulat font apparaître la présence de portlandite dont la quantité diminue avec l'âge : l'intensité des pics caractéristiques de ce minéral décroit avec le temps.

L'orientation des cristaux de portlandite, dans le plan de séparation augmente avec le temps (figure 5) ; l'indice d'orientation (défini précédement) est égal à 5,1 à 1 jour, puis atteint 5,8 à 7 jours et à 6,6 à 28 jours. Ces résultats sont comparables à ceux trouvés précédemment.(Grandet et Ollivier)

L'épaisseur de l'auréole demeure constante dans le temps et sa valeur est de l'ordre de $\delta_0 = 40\ \mu m$.

4.33 Etude de l'auréole de transition de la pâte pure de CPA + cendre silico-alumineuse CC
Voici l'évolution que présentent avec le temps, les diverses caractéristiques de l'auréole de transition granulat / pâte CPA + cendre silico-alumineuse.

A la surface du granulat apparaissent des dépôts plus étendus d'une épaisseur moyenne de 2 μm : seule la portlandite y est décelée.

Les valeurs de l'indice d'orientation sont plus basses que dans les pâtes de CPA pur et varient de 1,7 à 2,5 suivant l'age (figure 6).

Les effets de l'orientation sont faibles à l'interface, mais semblent se faire ressentir loin dans la pâte : la limite δ_0 est tout de même difficile à préciser et se situe dans la fourchette 60 - 100 μm.

A la portlandite est associée de l'ettringite, toujours bien cristallisée.

Dès 1 jour, il apparaît, en très faible quantité, un aluminate de calcium C_4 AHx.

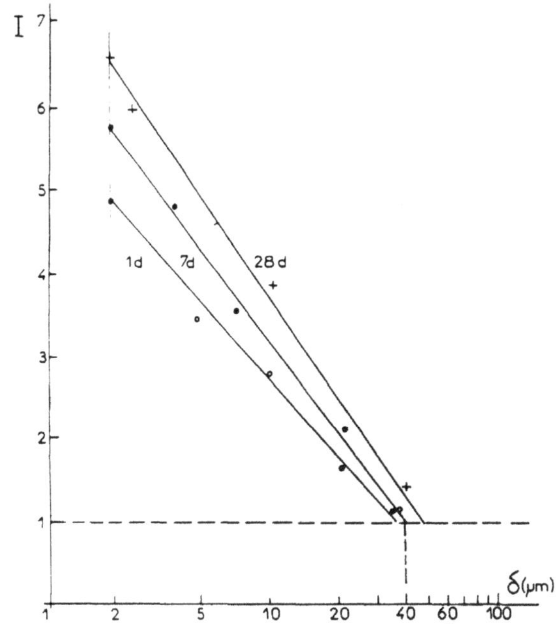

Fig.5. Orientation de la portlandite
dans une pâte de CPA 55 au contact d'un granulat calcaire

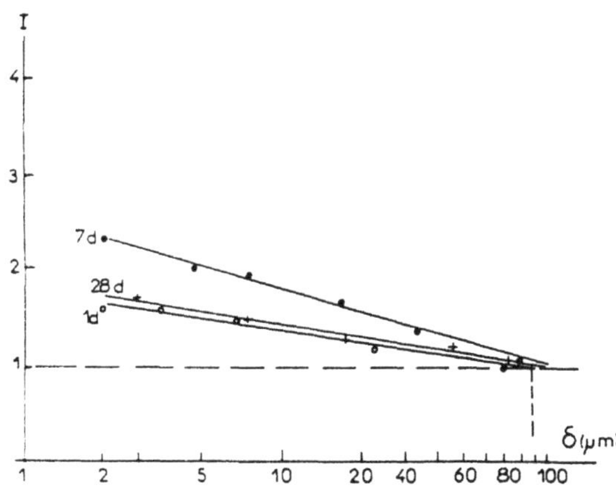

Fig.6. Orientation de la portlandite
dans une pâte de CPA 55 + cendre silico-alumineuse CC

4.34 Etude de l'auréole de transition de la pâte pure de CPA + cendre sulfitique CD.

L'influence de cette cendre est comparable à celle de la cendre CC.

Les dépôts restant adhérents au granulat sont constitués essentiellement de portlandite et leur épaisseur moyenne est de l'ordre de 2 à 3 μm. Par ailleurs (figure 7), l'indice d'orientation se maintient à des valeurs basses, voisines de 2.

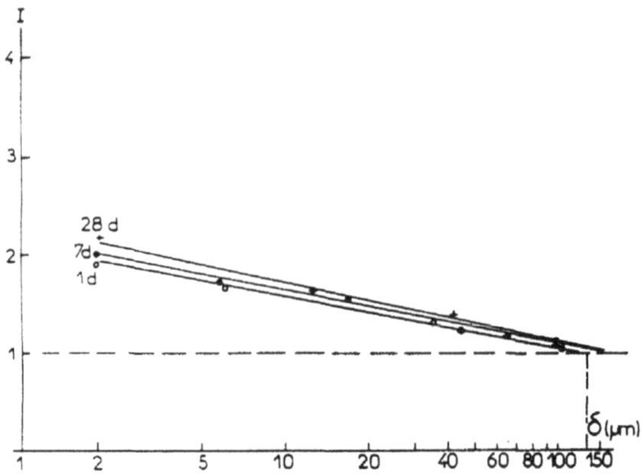

Fig.7. Orientation de la portlandite
dans une pâte de CPA 55 + cendre sulfitique CD

Par contre, l'effet d'orientation est perçu jusqu'à une distance évaluée à 100 - 150 μm. Comme dans les cas précédents, à la portlandite est associée de l'ettringite bien cristallisée. Dès 7 jours est présent de l'aluminate de calcium. Les sulfites demeurent stables avec le temps.

4.35 Etude de l'auréole de transition de la pâte de CPA + cendre LFC

Les différentes caractéristiques que présente l'auréole de transition granulat/pâte CPA + cendre LFC sont les suivantes :

* les dépots adhérents aux granulats diminuent avec le temps et sont constitués essentiellement de portlandite.

* les valeurs de l'indice d'orientation sont plus basses que dans le cas du CPA mais augmentent avec le temps I_r = 2,5 à 1 jour et 4 à 7 jours (figure 8).

* l'épaisseur de l'auréole peut être évaluée à 80 - 100 μm.

A coté de la portlandite, on note des quantités d'ettringite bien cristallisée supérieures à celle que forme une pâte de CPA pur : ceci dénote l'apport des sulfates de la cendre en plus de l'ettringite déjà existante dans la cendre elle-même. Tout le gypse présent dans l'auréole n'a pas été consommé à 7 jours.

5 Discussion des résultats

5.1 Indice d'orientation de la portlandite.

Notre discussion repose sur les explications admises jusqu'à présent, à savoir qu'un indice d'orientation élevé signifie que les cristaux de portlandite sont majoritairement parallèles à l'interface. Ceci se réalise lorsque :

* la surface du granulat, au moment de la formation de Ca $(OH)_2$, est parfaitement plane ;

* la quantité d'ions Ca^{2+} à l'interface est suffisamment élevée pour que se déroule le processus saturation-précipitation-cristallisation.

. pâte CPA 55 R pur.

La pâte est gâchée à un rapport E/C faible, égal à 0,30. Les études d'OLLIVIER ont montré que l'aluminate tricalcique (C_3A) s'hydrate en formant (au niveau de l'auréole) de l'ettringite en quantité importante : celle-ci freine alors la diffusion des ions Al^{3+}, ralentissant ainsi l'hydratation du C_3 A et donc la formation d'aluminate de calcium hydraté à l'interface.

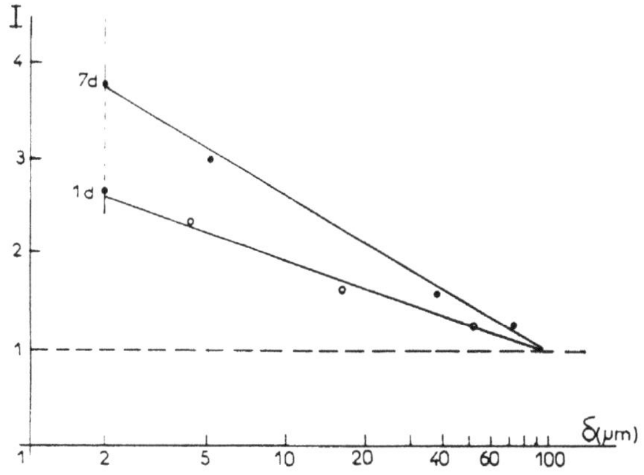

Fig.8. Orientation de la portlandite dans une pâte CPA + cendre LFC

La formation des cristaux de portlandite se fait sur une surface de granulat polie : les deux conditions énoncées ci-dessus sont alors réunies et nous obtenons un indice d'orientation élevé.

C'est effectivement ce que nous avons observé sur les éprouvettes mixtes de pâte pure où l'indice d'orientation est important. D'autre part, le C_3 A s'hydrate plus tardivement au contact du granulat de marbre pour former des aluminates qui attaquent le marbre ; nous assistons alors à la formation de carbo-aluminate. La réaction étant ralentie par la barrière d'ettringite, cela explique que ces carbo-aluminates sont formés en faible quantité à 28 jours.

. <u>Cas de la cendre silico-alumineuse CC.</u>

L'indice d'orientation de la portlandite est beaucoup plus faible que dans le cas de la pâte de CPA pur. Au vu des résultats précédents, nous pouvons considérer qu'au voisinage du granulat nous avons un rapport E/C plus élevé. Ceci entraîne la formation d'une barrière d'ettringite plus lâche ; les ions Al $^{3+}$ vont pouvoir migrer plus facilement et de l'aluminate de calcium hydraté C_4 AH_x se forme dès 1 jour au niveau du granulat. La portlandite se dépose alors sur une surface qui n'est plus parfaitement polie et son indice d'orientation sera alors plus faible.

. <u>Cas de la cendre sulfitique CD</u>

Cette cendre entraîne des modifications de l'indice d'orientation comparables à celles induites par la cendre silico-alumineuse.

Toutefois quelques modifications sont à noter par rapport à la cendre précédente du fait du caractère hydraulique de cette cendre. En effet la quantité d'aluminate tétracalcique hydraté est plus importante que dans le cas précédent : nous avons vu précédemment que lors de l'hydratation de la cendre sulfitique se forme de l'aluminate qui va venir s'ajouter à celui formé lors de l'hydratation du ciment. La pâte étant gâchée à un rapport eau/liant élevé (E/L = 0,42), de la même façon que précédemment, la barrière d'ettringite est assez lâche pour que les ions Al^{3+} puissent se déplacer. Cette formation d'aluminate modifie le faciès de la surface du granulat et empêche une orientation importante de la portlandite.

. <u>Cas de la cendre LFC.</u>

Les sulfates présents dans cette cendre jouent un rôle important dans la formation de l'auréole ; à 7 jours le gypse est encore présent et l'ettringite plus abondante. Il apparaît donc très rapidement une cristallisation de trisulfoaluminate qui bloque le C_3A

et prévient toute formation d'aluminate. La surface du granulat n'étant pas alors attaquée par des aluminates, comme pour les deux autres cendres, on conçoit qu'elle oriente de façon plus marquée la portlandite qui s'y dépose: en d'autres termes des trois cendres examinées, c'est celle qui modifie le moins la cristallisation de la portlandite dans l'auréole.

5.2 Epaisseur de l'auréole

La substitution d'une partie du ciment par un égal volume de cendre volante a dans tous les cas pour conséquence d'accroître l'épaisseur de l'auréole de transition. Toutefois cet accroissement n'est pas très significatif dans la mesure où la pente des droites représentant l'évolution de l'indice d'orientation en fonction de la distance au granulat est faible et où l'imprécision est donc plus grande.

6 Conclusion

Il est apparu que les échantillons de cendres volantes tels que nous les avons reçus ne modifient pas de façon différentielle l'auréole de transition qui apparaît dans les bétons hydrauliques puisque dans tous les cas on observe une diminution de l'indice d'orientation de la portlandite associée à un accroissement pas très significatif de l'épaisseur de cette auréole.

Pour arriver toutefois à une conclusion parfaitement incontestable que l'origine des cendres influe peu sur la structuration de l'auréole de transition des bétons hydrauliques qui en renferment, il faudra compléter ces premiers résultats par d'autres portant sur des prélévements bien représentatifs des fraîches productions industrielles, à savoir : cendre n'ayant pas subi de dépoussiérage préalable (comme il semble que ce soit le cas pour la cendre CD), ou d'humidification ayant conduit à une hydratation partielle (LFC).

Ces deux opérations ont, en effet, une répercussion sur la granularité du produit et, par voie de conséquence, la structuration de l'auréole en est très probablement affectée.

bibliographie :

Gokhale, A.J. et Burnet, G. (1988) By-products sulfur from the stabilization of coal solid wastes. **MRS**, vol.136, p55.

Grandet, J. et Ollivier, J.P. (1980) Orientation des hydrates au contact des granulats. **7e Congrès Int. Chim. Cim. Paris**, vol.III.

Grandet, J. et Ollivier, J.P. (1980) Nouvelle méthode d'étude des interfaces ciment-granulats. **7e Congrès Int. Chim. Cim. Paris**, vol.III.

Husson, B.(1991) Utilisation des cendres volantes à sulfite de calcium dans les bétons hydrauliques. Thèse de Doctorat.

Odler, I. et Zysk, K.H. (1987) Characterization of products of primary flue gas desulphurization. **MRS**, vol.113, p179.

6 EFFECTS OF POZZOLANIC AND NON-REACTIVE MICROFILLERS ON THE TRANSITION ZONE IN HIGH STRENGTH CONCRETES

A. GOLDMAN and A. BENTUR
Technion - ITT, Haifa, Israel

Abstract
This paper discusses the influence of the pozzolanic and the microfiller effects, provided by silica fume, on the microstructure and chemical composition of the transition zone in concretes of low water-cement ratio. In order to characterize these effects separately, carbon black, which has similar size and shape as silica fume, was introduced as a non-reactive microfiller. SEM observations, porosity and size measurements of the transition zone were carried out. The results demonstrated that carbon black was as effective in filling the spaces in the transition zone, as silica fume. At the early age of one day each of the two materials caused significant densification inside the transition zone. The microstructure had similar appearance in both cases, thus indicating microfiller effect which is independent of the reactivity of the filler. In time, up to two months, silica fume developed some additional densification, related to the pozzolanic effect. The width of the transition zone, evaluated at one day, did not depend on such factors as water-cement ratio, aggregate size and orientation.
Keywords: Transition Zone, Microfillers, Silica Fume, Microstructure.

1 Introduction

Previous studies of silica fume concretes led to the conclusion that the strengthening mechanism of silica fume in concretes is associated with densening of the transition zone, resulting in a strong aggregate-matrix interface (Bache (1981), Bentur et al (1988), Goldman and Bentur (1989), Sellevold (1987). This strong interface induced a "true" composite behavior in the silica fume concrete where the aggregate became an active reinforcing inclusion, which could account for the observation that in silica fume concrete the compressive strength of the concrete is higher than that of its paste matrix. This is just the opposite of the behavior of normal and high strength concretes without silica fume where the interface is porous and actually becomes a weak link (Hadley (1972), Monteiro (1985), Swamy (1987), Zimbelman (1985), leading to concrete strength which is smaller than that of its paste matrix.

The densening influence of the silica fume, on the transition zone, was attributed to pozzolanic and filler effects. The filler effect is assumed to be the consequence of the fine and spherical shape of the silica fume particles. As a result, silica fume particles eliminate bleeding and pack efficiently at the vicinity of the aggregate surface, thus preventing the formation of water-filled space around the aggregate, which is a major cause for formation of a porous transition zone in concretes without silica fume.

In order to resolve and separate between influences of filler effects and pozzolanic effects an extensive study was carried out, in which the performance of paste and

Interfaces in Cementitious Composites. Edited by J. C. Maso. © RILEM.
Published by E & FN Spon, 2 - 6 Boundary Row, London SE1 8HN. ISBN 0 419 18230 6.

concrete systems was evaluated, replacing silica fume with carbon black particles. Carbon black particles are of similar shape and size as silica fume, but they can be considered inert. It was shown in this study that carbon black exerted a similar influence on strength characteristics in concrete as silica fume: The strength of the concrete with carbon black was higher than that of its paste matrix, which is similar to the trend with silica fume systems, but different than the trend in cementitious systems without fillers (Table 1).

Table 1

28 days compressive strength of concretes with the same water/binder ratio of 0.40. The binder is portland cement (reference) or portland cement with 18% replacement of non-reactive filler (carbon black) and pozzolanic filler (silica fume).

	28 days compressive strength, MPa		
	Reference	Carbon Black	Silica Fume
Concrete	70.0	81.8	91.1
Paste Matrix	86.4	67.8	86.8

The present paper describes the results obtained in the study of the transition zone in such systems, to evaluate if, indeed, the carbon black densifies this zone, as is the case with silica fume, in spite of the fact that carbon black is non-reactive.

2 Experimental

The experimental work included the study of the properties of concretes and pastes extracted from the fresh concrete. To make extraction possible, a mix was developed with two fractions of aggregates, which consisted of uniformly sized 5 mm and 19 mm particles of crushed dolomite. Such a mix was labelled "model concrete". This way of obtaining the paste, directly from the fresh concrete, was preferred, since a corresponding paste, mixed independently of the concrete, may not accurately represent a paste matrix of that concrete, particularly when considering the effect of aggregates in the mixer in producing higher shear rates and more efficient dispersion of fine particles such as silica fume and carbon black.

The paste and concrete specimens were fractured for SEM observations. Polished impregnated surfaces were prepared for quantitative image analysis, with optical microscope. The technique used was based on QUANTIMET 970, an optical microscope equipped with a programmable image analyser, produced by Cambridge Instruments, UK. The image of a specimen consisted of 640x450 pixels, and the magnification was x160, which provided the resolution of 0.43 μm. Thus the system identified pores of approximately 0.5 μm and larger.

The specimens for the analysis were 70x33x14 mm slices, obtained by sawing the model concrete cubes. The top and the bottom of each slice were marked, indicating the direction of casting of the fresh concrete into the cube mold.

Scanning with the optical microscope and the quantitative imaging started at the aggregate edge, and was continued into the bulk paste, far beyond the transition zone. The measurements were taken at the aggregate top and bottom, in accordance with its orientation.

The properties of the materials used for making the model concretes were the following:

- Ordinary portland cement, corresponding to ASTM type I.
- Silica fume, a raw byproduct supplied by SKW, Canada, containing 92.8% SIO_2. Its specific surface was 18.3 m^2/g (N_2 absorption), and the bulk density was 2110 kg/m^3.
- Carbon black was a product of Cabot U.K. Its specific surface area was 28.4 m^2/g (N_2 adsorption) and the bulk density was 1740 kg/m^3.

These properties of carbon black, as well as its spherical particle shape, are similar to those of silica fume.

The model concretes were prepared with Naftalene sulphonate based superplasticizer. The composition of the concretes was as follows:

19 mm aggregate	-	1106 kg/m^3
5 mm aggregate	-	548 kg/m^3
Binder (cement, or cement with 18% fillers)	-	560 kg/m^3
Water	-	186 kg/m^3
Superplasticizer	-	9 kg/m^3

3 Results

3.1 Densification of the transition zone by silica fume and carbon black

The typical morphology of the transition zone at the age of one day in concrete without microfillers included large, continuous voids situated around coarse aggregate. This was observed in a variety of specimens, taken of this concrete. Similar observations were reported earlier by Bentur, Goldman and Cohen (1980), taken near a fine aggregate. The average width of the transition zone near coarse aggregate could be estimated as 10-20 μm, sililar to that reported in several other works (5-50 μm), also measured near fine aggregate.

The presence of carbon black resulted in considerable reduction in the porosity at the interface, which could clearly be observed at one day (Fig. 1). Bentur, Goldman and Cohen (1980) reported much the same results, obtained with silica fume. In both cases large crystals of calcium hydroxide could not be found. Since silica fume and carbon black had similar physical properties, like spherical particle shape, density and specific surface, the similarity of the two microstructures is not surprising, if it is assumed that filler effect plays a major role in the densification of the transition zone.

Thus, according to these results, a microfiller can be effective in densification of the transition zone, whether it is silica fume, or carbon black, or perhaps another material, identified as a microfiller of similar physical properties.

At the age of one day, the aggregate was already able to make a significant contribution to the increase in concrete strength (i.e. concrete strength higher than the paste matrix) as reported by Goldman (1987). The increase took place in the silica fume concrete, as well as in that containing carbon black, thus suggesting that the densification of the transition zone in both fillers, led to improved mechanical performance.

The observations presented in Fig. 2 for silica fume, and in Fig. 3 for carbon black, support this explanation. The massive concentrations of the microfiller appear as just an assemblage of tiny particles (Figs. 2a, 3a), at lower magnification, while higher ones revealed linkages, formed between the particles (Figs. 2b, 3b), part of them intercepted by the hydration products. Two types of interparticle linkages may therefore exist:

- Direct, possibly of chemical or physical origin.
- Indirect, created through the hydration products, formed on the particle surfaces, between adjacent particles.

The direct links can appear as demonstrated in Fig. 4, where the "chain" of carbon black particles is connecting between two cement grains. This SEM observation has been obtained on a dry mix of cement with carbon black. The latter suggestion refers to the known hypothesis on "nucleation sites", proposed by Grutzeck et al (1983). In such case, the neighboring particles could be connected by C–S–H gel, precipitated on their surfaces, or by other hydration products.

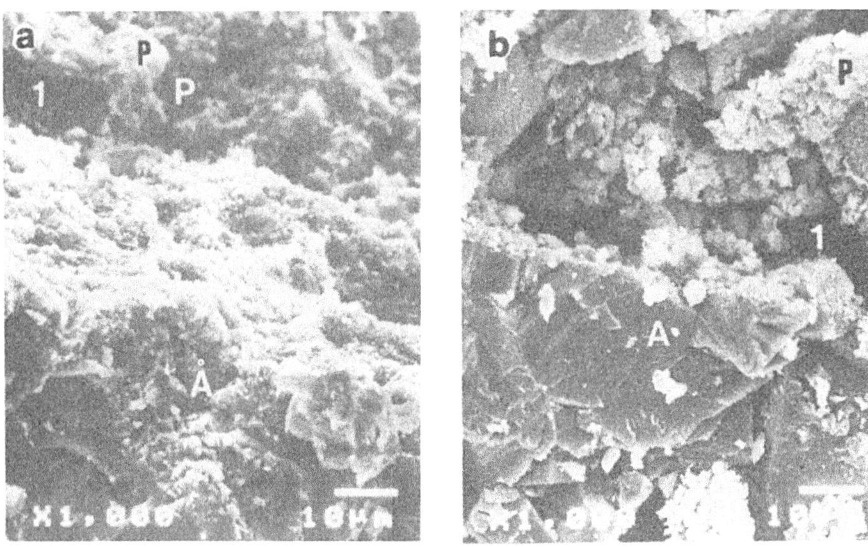

Fig. 1. Typical morphology of the transition zone in concrete with carbon black filler (a) and without filler (b) at one day (W/B=0.4):
A – aggregate; P – paste matrix; 1 – voids; 2 – carbon black.

Further investigation of the complexity of processes generated within the transition zone has not been included in this study. However, the results represented here showed that a microfiller alone has a considerable potential. It can provide a densification to a high degree at a very early stage, and has nothing to do with pozzolanic reaction. This pozzolanic activity provides silica fume with the advantage over carbon black, but this can only be realized at later stages.

The SEM observations on carbon black concrete at the age of 1–2 months, indicated an additional densification of the transition zone. Such observations may also support Grutzeck's hypothesis on the nucleation sites provided by microfiller.

3.2 Porosity and width of the transition zone

The scanning of the zones near the aggregate surface, in polished model concrete sections under optical microscope showed, as expected, differences depending on orientation: lower porosity above the aggregate and higher beneath it. The mode of

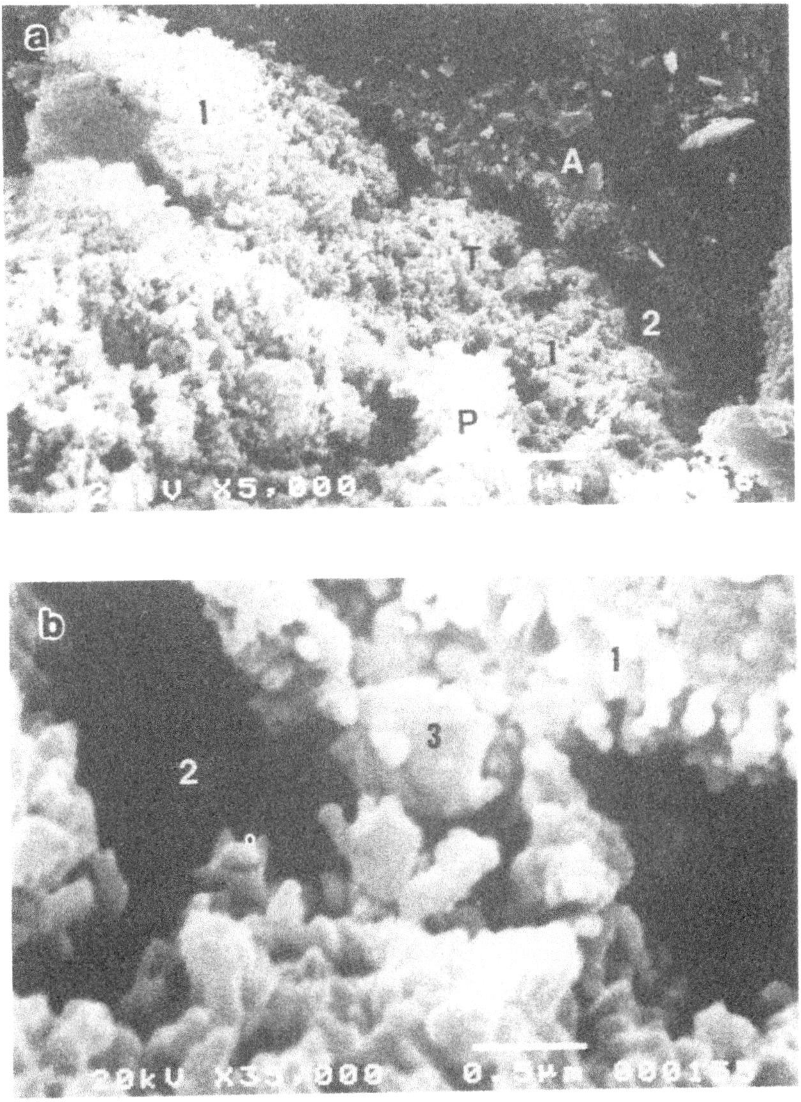

Fig. 2. The transition zone (T) between aggregate (A) and paste matrix (P) at one day, filled by silica fume (W/B=0.4):
a – silica fume inside the transition zone, 5000 X
b – microstructure in the same area, 35000 X
1 – silica fume; 2 – voids; 3 – hydration products.

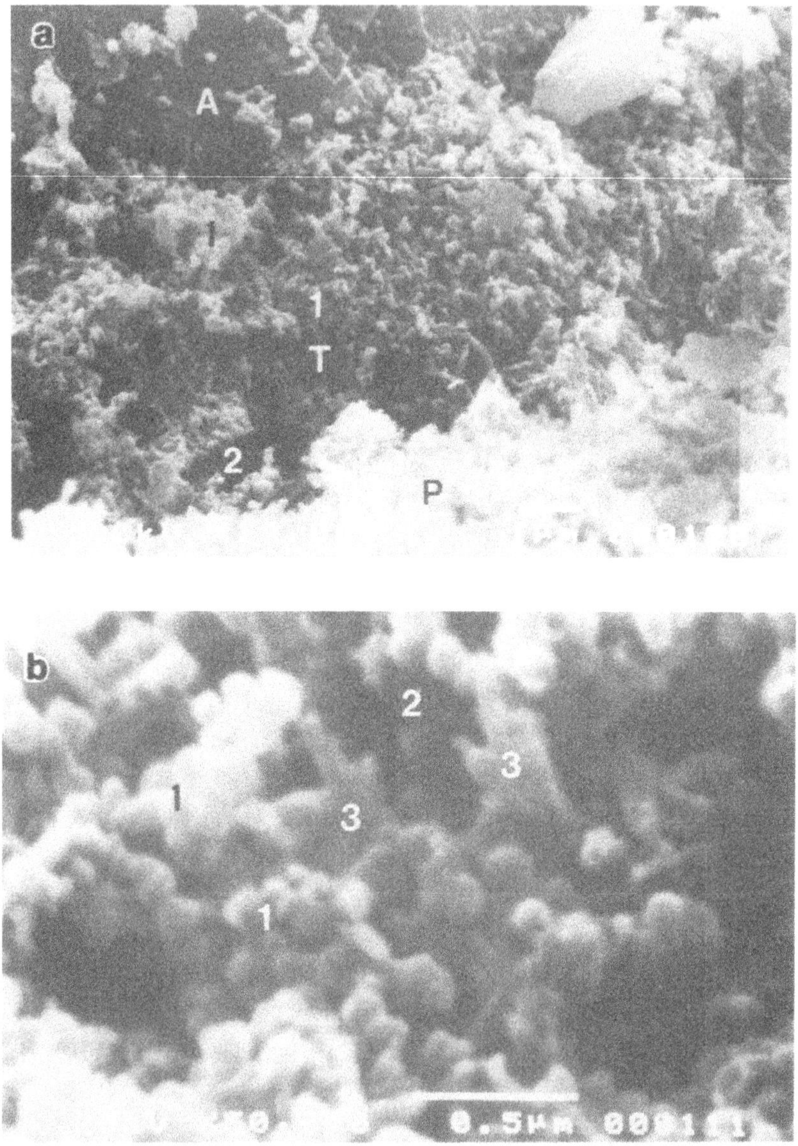

'ig. 3. The transition zone (T) between aggregate (A) and paste matrix (P) at one day, filled by carbon black (W/B=0.4):
a – carbon black inside the transition zone, 10000 X
b – microstructure in the same area, 50000 X
1 – carbon black; 2 – voids; 3 – hydration products.

Fig. 4. Clumps of carbon black (CB) between cement grains (PC).

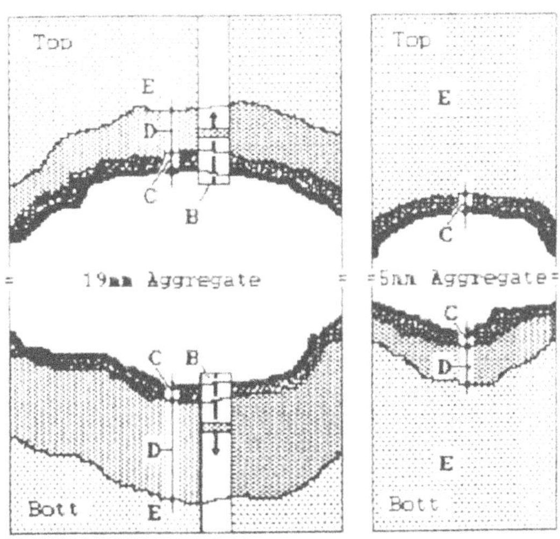

Fig. 5. Schematic description of the aggregate vicinity on polished concrete section. Zones of the scanning area at the aggregate top and bottom are shown.

scanning is shown in Fig. 5. In concrete without microfillers the transition zone could be observed, as a gap between the aggregate and the bulk of the paste matrix, consisting of continuous voids. In the presence of silica fume the void content was much lower at one day, but the transition zone of similar width still appeared. As for the carbon black concrete, similar reduction of the voids was observed.

Scanning of the paste matrix beyond the transition zone indicated that its bulk was not entirely uniform. Around the transition zone, the porosity was definitely higher than in deeper areas of the bulk. This zone of increased porosity was much larger than the transition zone itself. In all cases observed, it was wider at the bottom of aggregates, than at the top.

Taking this situation into account, the scanning areas were divided into zones B (aggregate edge), C (transition zone), D (the more porous part of the paste matrix), and E (the bulk), shown in Fig. 5.

Since the present paper deals with the transition zone (C), only relevant results are included. Another publication will represent the results, related to the bulk morphology.

Fig. 6 shows typical results, where the width and the relative porosity of the transition zone were measured near a 5 mm aggregate.

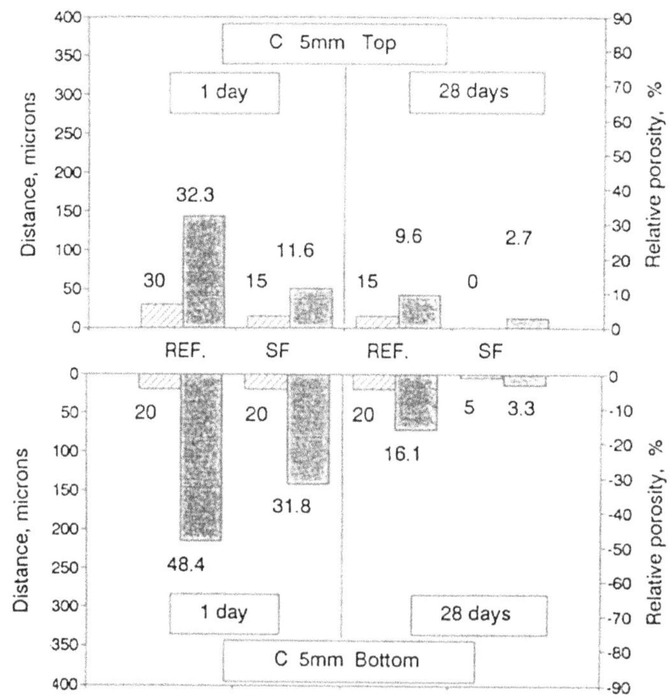

Fig. 6. The width and relative porosity in the transition zone near a 5 mm aggregate grain, at 1 and 28 days (W/B=0.2): – porosity. – width.

At one day, the porosity was considerably reduced in the presence of silica fume; however, the width of the transition zone remained the same as in the reference concrete. At 28 days, additional reduction of porosity took place, with and without silica fume. As can be seen, silica fume was effective at one day (due to the microfiller effect), as well as later. Qualitatively, the same results were obtained for a 19 mm aggregate, but the porosity values were higher.

For a 5 mm aggregate, the width of the transition zone was 15–30 μm at both ages (10–40 μm for a 19 mm aggregate), while it could not be identified at 28 days, in the presence of silica fume. The latter results is in agreement with conclusions derived in previous works, that silica fume can provide the transition zone with the microstructures which is as dense as the bulk of the paste.

The results obtained by the quantitative analysis are in good agreement with those reported from SEM observations. The microfiller effect could be identified by both techniques. Similar results were obtained for the width of the transition zone.

4 Summary

Carbon black, the non-reactive microfiller, was highly effective in densifying the transition zone in concrete. Silica fume was effective to a similar degree at the early stage, when its pozzolanic activity was not yet of significance. Both fillers enhanced the strength of the concrete, to levels higher than the paste matrix.

It is suggested that this similarity implies that the modification of the concrete strength and microstructure by the presence of silica fume, is due to a large extent to filler effect, rather than pozzolanic reactivity.

5 References

Bache, H.H. (1981) Densified Cement/Ultra-Fine Particle Based Materials. 2nd Int. Conf. on Superplasticizers in Concrete, Ottawa.

Bentur, A., Goldman, A. and Cohen, M.D. (1988) The contribution of the transition zone to the strength of high quality silica fume concrete. MRS Pros., Vol. 114, pp. 97-103.

Goldman, A. (1987) Properties of Concretes and Pastes with Microsilica, M.Sc. Thesis.

Goldman, A. and Bentur, A. (1989) Bond Effects in High Strength Silica Fume Concretes. ACI Materials J., Vol. 86, pp. 440-447.

Grutzeck, M.W., Atkinson, S., Roy, D.M. (1983) Mechanism of Hydration of Condensed Silica Fume in Calcium Hydroxide Solutions. ACI SP 79-33, pp. 643-664.

Handley, D.W. (1972) The Nature of the Paste-Aggregate Interface. Ph.D. Thesis, Purdue University.

Monteiro, P.J.M. (1985) Microstructure of Concrete and its Influence on the Mechanical Properties. Ph.D. Thesis, Univ. of California.

Sellevold, E.J. (1987) The Function of Condensed Silica Fume in High Strength Concrete. Proc., Util. of High Str. Concr., Stavanger, pp. 39-49.

Swamy, R.N. (1987) The Nature of the Strength of Concrete. Proc., Util. of High Str. Concr., Stavanger.

Zimbelmann, R. (1985) A Contribution to the Problem of Cement - Aggregate Bond. Cem. and Concr. Res., Vol. 15, pp. 801-808.

PART 1.2

INTERFACIAL MICROSTRUCTURE AND PROPERTIES: DEVELOPMENT AND NATURE OF INTERFACIAL MICROSTRUCTURE

(Microstructure et propriétés de l'auréole de transition: Formation de l'auréole)

7 INFLUENCE OF MINERAL ADMIXTURES ON THE TRANSITION ZONE IN CONCRETE

U. NILSEN, P. SANDBERG and K. FOLLIARD
Department of Civil Engineering, University of California, Berkeley, USA

Abstract
The influence of several mineral admixtures on the transition zone in concrete has been studied using the Grandet-Ollivier method. By substituting cement with either condensed silica fume, ground granulated blast furnace slag, or rice husk ash it is shown to that a smaller transition zone and a lower concentration of calcium hydroxide (CH) crystals in the transition zone. None of the mineral admixtures had a significant influence on the degree of preferred orientation of CH crystals in the transition zone.
Keywords: Condensed Silica Fume, Slag, Rice Husk Ash, Calcium Hydroxide, Preferred Orientation, Concentration Intensity.

1 Introduction

The substitution of cement in concrete by mineral admixtures (MA) is gaining popularity all over the world. The use of MA is cost saving, environmentally friendly, and in many cases results in more durable and stronger concrete. It has earlier been shown that mineral admixtures have a strong influence on the transition zone in concrete [1-3]. In this study the influence of several different mineral admixtures is compared.

2 Materials and procedures

Cement pastes incorporating different mineral admixtures were cast against polished granite aggregates, which are assumed to be inert. Following the Grandet - Ollivier method [4], the effect of condensed silica fume (CSF), rice husk ash (RHA), and granulated blast furnace slag (BFS) on the transition zone was investigated. A neat portland cement paste (PC) was used for reference purpose. In Table 1 the specific surface areas of the cementitious materials obtained by nitrogen absorption are shown. Four different cementitious mixtures were used: 90% PC and 10% CSF; 70% PC and 30% BSF; 80% and 20% RHA; and 60% PC, 30% BSF, and 10% CSF.

A preferred orientation of calcium hydroxide is detected in XRD patterns by a change in relative heights of the peaks at $2\theta=18.1^{o}$ (001 crystal plane) and $2\theta=34.1^{o}$ (101 crystal plane). For each of the paste - aggregate combinations, successive abrasions and X-ray scans of the 18.1^{o} (001 crystal plane) and 34.1^{o} (101 crystal plane) 2θ peaks were performed. This process was continued until a constant ratio between the heights of the 18.1^{o} and 34.1^{o} 2θ peaks was obtained. At random

Interfaces in Cementitious Composites. Edited by J. C. Maso. © RILEM.
Published by E & FN Spon, 2 - 6 Boundary Row, London SE1 8HN. ISBN 0 419 18230 6.

Table 1. Specific surface area of cement and mineral admixtures.

Material	Surface Area (m^2 / kg)
PC	350
CSF	21900
RHA	24400
BFS	830

orientation, the ratio between the heights of the 18.1° and 34.1° 2θ peaks is close to 0.74, (0.74 is the ratio obtained with a standard powder sample of CH).

Quantitative measurements of hydration products in the transition zone cannot be made using this method. However, relative concentration gradients can be obtained with fairly good accuracy simply by comparing peak heights from different depths in the transition zone. The height of the 34.1° 2θ peak represents the relative concentration of CH at a given distance from the interface. For comparison between samples, the peak heights were measured in counts per seconds. Relative comparison between samples was possible, since all the samples were of the same size and were cast, cured and tested the same way.

3 Results and discussion

In Table 2 the calculated thickness of the transition zones at 28 days is shown. The thickness of the transition zone is calculated at the point in the cement paste where the ratio of the 18.1° and 34.1° 2θ peaks becomes constant. A summary of other experimental results is given in Fig. 1a) and 1b) and Fig. 2a) and 2b).

As described by Grandet and Ollivier [5], the aggregate-cement paste transition zone is characterized by the formation of CH crystals with a preferred orientation. This degree of preferred orientation has been proven to decrease with distance from the interface, Grandet and Ollivier [5], until the bulk cement paste is reached. At this point the ratio of the 18.1° to 34.1° 2θ peaks heights is close to 0.74. The results of this study confirm both that the CH at the interface exhibits preferred orientation with the c-axis normal to the aggregate face, and that the amount of CH decreases with distance from the interface. This is in accordance with earlier work, [1,5]. It was further observed that the CH concentration increases with time, since the formation of hydration products increases with time.

Table 2. Thickness of the transition zone at 28 days

Mix Type (w/c = 0.35)	Estimated thickness of transition zone (μm)
100% PC	37
90% PC, 10% CSF	25
70% PC, 30% slag	45
60% PC,30% slag,10% CSF	40
80% PC, 20% RHA	35

From Fig. 2a) and 2b) it can be seen that the substitution of cement by CSF reduces the amount of CH drastically, but Fig. 1a) and 1b) show that the degree of preferred orientation does not change. For all ages there is still an accumulation of CH

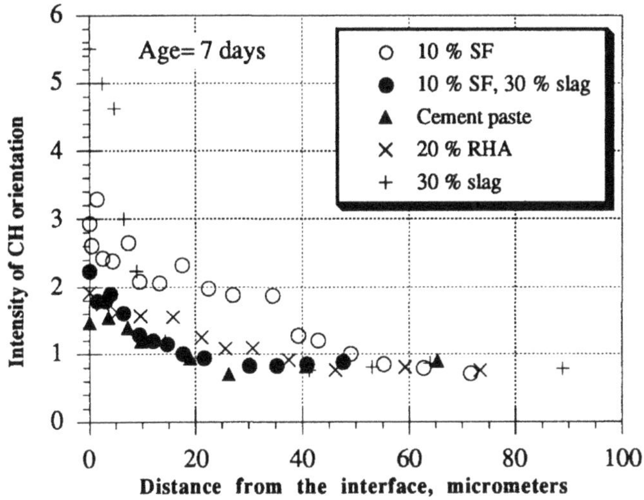

a) Age = 7 days.

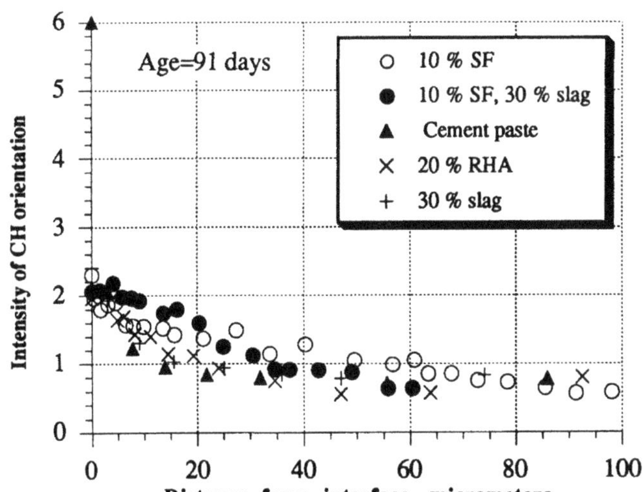

b) Age = 91 days.

Figure 1. a) -b) - Intensity of CH orientation at 7 and 91 days respectively.

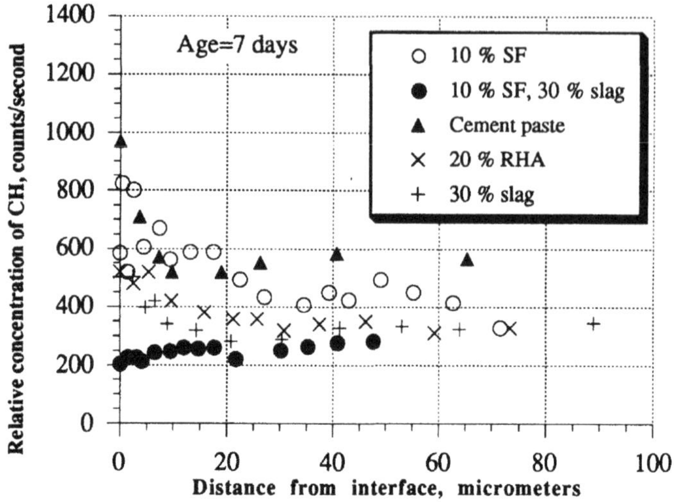

a) Age = 7 days.

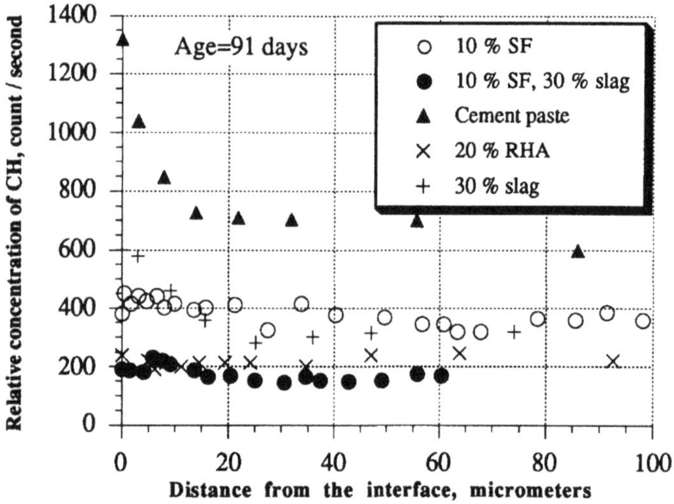

b) Age = 91 days.

Figure 2. a)- b) - Relative Concentration of CH at 7 and 91 Days Respectively.

oriented in a preferred manner, yet, in each case, the amount of CH is reduced considerably. This is attributed to the pozzolanic reaction between CH and silica forming C-S-H as the reaction product , thus resulting in a more uniform and less porous transition zone. These results agree with a study by Detwiler et al. [6], and Monteiro [1]. The addition of silica fume affects the transition zone in the following ways, Monteiro and Mehta [7]:

1. Less accumulation of free water at the interface.
2. Nucleation sites preventing formation of large CH crystals.
3. Pozzolanic reaction resulting in a general reduction of CH.

Fig. 2a) and 2b) show that the replacement of cement by RHA tends to decrease the relative concentration of CH in the transition zone to the extent that, after 91 days, the difference in the 34.1° 2ϴ peak height with increased distance from the interface is negligible. It is obvious from Fig. 1a) and 1b) that CH still forms in a preferred orientation at or near the interface. Interestingly, the degree of preferred orientation at a given distance from the interface is measurably lower for 91 days, compared to 7 days. Since RHA, is mostly comprised of amorphous silica, it readily reacts with CH to produce C-S-H and gives similar results to those of CSF substitution. There has been no published research detailing a similar study of RHA.

The addition of 30% slag does not prevent the preferred orientation of CH at or near the interface, as shown in Fig. 1a) and 1b). In fact, at 7 days, the ratio of the 18.1° peak to the 34.1° 2ϴ peak is measured to be 5.5, the highest degree of preferred orientation for any sample in this study. Even if this initial value is ignored, there still exists a preferred orientation for all ages. Though there are differences in the degree of preferred orientation near the interface, a constant value is obtained at about 25 µm from the interface for the different curing times. Thus, it can be seen that the improvement in the transition zone in this case manifests itself in the first 7 days, since later ages do not yield a smaller transition zone or decrease in CH concentration. Detwiler et al. [3] suggested that the effect of slag is mainly due to nucleation of numerous small and randomly oriented CH crystals on slag particles. The pozzolanic reaction seems only to have more of a secondary effect, since slag does not consume as much CH as compared to CSF and RHA.

The addition of 10% CSF and 30% slag has a similar effect on the transition zone, reducing the amount of CH while still exhibiting a preferred orientation of CH at or near the interface. This seems logical since the use of either 10% CSF or 30% slag alone reduces the amount of CH while having little effect on the preferred orientation.

4 Conclusions

The influence of several mineral admixtures on the transition zone in concrete has been studied. It is shown that the substitution of cement by either CSF, RHA and BFS to some extent, changes the thickness of the transition zone. The concentration of CH is greatly reduced while the introduction of mineral admixtures do not change the degree of preferred oriented CH crystals in the transition zone.

5 Acknowledgements

The authors are thankful for the advice given by Professors Mehta and Monteiro.

6 References

1. Monteiro, P. J. M., "Microstructure of Concrete and its Influence on the Mechanical Properties", Pd.D. Thesis, University of California at Berkeley, 1985.
2. Mehta, P. K. and Monteiro, P.J.M.,"Effect of Aggregate, Cement, and Mineral Admixtures on the Microstructure of the Transition Zone", Bonding in Cementitious Composites, MRS, V. 114, 1988, pp. 65-75.
3. Detwiler, R. J., Krishnan, K., Mehta, P.K., "Effect of Granulated Blast Furnace Slag on the Transition Zone in Concrete", American Concrete Institute, SP-100, Vol.1, 1987, pp. 63-72.
4. Grandet, J. and Ollivier, J. P., "Nouvelle Methode d'Etude des Interface Ciments-Granulats", 7th International Congress on the Chemistry of Cement, Vol. III, 1980, pp. VII.85-89.
5. Grandet, J. and Ollivier, J. P., "Orientation des Hydrates au Contact des Granulats", 7th International Congress on the Chemistry of Cement, Vol. II, 1980, pp. VII. 63-68.
6. Detwiler, R. J., Monteiro, P. J. M., Wenk, H.R. and Zhong, Z., "Texture of calcium hydroxide near the cement paste-aggregate interface", Cement and Concrete Research, Vol. 18, 1988, pp. 823-829.
7. Monteiro, P. J. M. and Mehta, P. K., "Improvement of the Aggregate-Cement Paste Transition Zone by Grain Refinement of Hydration Products", 8th International Congress on the Chemistry of Cement, Vol. III, 1986, pp. 433-437.

8 STUDIES OF THE INTERFACIAL BOND BETWEEN CEMENT PASTES AND EXPANDED SHALE

QI YUN and CHU QINGLI
Chongqing Institute of Architecture and Engineering,
Chongqing, China

Abstract
Expanded shale is a synthetic, porous, lightweight aggregate calcined at high temperature. The properties and characteristics of expanded shale and the concrete made with it were investigated experimentally, and the results have shown that there is a great deal of influence of shale strength and interfacial bond between the shale and cement pastes on the mechanical properties of this concrete. A new technique, called Enveloping Expanded Shale in Cement paste of low water-cement ratio, EESC, is given and some parameters, such as water-cement ratio in the enveloping layer (W_1/C_1) and the thickness proportion of the enveloping layer (C_1/C) etc, are defined in this paper. The influence of these parameters on concrete strength was investigated experimentally. Microhardness and pore size distributions of the interface layer were determined, and SEM was used successfully to observe the differences of interfacial bond. The experimental results of mechanical properties confirm that this new technique, EESC, can strengthen expanded shale and improve the interfacial bond. Thus an excellent concrete can be obtained by EESC.
Keywords: Expanded shale, Lightweight aggregate, Strength, Interface, Concrete.

1 Introduction

Expanded shale concrete has some special properties, for example, low density, good thermal conductivity and anti-seismic properties. With its applications in many fields, it is important to study the microstructure relative to the properties of the concrete. The strength of expanded shale concrete is often not satisfactory because of the porous structure and low strength of the shale aggregate. In order to get higher strength, it has been thought that strengthening the expanded shale and the interfacial bon between the shale and cement paste can improve the mechanical properties of this concrete (1,2). Using the new procedure for mixing fresh concrete given by the authors, the study was carried out through strengthening expanded shale and improving interfacial bond, thus the strength of concrete is increased.

Interfaces in Cementitious Composites. Edited by J. C. Maso. © RILEM.
Published by E & FN Spon, 2 - 6 Boundary Row, London SE1 8HN. ISBN 0 419 18230 6.

W_1 = weight of water used in enveloping layers
C_1 = weight of cement used in enveloping layer
C = weight of total cement used in the concrete
Q = weight of expanded shale used in the concrete

2 Experiments

2.1 Materials

The expanded shale used was produced in Zigong of China from raw material of a red clay. The clay was first ground and mixed with certain additives, then shaped in a machine, and finally put in a rotary kiln and calcined at high temperature.

The shape of the expanded clay generally is elliptical, as shown in Fig. 1. The grain grading of the shale, Fig. 2, is suitable. The cylinder strength is 3.3 MPa, loose density 620 kg/m^3, water absorption at one hour is 10.5%, and porosity 58.4%.

With the help of SEM we observed many link holes and blind holes in the section of the shale as shown in Fig. 3(a), and the range of pore size varied greatly from several millimetres down to molecular sizes. As a result of the porosity, it has a lower density. Fig. 3(b) shows the shale surface which is very rough, even large opening pores which have a strong influence on the water absorption of the shale because water will enter inside rapidly from the opening pores when the shale is immersed in water. By our experiments, the water absorption after one minute is 67.6% of that after one hour. So compared with dense aggregate, the expanded shale used has the properties of light weight, high porosity and high water absorption.

Portland blastfurnace cement, and river sand from Jianyang with Mx = 2.32 were used.

Fig.1 Expanded shale of Zigong

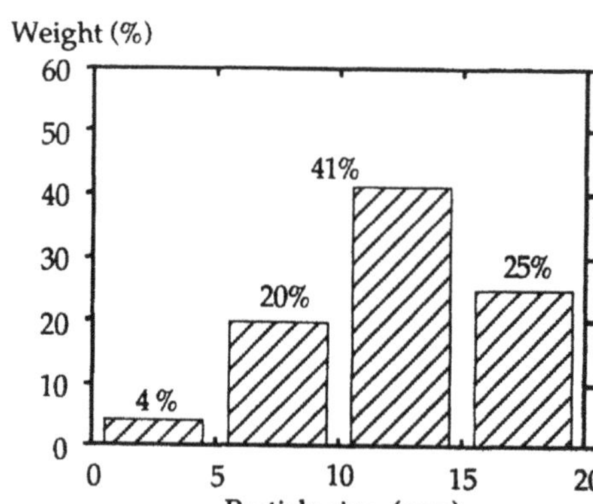

Fig. 2 Size distribution of the shale

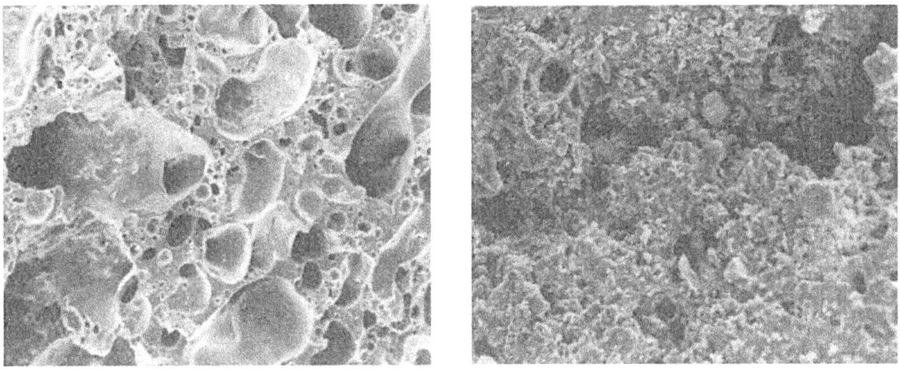

(a) Fracture section X100 (B) Surface X150

Fig.3 SEM micrographs of expanded shale

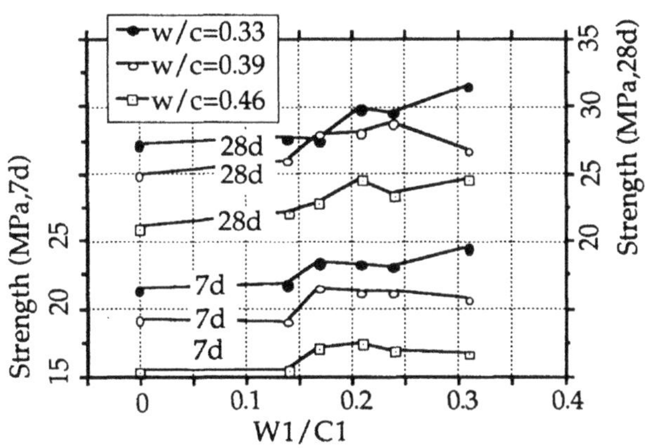

Fig.4 Effect of W1/C1 on strength of the concrete

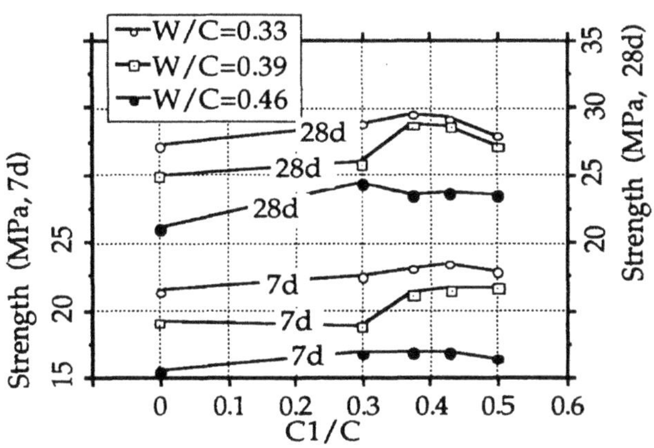

Fig.5 Effect of C1/C on strength of the concrete

2.2 Mixing procedure

First, part of the water (W_1) was sprinkled on the surface of the shale, then it was mixed with part of the cement (C_1). This procedure resulted in the shale being enveloped by a layer of cement paste with a low water-cement ratio. Some minutes later, sand and the rest of the water and cement were mixed with the enveloped shale and became fresh EESC concrete. The samples for the measurement of microhardness, SEM and pore size distribution did not contain sand, in order to get good precision.

3 Experimental results and discussion

3.1 Strength of concrete

The effect of the three parameters, W_1/C_1, C_1/C and C/Q, on the strength of the concrete is shown in Figs 4 - 6. The strength of EESC concrete is higher than that made with an ordinary mixing procedure. Higher water-cement ratio was better for

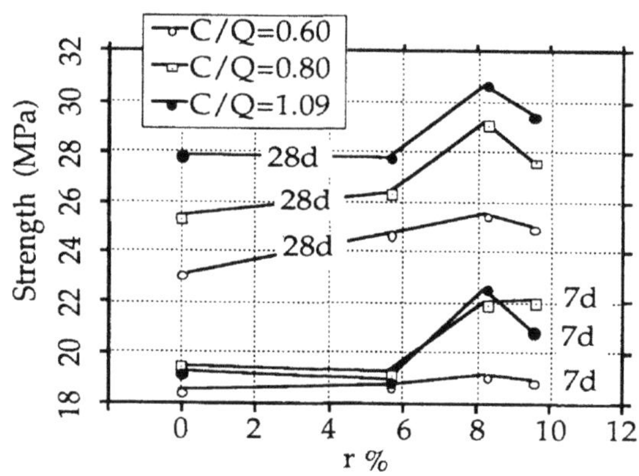

Fig.6 Effect of C/Q on strength of concrete

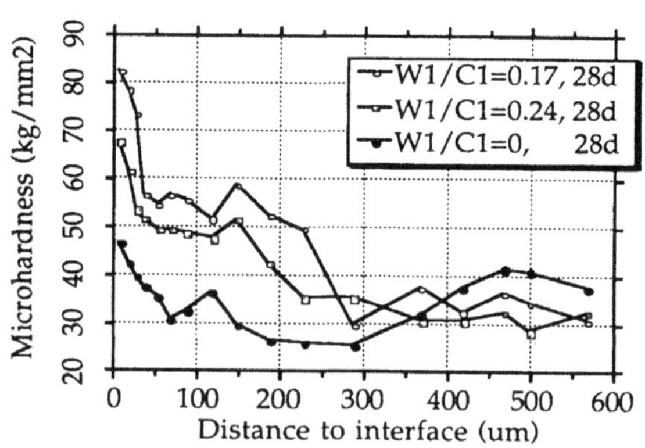

Fig.7 Microhardness near interface

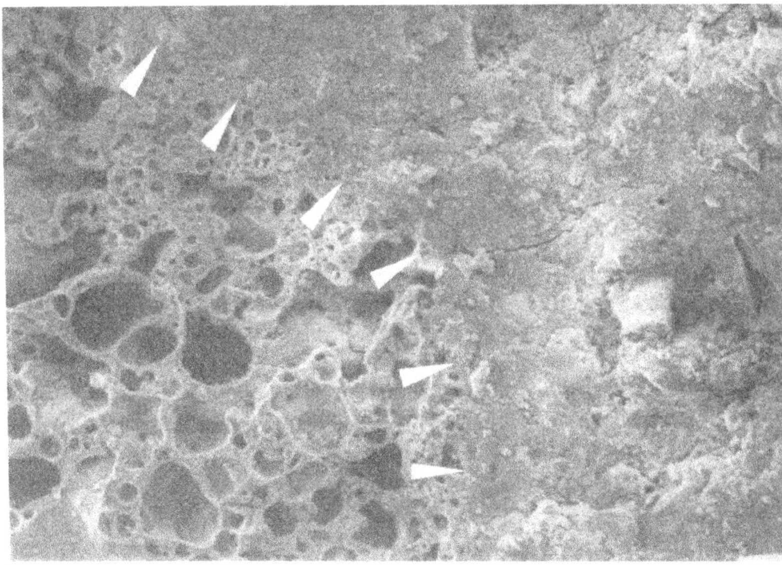

(a) EESC sample X100

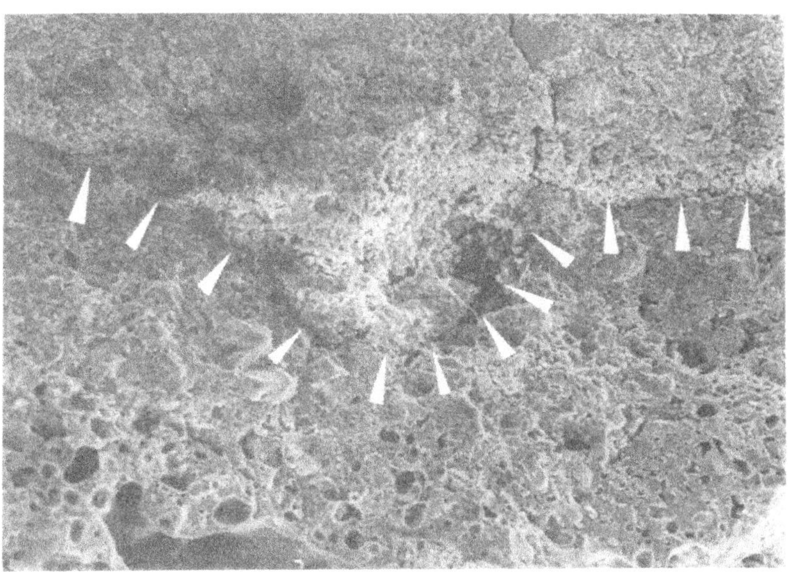

(b) Ordinary sample X75

Fig.8 SEM macrocraphs near the interface

using EESC procedures and when W_1/C_1 was 0.21 - 0.31, and C_1/C was 0.375 - 0.43, the strength development was more satisfactory.

3.2 Microhardness

Microhardness was measured by a microhardness meter attached to an Orthoplan microscope, and the results are shown in Fig. 7. Near the interface, the microhardness of EESC samples with enveloped shale was obviously higher than that of the ordinary samples with the shale non-enveloped, especially in the range of 0 - 50 micrometers near the cement-aggregate interface.

3.3 SEM observation

SEM was used to observe the interfaces between the shale, enveloped and non-enveloped, and cement paste. With EESC samples, it was found that there was a denser layer of hardened cement paste and the shale was enveloped by it. The thickness of the layer is about 200 μm and this is in keeping with what the hardness confirmed, From Figs 8 (a) and (b), we could find the structure differences near interfaces which are in front of the arrow heads. This structure difference probably results from the differences of water absorption between the shale enveloped and non-enveloped. We consider that the important reason for strengthening the concrete with the EESC procedure is the dense layer which is like a hard walnut shell.

3.4 Pore size distribution near interface

Mercury porosimetry was used to measure the pore size distribution near the interface of the two kinds of shale (enveloped and non-enveloped) and the cement paste. Although it was very difficult to separate hardened cement paste from the surface of the shales selected especially, interestingly Fig. 9 still shows that the curve of the EESC sample was situated more or less below that of the sample with non-enveloped shale, and this means more particles of cement or hydrated cement in the interface region.

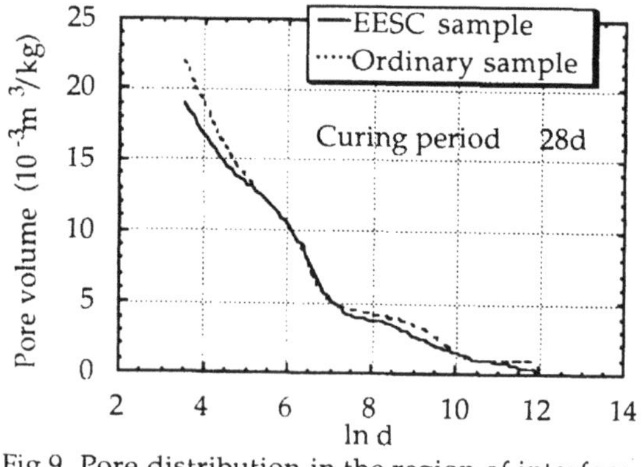

Fig.9 Pore distribution in the region of interface

4 Conclusions

Expanded shale is porous lightweight aggregate, which has high porosity and water absorption, and low strength. Compared with dense aggregate concrete, expanded shale concrete has some special structural characteristics.

The new mixing procedure, EESC, can increase the strength of expanded shale concrete by about 20%. It has the effect of increasing the strength of the expanded shale concrete when the density of the concrete is kept at a certain value. That is because expanded shale is reinforced by the layer of hardened cement paste and the new procedure makes the structure closer near the cement-aggregate interface.

The water-cement ratio in the enveloping layer is defined as W_1/C_1, and the thickness proportion of the enveloping layer as C_1/C. When W_1/C_1 is in the range 0.21 - 0.31, and C_1/C is in the range 0.375 to 0.43, the strength of EESC concrete is more satisfactory.

5 References

Feng, N. Structure characteristics and strength of synthetic light-aggregate concrete. **J. Concrete and Cement Products** (China), No. 1, 1981.

Zhao, R. Studies on the failure mechanism of expanded shale concrete. J. **Concrete and Cement Products.** (China), No. 4, 1983.

9 ETUDE DES INTERFACES CIMENT-FIBRES ORGANIQUES

(Interfaces between synthetic fibers and cement paste)

V. HOUGET, M. CHABANNET and J. PERA
Laboratoire des Matériaux Minéraux, INSA de Lyon, Villeurbanne, France

Résumé *(English abstract, page 307)*
Ce travail concerne l'étude systématique des interfaces ciment hydraté-fibres organiques. Les fibres retenues sont celles de polypropylène, polyamide, polyester, polyacrylonitrile, polyvinylalcool, polyaramide et carbone. Plusieurs échéances d'hydratation de la matrice cimentaire ont été choisies : 7, 28, 90 et 180 jours. L'évaluation de l'énergie de rupture des composites à ces différentes échéances couplée à une observation microscopique électronique à balayage a permis de préciser certains phénomènes régissant le comportement mécanique des composites.
Mots clés : Composites, Ciment, Fibres Organiques, Energie de Rupture, Interfaces, Microscopie Electronique à Balayage.

1 Introduction

Depuis ces dernières années, on note un intérêt croissant des industriels du béton pour les fibres organiques, certainement dû au fait que, calculé sur une base volumique et non pondérale, leur prix est inférieur à celui des fibres d'acier ou de verre (Krenchel, 1986). Si leur utilisation présente de nombreux avantages comme la diminution de la perméabilité ou la limitation des retraits des bétons aux premiers âges (Swamy, 1979), elle pose de gros problèmes tels que leur mauvaise adhérence au béton et un module relativement faible (Krenchel, 1986).

Comme les propriétés macroscopiques des composites sont souvent gouvernées par les interfaces fibres-matrice, une étude systématique des interfaces fibre organique-ciment hydraté a été entreprise. Sept fibres ont été choisies : polypropylène, polyamide, polyester, polyacrylonitrile, polyvinylalcool, polyaramide et carbone.

2 Expérimentation

2.1. Propriétés des fibres retenues

Interfaces in Cementitious Composites. Edited by J. C. Maso. © RILEM.
Published by E & FN Spon, 2 - 6 Boundary Row, London SE1 8HN. ISBN 0 419 18230 6.

Elles sont rassemblées dans le tableau 1.

Tableau 1. Propriétés des fibres retenues

Fibre	Nom commercial et Société	Densité	Module (GPa)	Allongement à rupture (%)	Longueur (mm)	Diamètre (μm)
Polypropylène	Marduel	0,92	6,4	16	10	fibre plate
Polyamide	Rhone-Poulenc	1,14	4	20-30	20	20
Polyester	Hi-Tech Chryso	1,39	6	12-30	20	65
Polyvinylalcool	Kwaway	1,30	20-40	7-10	15	40
Polyacrylonitrile	Dolanit Hoescht	1,18	17-20	6-11	18	52
Polyaramide	Twaron Enka Akzo	1,44	80	3,3	6	12
Carbone	Tenax Enka Akzo	1,78	238	1,4	6	7

Le pourcentage de fibres utilisé a été déterminé de façon à ce que le prix de revient de chaque composite soit égal à celui d'un composite renforcé à raison de 2,5 % de fibres de verre alcali-résistant, pourcentage habituellement utilisé pour la réalisation de composites ciment-verre par premix (Majumdar, 1991). Les résultats obtenus sont présentés dans le tableau 2.

Tableau 2. Pourcentages de fibres utilisés

Coût et pourcentage	Polypropylène	Polyamide	Polyester	Polyvinylalcool	Polyacrylonitrile	Polyaramide	Carbone
Coût (FF/kg)	60	60	30	80	40	150	220
Pourcentage (% en masse)	1,25	1,25	1,50	0,95	1,10	0,50	0,35

Les pourcentages de fibres de polyester et de polyacrylonitrile ont également été limités par les conditions d'ouvrabilité du mortier.

2.2. Réalisation des composites ciment-fibres (C.C.F.)

Pour chaque fibre, des plaques d'épaisseur 15 mm, de largeur 80 mm et de longueur 200 mm, ont été réalisées à partir des constituants suivants :

- ciment : CPA HP des Ciments d'Origny,
- sable : silice pure de Silice et Kaolin, bien calibrée d'un point de vue granulométrique (Farha, 1990), le rapport pondéral sable-ciment est pris égal à 1,

. eau : rapport eau-ciment égal à 0,45 sauf lors de l'utilisation des fibres de carbone où il a été abaissé à 0,38,

. fibres : pourcentage défini dans le tableau 2.

La mise en place des composites dans les moules est faite par vibration pendant 2 minutes. Les plaques sont conservées en sacs étanches à 20°C jusqu'aux dates des essais mécaniques.

2.3. Analyses effectuées

Des essais de flexion trois points ont été réalisés aux échéances de 7, 28, 90 et 180 jours, sur une presse universelle ADAMEL LHOMARGY asservie en déplacement (1 mm/mn) et de capacité maximale 20 kN.

A ces mêmes échéances, les surfaces de rupture des éprouvettes ont été observées en microscopie électronique à balayage couplée à une analyse élémentaire en dispersion d'énergie (MEB JEOL 840 ALGS).

3 Résultats et discussion

3.1. Comportement mécanique des composites

L'allure générale des courbes charge-flèche obtenues après 180 jours d'hydratation est donnée sur les figures 1 et 2.

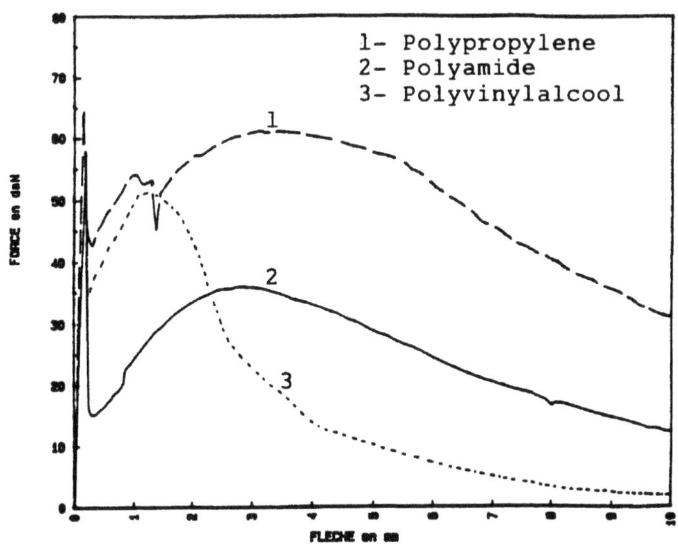

Fig. 1. Courbes charge-flèche. Fibres de polypropylène et polyvinylalcool.

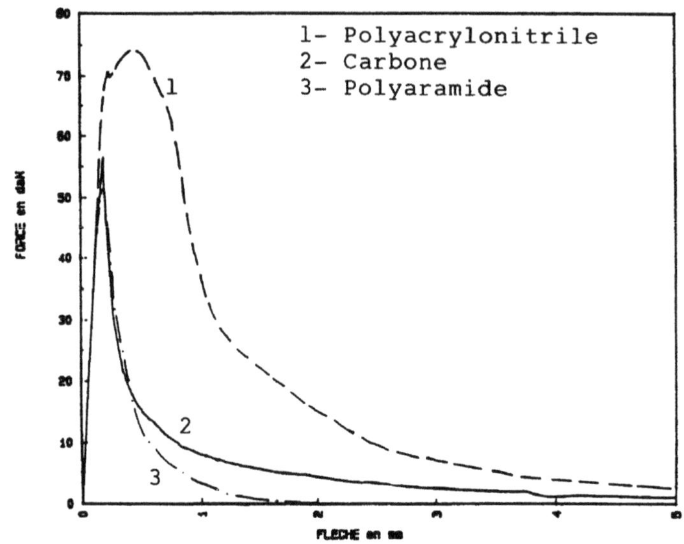

Fig. 2. Courbes charge-flèche. Fibres de polyacrylonitrile carbone et polyaramide.

Pour les fibres de polyester, l'évolution du comportement mécanique au cours du temps, est représentée sur la figure 3. Après 180 jours d'hydratation, les propriétés du composite sont équivalentes à celles d'un mortier non fibré.

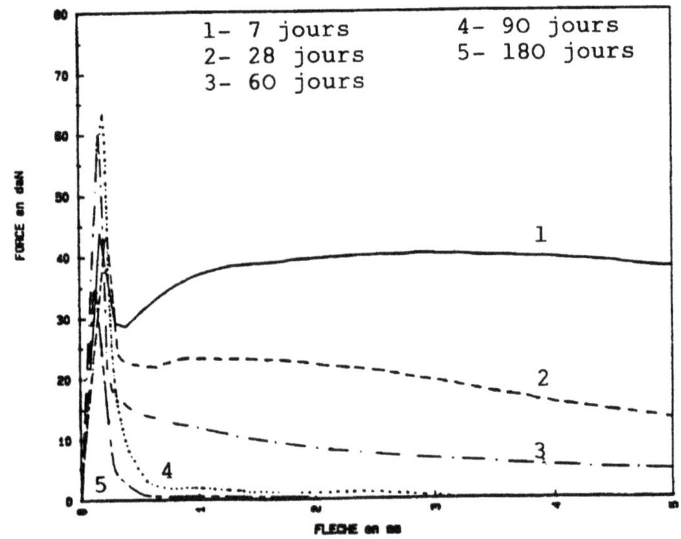

Fig. 3. Courbes charge-flèche. Fibres de polyester.

Les énergies de rupture de chaque composite ont été calculées pour une valeur de la flèche égale à 12 mm pour les fibres de polypropylène, polyamide, polyvinylalcool et polyester et pour la flèche à rupture dans le cas des autres fibres. Leur évolution au cours du temps est représentée sur les figures 4 à 7.

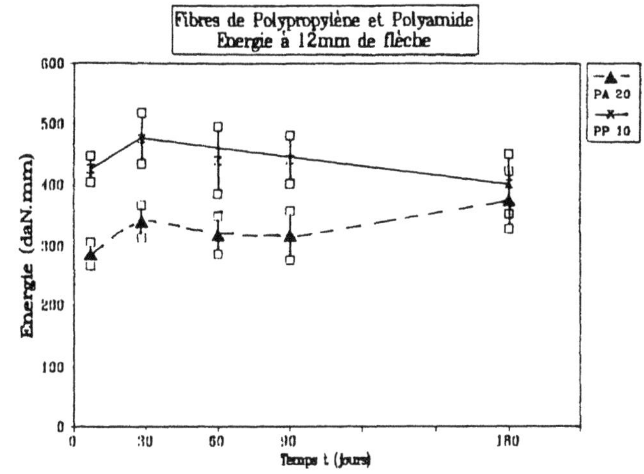

Fig. 4. Evolution des énergies de rupture. Fibres de polypropylène et de polyamide.

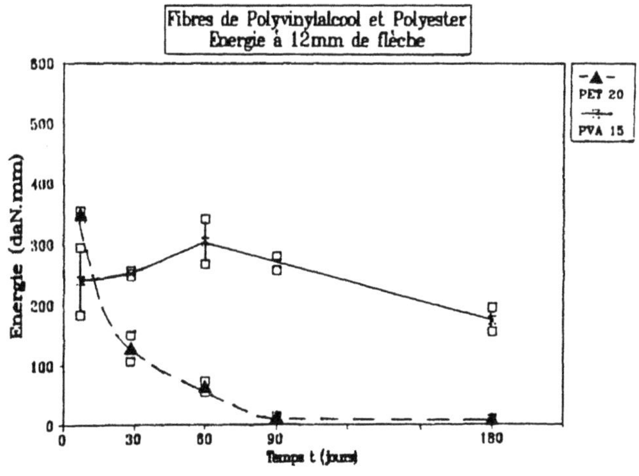

Fig. 5. Evolution des énergies de rupture. Fibres de polyvinylalcool et de polyester.

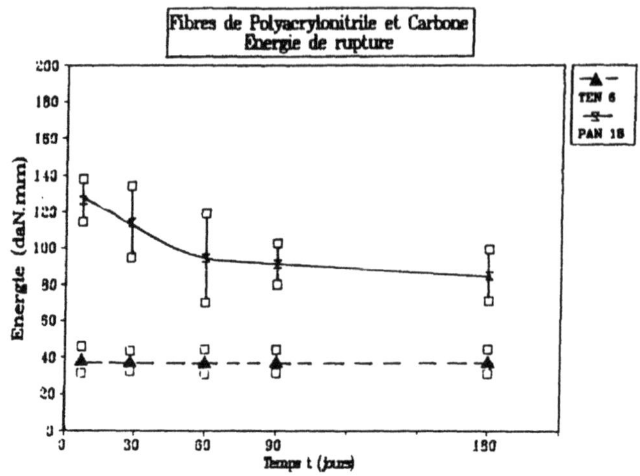

Fig. 6. Evolution des énergies de rupture. Fibres de polyacrylonitrile et de carbone.

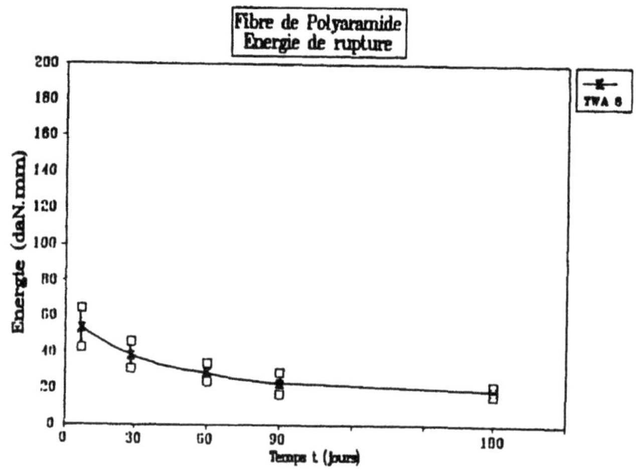

Fig. 7. Evolution des énergies de rupture. Fibres de polyaramide.

A partir de ces résultats, il se dégage plusieurs types de comportement :

- énergie de rupture élevée et constante dans le temps pour les fibres de polypropylène et de polyamide ;
- énergie de rupture élevée et diminuant dans le temps pour les fibres de polyvinylalcool et de polyester ;
- énergie de rupture faible et constante dans le temps pour les fibres de carbone ;
- énergie de rupture faible et diminuant dans le temps pour les fibres de polyacrylonitrile et de polyaramide.

3.2. Etude des interface fibre-matrice

Les fibres de polypropylène extraites de la matrice à 90 jours présentent une surface très striée (figure 8) due à un bon accrochage de la fibre dans la matrice (figure 9), ce qui explique le bon comportement mécanique du composite. Les interfaces sont constituées de CSH et de portlandite. Aucun cristal de portlandite ne pénètre cependant la fibre.

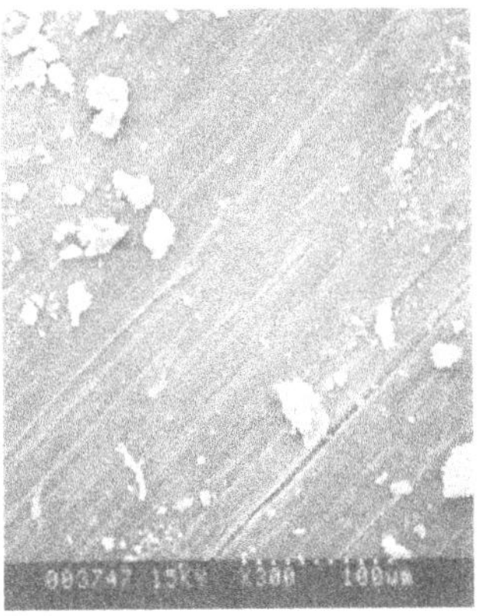

Fig. 8 . Fibre de polypropylène extraite de la matrice. 90 jours d'hydratation.

Fig. 9. Interface fibre de polypropylène-matrice. 90 jours d'hydratation.

Les fibres de polyamide ne subissent aucune dégradation (figure 10) et semblent glisser dans la matrice (figure 11). La portlandite est le principal hydrate présent à l'interface. Ce glissement de la fibre permet d'obtenir une bonne ductilité au niveau du composite.

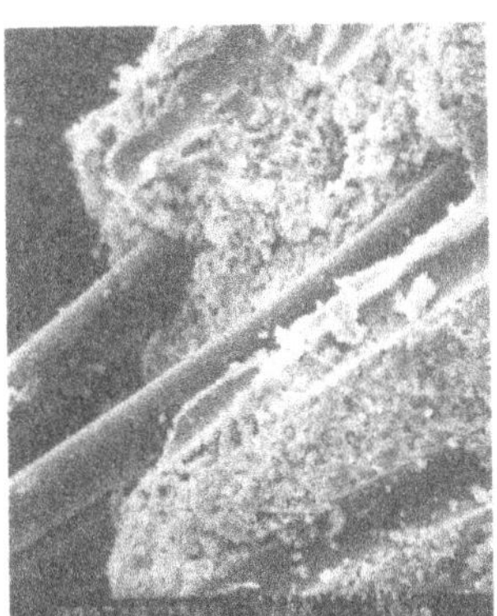

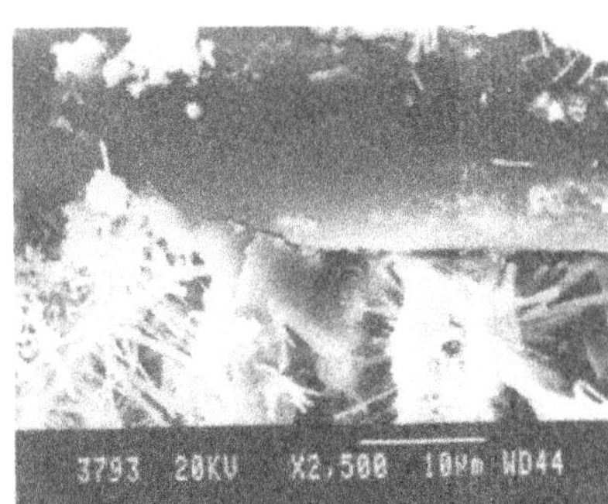

Fig. 10. Fibre de polyamide 90 jours d'hydratation.

Fig. 11. Interface fibre de polyamide-matrice. 180 jours d'hydratation.

L'examen microscopique montre que les fibres de polyvinylalcool sont légèrement dégradées après 90 jours d'hydratation (figure 12) et bien bloquées dans la matrice par des cristaux de portlandite (figure 13). Cette attaque superficielle accompagnée de l'attaque physique des fibres par la portlandite explique la perte de ductilité du composite au cours du temps.

Fig. 12. Fibre de polyvinylalcool. 90 jours d'hydratation

Fig. 13. Interface fibre de polyvinylalcool-matrice. 90 jours d'hydratation.

Les fibres de polyester sont très dégradées à 90 jours et recouvertes d'une pellicule de cristaux de térephtalate de calcium, comme le montre l'analyse réalisée en spectrométrie infra-rouge (figures 14 et 15). La fibre a donc réagi avec la portlandite libérée lors de l'hydratation du ciment.

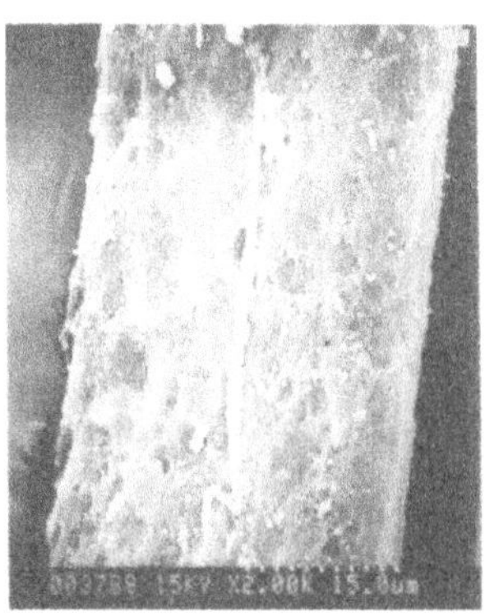

Fig. 14 . Fibre de polyester 90 jours d'hydratation.

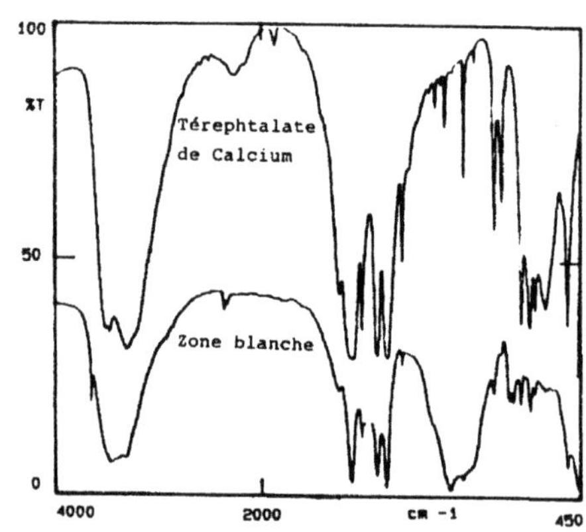

Fig. 15. Produit de réaction entre la fibre de polyester et la portlandite.

Dès 28 jours d'hydratation, on note une incrustation des cristaux de portlandite dans les fibres de polyacrylonitrile (figures 16 et 17) ce qui provoque la fragilisation du composite.

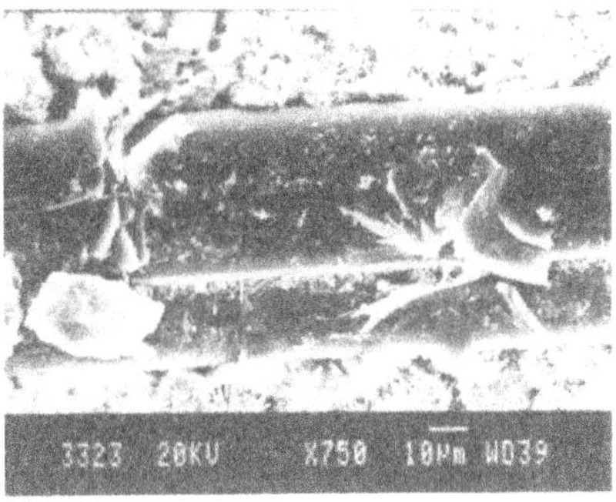

Fig. 16. Interface fibre de polyacrylonitrile-matrice. 28 jours d'hydratation

Fig. 17. Interface fibre de polyacrylonitrile-matrice. 90 jours d'hydratation.

Les fibres de carbone sont mal distribuées dans la matrice, l'interface est constituée de cristaux de portlandite qui ne pénètrent pas la fibre (figure 18).

Fig. 18. Empreinte d'une fibre de carbone dans la matrice . 90 jours d'hydratation.

Fig. 19. Fibre de polyaramide. 9o jours d'hydratation.

Les fibres de polyaramide sont soit "effilochées" soit cassées au niveau de la surface de rupture (figure 19). Cette fragilisation des fibres est due à une perte de résistance en traction en milieu alcalin.

4 Conclusion

La portlandite est le principal hydrate présent à l'interface fibres organiques-ciment hydraté. Dans certains cas, elle entre en réaction avec la fibre (polyester par exemple) pour donner de nouvelles espèces chimiques (térephtalate de calcium) et conduire à une fragilisation du composite au cours du temps. Dans d'autres cas, elle s'incruste dans la fibre et empêche tout glissement, provoquant une diminution de la ductilité du composite (polyvinylalcool, polyacrylonitrile). Quant elle ne pénètre pas dans la fibre, la ductilité des composites est conservée dans le temps (polypropylène, polyamide).

D'autres phénomènes interviennent également pour expliquer le comportement mécanique des composites : solubilisation partielle du polyvinylalcool, baisse de résistance en traction des fibres de polyaramide en milieu alcalin.

Les fibres de polyaramide et de carbone, utilisées avec les pourcentages retenus dans l'étude, donnent des composites fragiles alors que les fibres de polyamide et de polypropylène conduisent à des composites ductiles.

5 Références

Farha, M. (1990). Conception de composites ciment-verre à porosité minimale. **Thèse de doctorat.** Institut National des Sciences Appliquées de Lyon. 138 p.

Jelidi, A., Chabannet, M., Ambroise, J. and Péra, J. (1991) Development of composites reinforced by polyester fibres. Proceedings **Symposium Materials Research Society** (eds S. Mindess and J. Skalny), Vol 211, pp. 275-281.

Krenchel, H. and Shah, S.P. Synthetic fibres for tough and durable concrete, in **Developments in Fibre Reinforced Cement and Concrete**, FRC 86, RILEM Symposium, (eds R.N. Swamy, R.L. Wagstaffe, D.R. Oakley), RILEM Technical Committee 49-TFR, Vol. 1., Section 4-7.

Majumdar, A.J. and Laws, V. (1991). **Glass Fibre Reinforced Cement.** Published on behalf of the Building Research Establishment, BSP Profesional Books, Oxford, pp. 61-62.

Mark, H.F., Bikales, N.M., Overberger, C.G. and Menges, G. (1989). **Encyclopedia of polymer science and engineering.** John Wiley and sons, Inc. Vol. 17, Vinyl Alcohol Polymers, pp. 167-168.

Swamy, F.N. and Stavrides, H. (1979). Influence of fiber reinforcement on restrained shrinkage and cracking. **Journal of the American Concrete Institute**, March 1979, pp. 443-460.

10 ADHESION OF BONDED MORTAR TO POLYMER-CEMENT PASTE COATED MORTAR SUBSTRATES

S. N. PAREEK, Y. OHAMA and K. DEMURA
College of Engineering, Nihon University, Koriyama, Japan

Abstract
Mortar substrates are prepared, and coated with three types of polymer-cement pastes with polymer-cement ratios of 0, 5, 10 and 20%, and subjected to open times of 1 hour, 1 day and 7 days. In order to study the influence of the coating thickness of the polymer-cement pastes, the mortar substrates are coated with coverage rates of 500, 700 and 900 g/m^2. Specimens are prepared by placing the fresh mortar onto the polymer-cement paste coated mortar substrates after various open times, and subjected to a combined moist, water and dry cure. The adhesion of the bonded mortar to the polymer-cement paste coated mortar substrates is determined by direct tensile tests, and examined for the failure modes of the failed cross-sections. The microstructures of the interfaces are investigated by scanning electron microscopy. From the extensive test results, optimum adhesion is obtained for the bonded mortar to the polymer-cement paste coated mortar substrates with an open time of 1 hour regardless of the type of polymer dispersion, polymer-cement ratio and coverage rate, and is about 1.5 to 2.0 times the adhesion of the bonded mortar to unmodified cement paste coated mortar substrates.

1 Introduction

Use of polymer dispersions (latexes or emulsions) or polymer-cement pastes with them, as bonding agents for ordinary cement mortar is a widespread practice for troweling works in the construction industry[1]. The polymer dispersions or polymer-cement pastes contribute to the adhesion improvement of bonding mortar to concrete or mortar substrates. This research work deals with the application of the polymer-

Interfaces in Cementitious Composites. Edited by J. C. Maso. © RILEM.
Published by E & FN Spon, 2 - 6 Boundary Row, London SE1 8HN. ISBN 0 419 18230 6.

cement pastes as bonding agents for troweling works.

In the present research work, mortar substrates are prepared, coated with three types of polymer-cement pastes with various polymer-cement ratios at various coverage rates, and then given suitable open times. Specimens are prepared by jointing the fresh mortar onto the coated mortar substrates, and tested for adhesion in tension. From the test results, the effects of the type of polymer dispersion, polymer-cement ratio, coverage rate and open time on the adhesion in tension are discussed.

2 Materials

2.1 Cement and fine aggregate

Ordinary portland cement and Toyoura standard sand were used for mortar substrates and bonding mortars. The ordinary portland cement was also used for polymer-cement pastes.

2.2 Polymer dispersions for polymer-cement pastes

Commercially available polymer dispersions used as cement modifiers for polymer-cement pastes were a styrene-butadiene rubber (SBR) latex, an ethylene-vinyl acetate (EVA) emulsion and a polyacrylic ester (PAE) emulsion. Their properties are listed in Table 1.

Table 1 Properties of Polymer Dispersions.

Type of Polymer Dispersion	Specific Gravity (20°C)	pH (20°C)	Viscosity (20 °C)	Total Solids (%)
SBR	1.02	9.1	135	45.1
EVA	1.06	5.0	1000	44.5
PAE	1.08	9.3	43	44.7

3 Testing procedures

3.1 Preparation of mortar substrates

Ordinary cement mortar was mixed according to JIS R 5201(Physical Testing Methods for Cement) with a weight ratio of cement to standard sand 1:2 and a water-cement ratio of 65%. Mortar substrates 70x70x20mm were molded, and given a 1-day-20°C-80%R.H.-moist plus 1-day-70°C-water plus 5-day-20°C-50%R.H.-dry cure. Then the bonding surfaces of the mortar substrates were treated with the AA-150 abrasive papers specified in JIS R 6252 (Abrasive Papers), and blown with compressed air to remove free dust particles before coating the polymer-cement pastes.

3.2 Preparation of specimens

Mortar substrates were coated with polymer-cement pastes having a water-cement ratio of 40% and polymer-cement ratios(P/C) of 0, 5, 10 and 20% at the respective coverage rates of 500, 700 and 900g/m^2, and the coated mortar substrates were stored at 20°C and 50%R.H. for open times of 0, 1 hour, 1 day and 7 days before placing the bonding mortar. The bonding mortar was mixed with a weight ratio of cement to standard sand 1:3 and a water-cement ratio of 72%. Specimens were made by placing the bonding mortar on the polymer-cement paste coated surfaces of the mortar substrates as shown in Fig. 1, and subjected to a 1-day-20°C-80%R.H.-moist plus 6-day-20°C-water plus 21-day-20°C-50%R.H.-dry cure.

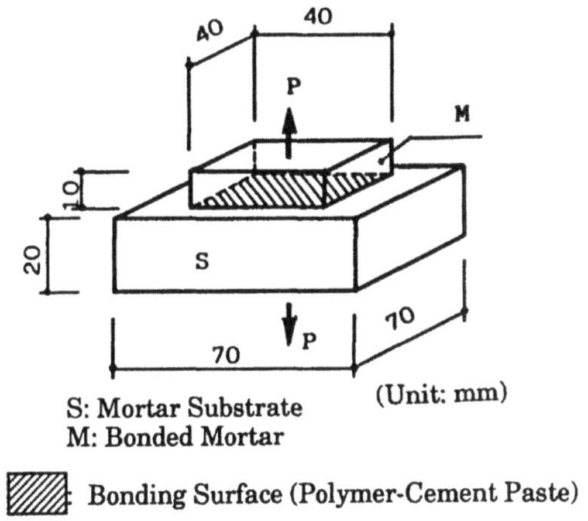

Fig. 1 Specimen for Adhesion Test.

3.4 Adhesion test

The cured specimens were tested for adhesion in tension by using a manually operated pull-gage in accordance with JIS A 6915 (Wall Coatings for Thick Textured Finishes). The adhesion in tension of the specimens was calculated as follows:

$$\text{Adhesion in tension (kgf/cm}^2\text{)} = P/A$$

where P is the maximum load (kgf), and A is the area (cm^2) of the bonded surfaces of the specimens.

After the adhesion test, the failed cross-sections of the specimens were observed for failure modes, which were classified into the following four types:

M : Cohesive failure in the bonded mortar
AP : Adhesive failure (failure in the polymer-cement paste)
AP_M : Adhesive failure at the interface between the polymer-cement paste and bonded mortar
AP_S : Adhesive failure at the interface between the polymer-cement

paste and mortar substrate

The total area of the bonded surfaces was supposed to be 10, and the respective approximate rates of M, AP, AP_M and AP_S areas on the failed cross-sections were expressed as suffixes for M, AP, AP_M and AP_S.

The adhesive interfaces between bonded mortar and polymer-cement paste coated mortar substrates were observed by a scanning electron microscope on the samples prepared from the specimens after adhesion test.

4 Test results and discussion

4.1 Adhesion and failure modes

Fig. 2 represents the adhesion in tension of bonded mortar to the polymer-cement paste coated mortar substrates subjected to open times of 0, 1 hour, 1 and 7 days. In general, the mortar substrates coated with the polymer-cement pastes provide the maximum adhesion of the bonded mortar at an open time of 1 hour irrespective of the type of polymer dispersion, polymer-cement ratio and coverage rate. The adhesion in tension of the bonded mortar to the polymer-cement paste coated mortar substrates is larger than that to unmodified cement paste coated mortar substrates. A significant improvement in the adhesion in tension of the bonded mortar is achieved by polymer-cement paste coatings to the mortar substrates. The improved adhesion in tension of the bonded mortar to the polymer-cement paste coated mortar substrates is 1.5 to 2.0 times that to the unmodified cement paste coated mortar substrates irrespective of the type of polymer dispersion, polymer-cement ratio and coverage rate. The poor adhesion of the bonded mortar to the mortar substrates coated with the polymer-cement pastes with an open time of 0 is due to an insufficient grab with the substrates because of the rapid diffusion of the pastes into the fresh bonded mortar. On the other hand, a prolonged open time of 7 days for the polymer-cement paste coated mortar substrates provides the poor adhesion of the bonded mortar because of the combined effect of drying shrinkage and dry-out phenomenon. Although not significant, similar factors also influence the adhesion of the bonded mortar to the unmodified cement paste coated mortar substrates.

Fig. 3 illustrates the effect of the coverage rate of polymer-cement pastes on the adhesion in tension of bonded mortar to polymer-cement paste coated mortar substrates with an open time of 1 hour. With an increase in the coverage rate, the adhesion in tension of the bonded mortar to the polymer-cement paste and unmodified paste coated mortar

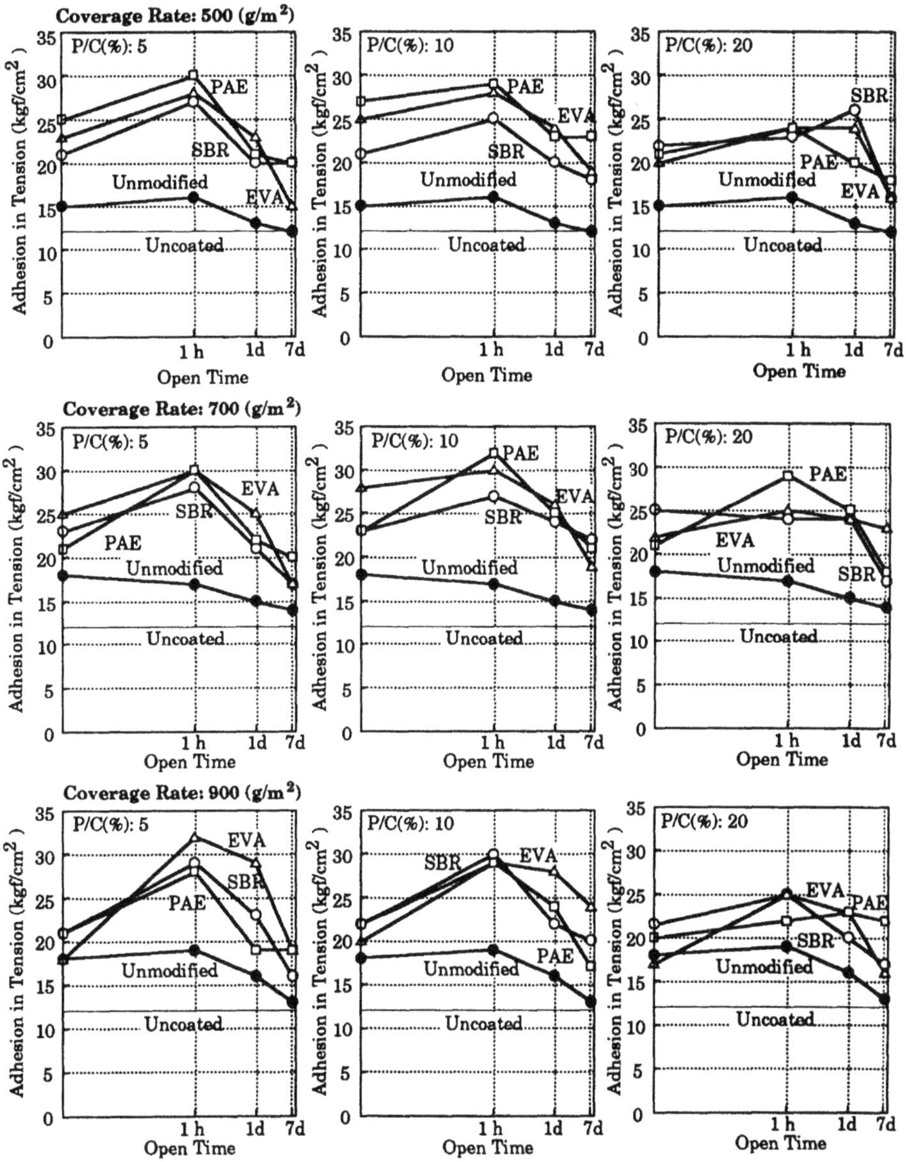

Fig. 2 Adhesion in Tension of Bonded Mortar to Polymer-Cement Paste Coated Mortar Substrates Subjected to Open Times of 0, 1 Hour, 1 and 7 days.

substrates tends to increase or reach a maximum at a coverage rate of 700 g/m^2. It seems that the thickness of the polymer-cement pastes varies depending on the coverage rate and determines the adhesion of the bonded mortar. Accordingly, it is considered that Bikerman's theory [2] can not be applied in the coverage rate range of 500 to 900 g/m^2.

Fig. 4 exhibits the effect of the polymer-cement ratio of polymer-cement pastes on the adhesion in tension of bonded mortar to polymer-

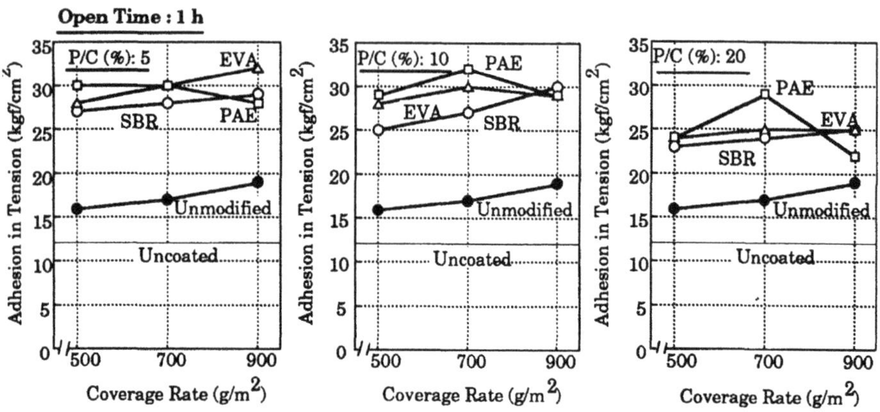

Fig. 3 Effect of Coverage Rate on Adhesion in Tension of Bonded Mortar to Polymer-Cement Paste Coated Mortar Substrates.

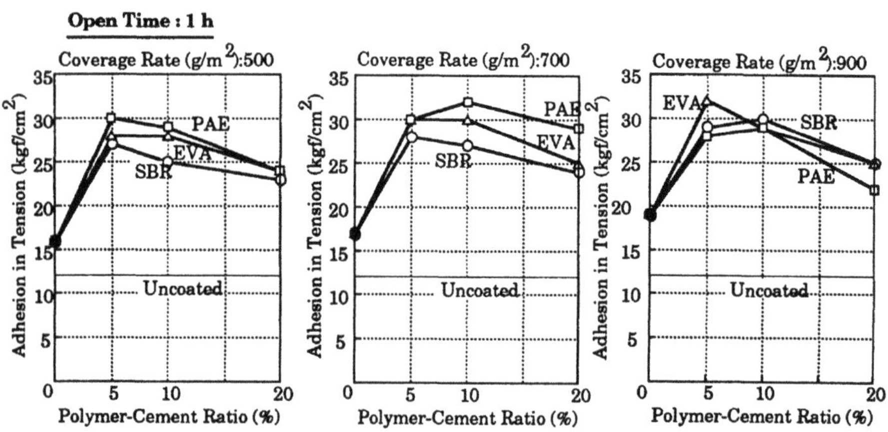

Fig. 4 Effect of Polymer-Cement Ratio on Adhesion in Tension of Bonded Mortar to Polymer-Cement Paste Coated Mortar Substrates.

cement paste coated mortar substrates. The adhesion in tension of the bonded mortar to the polymer-cement paste coated mortar substrates increases with increasing polymer-cement ratio of the polymer-cement pastes, and reaches the maximum at polymer-cement ratios of about 5 to 10 %.

Fig. 5 represents the failure modes in the adhesion tests of bonded mortar to polymer-cement paste coated mortar substrates with a coverage rate of 700 g/m^2. The occurrence of cohesive failure in the bonded mortar is dominant at open times of 0, 1 hour, and 1 day. The ratio of cohesive

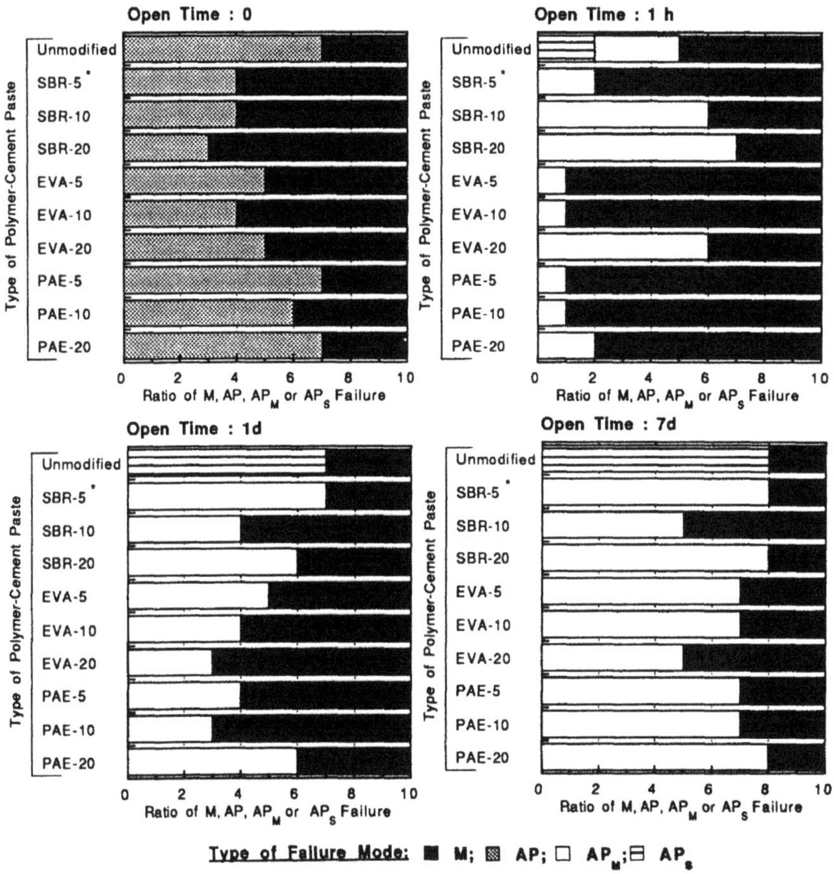

Fig. 5 Failure Modes in Adhesion Tests of Bonded Mortar to Polymer-Cement Paste Coated Mortar Substrates.

failure in the bonded mortar is 8 or higher at polymer-cement ratios of 5 and 10% and an open time of 1 hour, demonstrating a good bond to the polymer-cement paste coated mortar substrates. The occurrence of AP_M failure (failure at the interface between the bonded mortar and polymer-cement paste) increases in the bonded mortar to the polymer-cement paste coated mortar substrates with an open time of 7 days, revealing the poor bond of the bonded mortar. Such AP_M failure is found to be due to differences in the drying shrinkage between the polymer-cement pastes and bonded mortar during dry cure. However, AP_S failure (failure at the interface between the polymer-cement paste and mortar substrate) in the bonded mortar to unmodified paste coated mortar substrate is attributed to the inferior adhesion of the unmodified cement paste to the mortar substrate as a result of dry-out phenomenon.

4.2 Microstructural analysis of interfaces between bonded mortar and polymer-cement paste coated mortar substrates

Fig.6 shows the microstructures of the interfaces between bonded mortar and polymer-cement paste coated mortar substrates with a polymer-cement ratio of 20%, a coverage rate of 700g/m^2 and an open time of 1 hour. The presence of the polymer-cement paste layer's with the diffused edges facing the bonded mortar is distinctly seen. The diffusion of the polymer-cement pastes in the bonded mortar signifies the formation of a monolithic bond formation between the bonded mortar and polymer-cement pastes. Moreover, the very workable polymer-cement pastes fill the nonuniform mortar substrates, and develop a good bond by micromechanical interlocking mechanisms like polymer dispersion-coated mortar substrates [1].

From the test results, the influential factors affecting the adhesion between bonded mortar and polymer-cement paste coated mortar substrates are as follows:

(1) The formation of a monolithic bond between the bonded mortar and the polymer-cement paste coated mortar substrates.

(2) The workable polymer-cement pastes fill the nonuniform mortar substrates, reinforce them and provide a mechanical bond with the substrates apart from a physico-chemical bond.

(3) The poor adhesion of the bonded mortar to the polymer-cement paste coated mortar substrates with an open time of 7 days is chiefly attributed to differences in the drying shrinkage stresses between the bonded mortar and polymer-cement pastes, leading to the high ratio of AP_M failure as shown in Fig. 5.

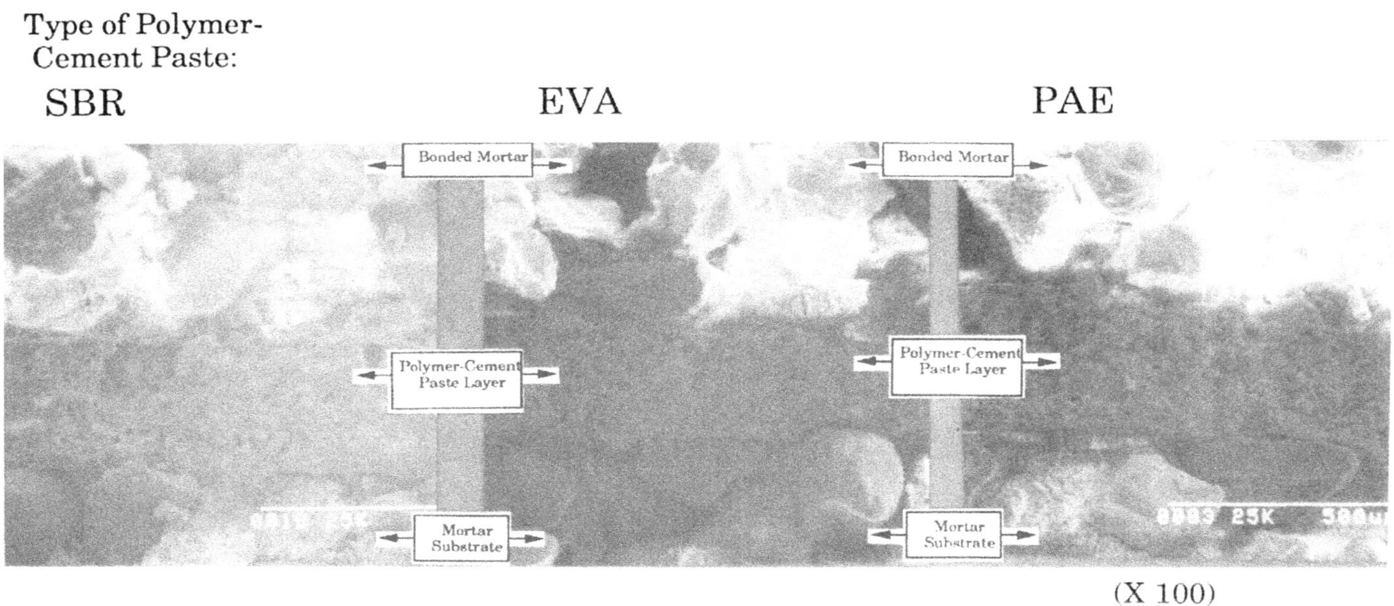

Fig. 6 Microstructure of Interfaces between Bonded Mortar and Polymer-Cement Paste Coated Mortar Substrates with Open Time of 1 hour, Polymer-Cement Ratio of 20% and Coverage Rate of 700 g/m^2.

5 Conclusions

(1) The adhesion in tension of bonded mortar to polymer-cement paste coated mortar substrates is increased to about 1.5 to 2.0 times that to unmodified cement paste coated mortar substrates.

(2) Optimum adhesion is attained for bonded mortar to polymer-cement paste coated mortar substrates with an open time of 1 hour irrespective of the type of polymer dispersion, the polymer-cement ratio and coverage rate of polymer-cement paste.

(3) The occurrence of M failure (cohesive failure in bonded mortar) of bonded mortar to polymer-cement paste coated mortar substrates with an open time of 1 hour, proves an adequate bond between the bonded mortar and polymer-cement pastes. AP_M failure (failure at the interface between the bonded mortar and polymer-cement paste) in the bonded mortar to the polymer-cement paste coated mortar substrates with an open time of 7 days is due to differences in the drying shrinkage between the bonded mortar and polymer-cement pastes during dry cure.

Pareek, S.N. Ohama, Y. and Demura, K. (1990) Adhesion mechanism of ordinary cement mortar to polymer dispersion coated mortar substrates, in **Proceedings of the Sixth International Congress on Polymers in Concrete**, (eds Y. Huang, K. Wu and Z. Chen), International Academic Publishers, Beijing, China, pp.442-449.

Bikerman, J.J. (1941) Strength and thinness of adhesive joints, in **J. of Society of Chemical Industry**, 60, pp. 23-24.

11 THE INTERFACE AND ADHESIVE STRENGTH BETWEEN REACTIVE RESIN AND CONCRETE

S.-W. CHEN
Chongqing Institute of Architecture and Engineering, Chongqing, P.R. China; at present TU Vienna, Austria
W. PERBIX and K. P. GROSSKURTH
Institute for Building Materials, Constructions and Fire Resistance in Civil Engineering, Technical University of Braunschweig, Germany

Abstract
Cracked building elements are increasingly being repaired by the injection of reactive resins. This includes the non-positive injection with epoxy resins for the protection of corrosion of the reinforcement and to restore simultaneously the stiffness of an uncracked construction by glueing the crack surfaces. In the tests the influence of different loading and climatic conditions on the morphology of the interface between several epoxy systems and crack surfaces was investigated by means of scanning electron microscopy. The bond strength between resin and concrete was measured in mechanical tests of concrete elements which were injected under dry and wet conditioning. The results of further tests with water-modified epoxy systems showed the influence of different water contents on the bond behaviour of the reactive resins.
Keywords: Interface, Morphology, Bond Strength, Crack Injection, Reactive Resin, Adhesive Behaviour.

1 Introduction

At present more and more attention has been paid to the protection and restoration of buildings in the world. In Europe the amount of the restoration for buildings takes up to 50 % of the amount of the investment for new buildings.

The buildings suffer from chemical, physical and mechanical attacks which can initiate cracks. The cracks often impair the appearence of the building and its usability, moreover, the stability of the building is endangered. In these case the cracks must be filled with cement lime or reactive resins.

Crack injection is usually divided into the positive and non-positive injection according to the aim of the restoration. Using a positive injection, only the protection against corrosion of the reinforcement is guarenteed, whereas a non-positive connection of the crack surfaces is advantageous in so far as the behaviour of the building element is additionally restored to its former condition.

Interfaces in Cementitious Composites. Edited by J. C. Maso. © RILEM.
Published by E & FN Spon, 2 - 6 Boundary Row, London SE1 8HN. ISBN 0 419 18230 6.

2 Aim of the investigations

At present epoxy resins are mostly used for the non-positive injection because of their better mechanical properties and small shrinkage compared with other reactive resins and cement lime.

The success of a non-positive crack injection is mainly determined by the bond strength between the injected epoxy resin and the crack surface because the cohesive strength of the cured thermosetting resin generally surpasses the average concrete tensile strength.

Therefore the cause of premature bond failure must be sought in the adhesive area and is scarcely found with the aid of macroscopic mechanical tests. Scanning electron microscopy permitts the investigation of the microstructure of the adhesion zone in order to correlate the morphology of the interface and the mechanism of the adhesion between resin and concrete.

3 Morphology of the interface between resin and concrete

The adhesion between epoxy resin and concrete does not only depend on the composition and the properties of the resin itself, but also on the boundary conditions, especially on moisture, temperature and loading of the crack surfaces during the injection and the curing of the epoxy resins.

The influence of moisture and loading state (static and dynamic loads) on the quality of the adhesion zone was preliminarily investigated by scanning electron microscopy. Some representative results are shown in figures 1-3.

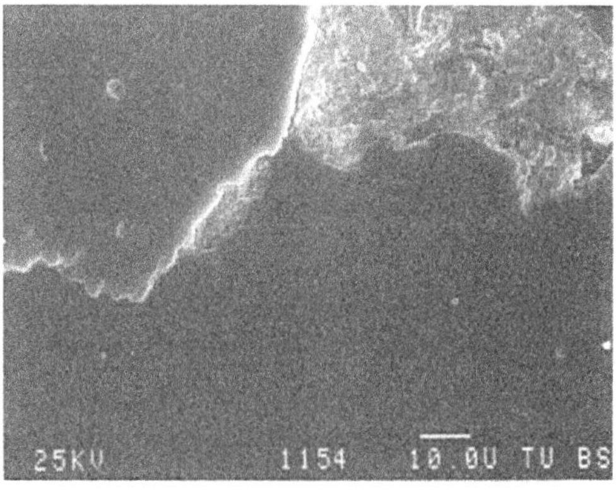

Fig. 1. dry crack, static load
(above: concrete, below: resin)

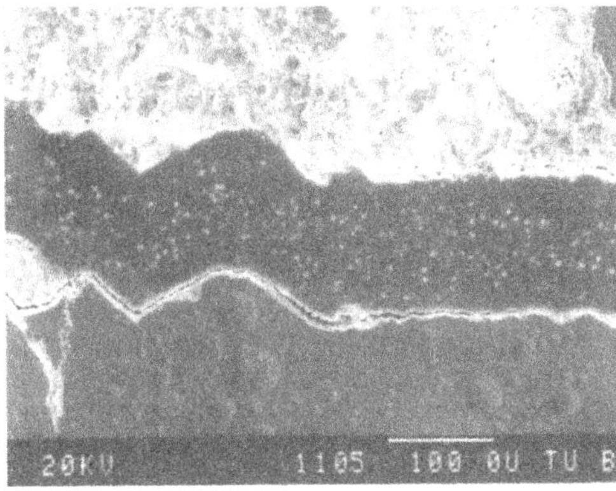

Fig. 2. wet crack, static load
(middle: resin)

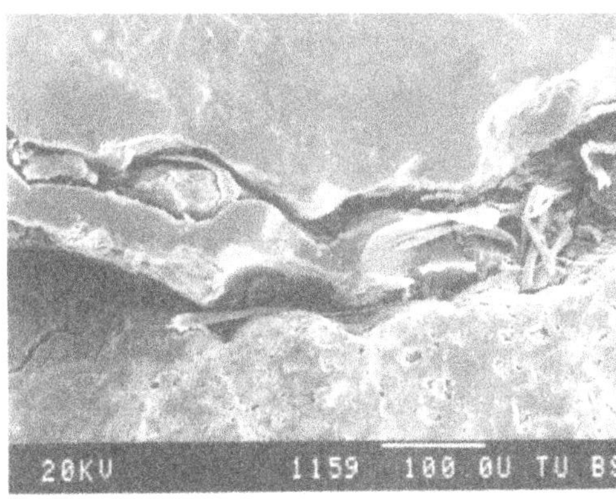

Fig. 3. wet crack, dynamic load
(middle: resin)

The results show that moisture influences tremendously the adhesion zone between resin and concrete if the resin does not possess a sufficient moisture stability. Under static loading the epoxy resins adhere strongly to the dry crack surface, and the interface shows an inseparable combination without obvious defects (Fig. 1).

In case of wet conditioning a water film remains between the resin and concrete which harms the adhesion and causes the interfacial cracking. The obvious adhesive defects occur particulary at the wet aggregate surface whereas the cured epoxy resin has a lot of bubbles and voids in the matrix (Fig. 2).

In worst cases there is no bonding between resin and wet crack surfaces under dynamic loading. Obviously the resin is not able to cure to a homogeneous matrix (Fig. 3).

Considering these defects an acceptable bond strength between resin and wet crack surface can not be expected.

4 Influence of moisture on the adhesive strength of epoxy resins

4.1 Adhesive strength between epoxy resin and concrete

Considering the results of the morphological investigations the chemical industry developed new epoxy resins for the non-positive injection of wet cracks.

For testing the bond strength these epoxy resins were injected into concrete cracks with a width of 0.5 mm and dry or wet surfaces. In addition the resins were injected in cracks filled with water. After 7 days under normal temperature and humidity conditioning (23 °C/50 % r.h.) the injected cubes (150 x 150 x 150 mm) were sawed into prisms (40 x 40 x 150 mm). The bond strength of the injected resins was measured with the test arrangement as shown in Fig. 4.

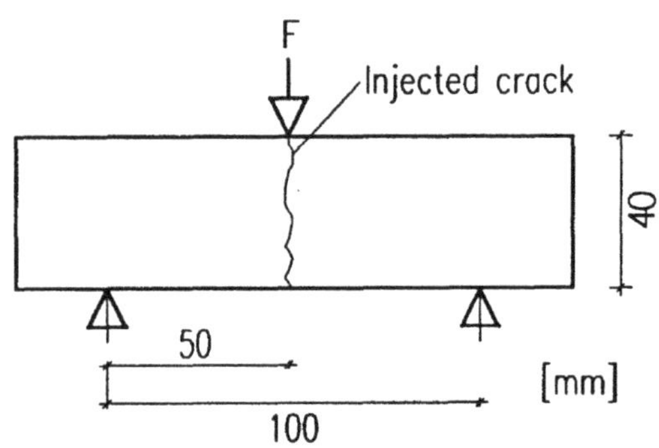

Fig. 4. Experimental arrangement for testing the adhesive strength of injected concrete specimen (bending test)

After overloading the prisms the adhesive areas of the specimen with a premature bond failure were observed visually. The tensile bending strength of the injected prisms are shown in Fig. 5. Table 1 gives an overview of the zone of failure in the overloading tests.

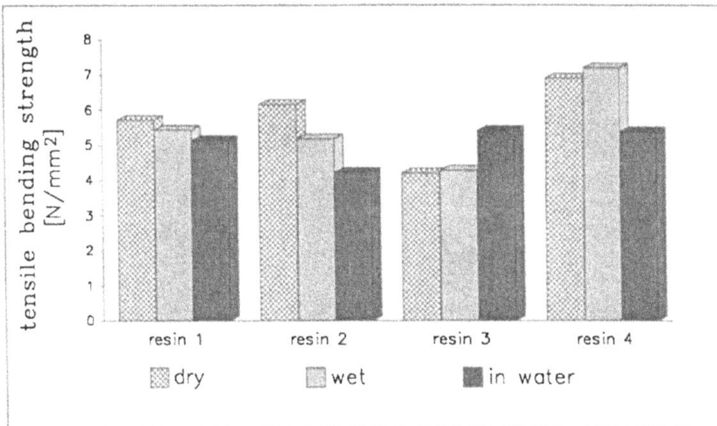

Fig. 5. Tensile bending strength of the injected epoxy resins as a function of the crack conditioning

Table 1. Zone of failure in the overloading test of the injected concrete specimen as a function of the crack conditioning

Resin	Crack Conditioning	Zone of Failure
Resin 1	dry	concrete
	wet	concrete
	water-filled	concrete
Resin 2	dry	concrete
	wet	adhesion zone
	water-filled	adhesion zone
Resin 3	dry	concrete
	wet	concrete
	water-filled	concrete
Resin 4	dry	concrete
	wet	concrete
	water-filled	adhesion zone

The experimental results indicate that moisture has a negative influence on the adhesion strength between the epoxy resin 2 respectively 4 and the concrete (table 1). The more is the moisture content of the crack surface the lower is the adhesive strength in general and the specimen failured in the adhesion area.

Looking at the results of resin 1 and 3 figure 5 and table 1 reveal that the bond strength between the resins and the wet crack surfaces is stronger than the expected tensile strength of concrete even the cracks have been filled with water.

4.2 Adhesive strength of water modified epoxy resins
For determining the influence of different water contents on the bond strength the epoxy resins were mixed with water and the adhesion strength was measured in tensile tests with the experimental arrangement as shown in Fig. 6. The results of these tests are listed in figure 7. The specified kind of failure in table 2 discribes the characteristics of the fracture surface after the overloading tests.

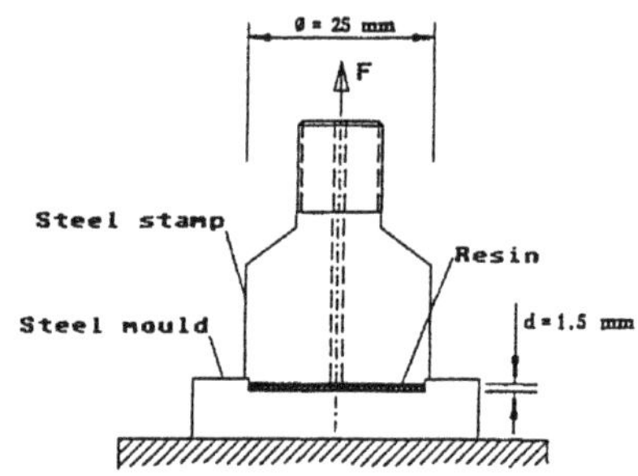

Fig. 6. Experimental arrangement for testing the adhesive strength (tensile test)

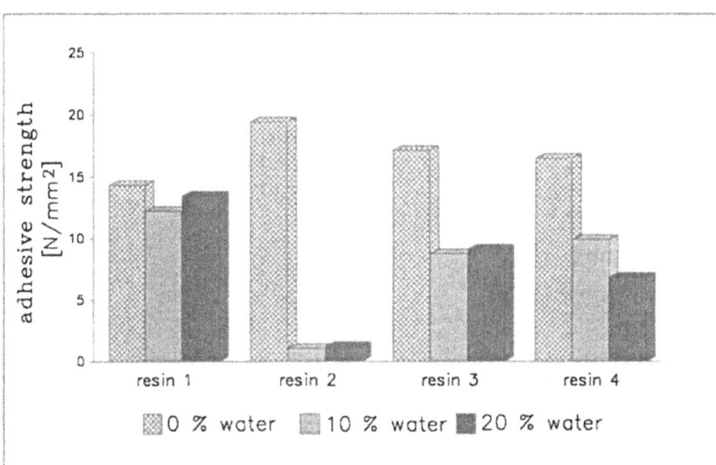

Fig. 7. Adhesive strength as a function of water content of the epoxy resins

Table 2. Kind of failure as a function of water content of the epoxy resins

Resin	Water Content (weight-%)	Kind of Failure
	0 %	95 % cohesive
Resin 1	10 %	63 % cohesive
	20 %	77 % cohesive
	0 %	28 % cohesive
Resin 2	10 %	100 % adhesive
	20 %	100 % adhesive
	0 %	93 % cohesive
Resin 3	10 %	100 % adhesive
	20 %	100 % adhesive
	0 %	81 % cohesive
Resin 4	10 %	100 % adhesive
	20 %	100 % adhesive

The results of the tensile tests shows the correlation between water content of the epoxy resins and their adhesion strength. Analogeous to the results of the bending tests with injected concrete elements resin 1 shows a high moisture stability whereas resin 2 failured in a critical way. Resin 3 and 4 show a significant reduction of their adhesive strength too but even with a water content of 20 % the strength is higher than the average tensile strength of concrete. Regarding the kind of failure in table 2 the water modified epoxy systems failured in the adhesive zone.

5 Conclusion

Investigations of the microregions of the adhesive area between resin and concrete with scanning electron microscopie show that the main reason for a premature bond failure must be sought in the interface between resin and crack surface because the bond behaviour of the injected resins depends on the moisture state of the crack surfaces.

The results of bending tests of concrete elements injected with special epoxy resins with an improved moisture stability and measurements of the adhesive strength of the water modified systems show that the bond strength between the cured resin and wet crack surfaces under static loading reached a level above the concrete tensile strength in general. The investigations of the microstructures of the adhesion zones will be the aim of further tests with these new epoxy systems.

6 Acknowledgement

The presented experiments are a part of a research project which was carried out at the Institute of Building Materials, Constructions and Fire Resistance in Civil Engineering at the University of Braunschweig, Germany. We wish to thank all colleagues at this laboratory.

7 References

Bundesminister für Verkehr, Abteilung Straßenbau (1988) Zusätzliche Technische Vorschriften und Richtlinen für das Füllen von Rissen in Betonbauteilen (ZTV-RISS 88).

Deutscher Betonverein (1981) Anwendung von Reaktionsharzen im Betonbau. (Fassung August 1981). Germany

Engelke, P. and Iványi, G. (1985) Kraftschlüssiges Verpressen von Rissen in Überbauten von Massivbrücken mit Epoxidharzen. **Beton- und Stahlbetonbau**, Heft 2/1985, 29-35,Heft 3/1985 79-82.

Grosskurth, K.P. and Perbix, W. (1986) Improvement in the durability of cracked concrete elements injected with synthetic resin by optimizing the bond behaviour. **Int. RILEM Symp. "Adhesion between Polymers and Concrete"**, France, Proceedings 403-409.

Grosskurth, K.P. and Perbix, W. (1990) Untersuchungen an Reaktionsharzen zur Bauwerksinstandsetzung. **Adhäsion 34.** (1990), Heft 3, 33-37, Germany.

Kern, E. (1980) Dichten von Rissen und Fehlstellen im Beton durch Injektion von Kunststoffen. **VDI-Berichte Nr. 384** (1980), Germany.

12 COMPUTER MODELLING OF THE INTERFACIAL ZONE IN CONCRETE

D. P. BENTZ, E. J. GARBOCZI and P. E. STUTZMAN
National Institute of Standards and Technology,
Gaithersburg, Maryland, USA

Abstract
A digital-image-based microstructural model is applied to simulating cement paste-aggregate interfacial zone microstructural development in a variety of concretes. A starting model system consists of a single square, flat plate, or cubic aggregate surrounded by circular or spherical cement and mineral admixture particles. A cement hydration model based on a cyclic process of dissolution, diffusion, and reaction is utilized to hydrate starting microstructures. In addition to supporting the well known wall effect, the model has suggested that a "one-sided growth" effect also contributes to the microstructural features developed in interfacial zones. The effects of mineral admixture particle size and reactivity and those of aggregate reactivity and water absorptivity on interfacial microstructure are explored via simulation. In many cases, it appears that improvements in the uniformity and density of the interfacial zone microstructure can be achieved by controlling these properties. Quantitative validation of the model is obtained by comparing model systems to ones prepared in the laboratory for concretes containing 0, 10, and 20% silica fume. Good agreement between the two systems is observed.
Keywords: Aggregate, Cement Hydration, Clinker, Computer Modelling, Interfacial Zone, Microstructure, Mineral Admixture, Simulation.

1 Introduction

As a composite material, concrete's performance can be expected to depend on the properties of the interfaces between its two major components, aggregate and cement paste. Bentur (1991) has reviewed the dependence of mechanical properties on interfacial microstructure. Durability will also be strongly influenced by interfacial zone microstructure and its effects on transport properties. Indeed, increases in electrical conductivity and permeability due to interfacial zones have been observed by Ping et al. (1991) and Costa et al. (1990) respectively, both using simplified specimen geometries.

Over the years, the general microstructural features of interfacial zones in ordinary portland cement concrete have been explored by a number of researchers including Zimbelmann (1985), Uchikawa (1988), and Scrivener and Gartner (1988). The general consensus reached is that the interfacial zone is a region, up to 50 μm wide around each aggregate, containing more porosity and larger pores than the bulk cement paste, and often containing large calcium hydroxide crystals. Due to these features, the interfacial zone is often considered to be the weak link in concrete with respect to mechanical performance and durability.

In this paper, a digital-image-based cement microstructural model is applied to simulate interfacial zone microstructural development. Reasons for the differences between interfacial zone and bulk paste microstructure will be developed and methods for improving the structure of this weak link will be explored.

Interfaces in Cementitious Composites. Edited by J. C. Maso. © RILEM.
Published by E & FN Spon, 2 - 6 Boundary Row, London SE1 8HN. ISBN 0 419 18230 6.

2 Model description

The microstructural model used to investigate interfacial zone microstructure was developed by Bentz and Garboczi (1990). The model has been described in detail by these authors (1991a and 1991b) and is outlined briefly here. In the model, two-dimensional (2-D) area or three-dimensional (3-D) volume is represented by a set of pixels each identified as a particular phase of concrete. For this paper, relevant phases are anhydrous cement (assumed to be pure C_3S), calcium hydroxide (CH), calcium silicate hydrate (C-S-H), porosity, aggregate, and mineral admixtures. Thus a cement particle is represented as a collection of contiguous pixels assigned to be C_3S. In this manner, real cement particle shapes may be utilized, as well as circles and spheres, in generating simulated cement microstructures. Figure 1 is an image of real cement particles in 2-D (obtained using a scanning electron microscope and image analysis) placed randomly around an idealized square aggregate and the same system after 77% hydration using the model. In the hydrated system, cement particles are white, CH is bright grey, C-S-H is dark grey, and porosity is black.

Hydration is simulated by operating on all the pixels present in the cement paste volume or area. The two reactions considered are the hydration of C_3S to form CH and C-S-H and the pozzolanic reaction between CH and silica (found in silica fume or fly ash) to form secondary or pozzolanic C-S-H. The assumed reaction stoichiometry, based on the work of Young and Hansen (1987), is as follows:

$$C_3S + 5.3H \rightarrow C_{1.7}SH_{4.0} + 1.3CH$$
$$S + 1.7CH + 2.3H \rightarrow C_{1.7}SH_{4.0}$$

In terms of volumes or areas, each unit of dissolved C_3S produces 1.7 units of C-S-H and 0.61 units of CH. Regarding the pozzolanic reaction, each unit volume of silica is capable of reacting with 2.08 units of CH to produce 4.6 units of pozzolanic C-S-H.

Hydration is executed in discrete cycles consisting of dissolution, diffusion, and reaction phases. During dissolution, all C_3S pixels in contact with water-filled porosity are given a chance to dissolve and produce diffusing CH and C-S-H species. The dissolution probability is based on the amount of C_3S surface in contact with water. During diffusion, the C-S-H and CH diffusing species execute random walks within the available pore space until reaction occurs. C-S-H forms on the surfaces of the original cement particles or on previously formed solid C-S-H. CH forms crystals by a nucleation and growth mechanism within the pore space. Additionally, if amorphous silica is present, the CH diffusing species react at silica surfaces to form pozzolanic C-S-H. When all diffusing species generated from one dissolution have reacted, a cycle is complete and a new dissolution is begun. By monitoring how much cement remains after any number of hydration cycles, the degree of hydration, α, of the system can be determined. As in real systems, the hydration is ultimately self-limiting as the surfaces of the remaining cement become totally surrounded by hydration product, preventing further dissolution. However, typically 80-90% hydration can be achieved before this situation occurs.

To simulate an interfacial zone in concrete, a single aggregate is first placed in the hydration volume (area) and then the cement particles are randomly placed such that no particles overlap. To simulate a lightweight absorptive aggregate, this original distribution of particles is modified by moving all particles toward the aggregate surface to simulate water absorption by the (dry) aggregate. This results in the aggregate acting somewhat as a filter, drawing in water and pulling cement particles towards its surface, as suggested by Fagerlund (1978). Additionally, the use of cement clinker as

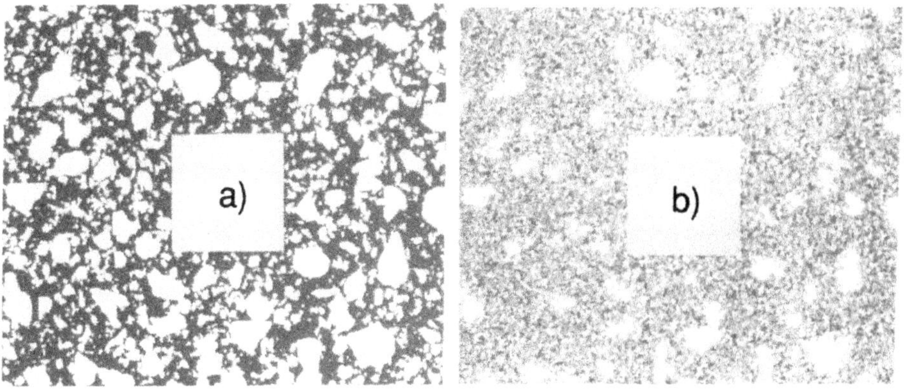

Fig 1. Cement particles placed around square aggregate (w/c=0.47) a) before and b) after 77% hydration.

aggregate may be simulated by assigning the aggregate to be C_3S. Due to its small surface-to-volume ratio, only a thin layer on the outer surface of the clinker aggregate will undergo significant hydration. Except where indicated otherwise, all simulations described below were performed using the two-dimensional version of the microstructural model. All 2-D results are the average of 5 to 10 separate configurations of a given system.

3 Interfacial zone microstructure

The application of the microstructural model to simulate interfacial zones has been described by Garboczi and Bentz (1991). The results of a simulation similar to the system in Fig. 1, but with circular cement particles, indicated that the interfacial zones contain less cement and C-S-H and more porosity and CH than the bulk paste, in accord with what has been observed in real concretes. Although particle packing effects obviously contributed to these characteristic features of the interfacial zones, a secondary one-sided growth mechanism also plays a role. In the interfacial zone, available porosity is filled with hydration products growing from one direction only, unlike the bulk paste where products grow inward from all directions. Using simulation, it was possible to separate these two effects. By first placing the cement particles throughout the entire system and then overlaying the aggregate and eliminating all portions of cement particles underneath the aggregate, it was possible to remove the effects of the initial particle packing (i.e. the wall effect) and study only the one-sided growth effect. This system could then be contrasted against one in which the aggregate is placed first followed by the cement particles, where both the wall effect and one-sided growth mechanism would be present.

Figure 2 shows the results of this simulation in terms of porosity as a function of distance from the aggregate surface for the initial and fully hydrated systems for both system configurations mentioned above. The one-sided growth mechanism appears to be a secondary but significant contributor to the increased porosity in the interfacial zone. The cement particles used in this simulation were monosize and 21 pixels in diameter, so that the thickness of the interfacial zone is about the same as the diameter of the cement particles.

To further investigate the link between cement particle size and interfacial zone thickness, a continuum three-dimensional spherical

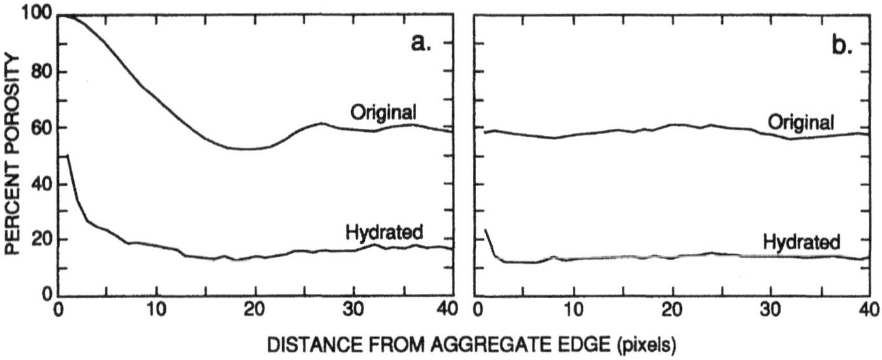

Fig. 2. Porosity vs. distance from aggregate surface before and after hydration for a) wall effect and one-sided growth mechanism and b) one-sided growth mechanism only.

particle parking program was implemented. Spherical cement particles of a known size distribution (from a sieve analysis) are placed randomly in order of largest to smallest in a cube 200 μm on a side such that no particles overlap. Rigid walls are present on two opposite faces of the box to represent aggregates while periodic boundary conditions are maintained on the other four sides. Volume fractions of cement and porosity are assessed using point sampling techniques. Using sieve size distributions presented in van Breugel (1991), results are shown in Figure 3 for two cements at various w/c ratios. Cement A1 has a median particle size, on a mass basis, of about 28 μm while the median particle size of cement A7 is about 11 μm. The thicknesses of the initial interfacial zones (i.e. the distance at which the porosity first approaches its "average" value) correspond closely to these median particle diameters. Conversely, w/c ratio has little effect on interfacial zone thickness, although it does effect the gradient in cement content as the aggregate surface is approached. Although hydration may decrease these interfacial zone thicknesses as seen in Fig. 2, these results suggest cement fineness to be one variable which strongly influences interfacial zone

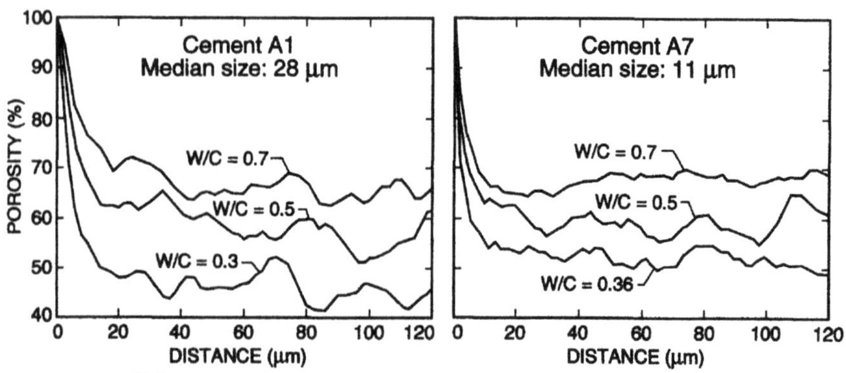

Fig. 3. Porosity fraction near aggregate before any hydration.

characteristics as the interfacial zone thickness is expected to be proportional to the median cement particle size.

Having outlined two causes of interfacial zone microstructural features, namely the wall and one-sided growth effects, we now explore various methods for improving interfacial zone microstructures.

4 Modifications of interfacial zone microstructure

4.1 Use of mineral admixtures

The modification of interfacial zone microstructure due to the addition of silica fume has been well documented by Scrivener et al. (1988). When properly prepared, concrete containing silica fume exhibits interfacial zones which are nearly as dense as the bulk paste. This is due to both the small size of silica fume particles which allows them to pack more efficiently near the aggregate surface and their high pozzolanic activity which converts CH into more C-S-H.

The simulations described in detail by Bentz and Garboczi (1991b) allow for both the size and reactivity of mineral admixtures such as silica fume and fly ash to be varied. Since in the 3-D version of the model, a pixel is typically 1 μm on a side, well dispersed silica fume can be represented by small particles one pixel in size. Conversely, fly ash or agglomerated silica fume is "large", and is assigned the same size as cement particles. Pozzolanic activity is varied by specifying how many units of CH each unit of mineral admixture may consume via the pozzolanic reaction. According to the reaction given earlier, pure silica would be able to consume 2.08 units of CH. Fly ash must be assigned a lower reactivity as it is typically only about 50% silica. Thus 1.04 might be a reasonable maximum activity to be expected for fly ash, while 0.47 appears to be a more likely value to be achieved in practice based on data from Hooton (1986).

Since increased strength has provided a major impetus for the use of silica fume in concrete, results will be summarized in terms of the total phase fraction of cement + C-S-H + mineral admixture present, assuming these to be the major load bearing phases in the cement paste matrix. Figure 4 shows the distribution of these phase fractions near an aggregate for a variety of concretes at a constant water-to-solids (w/s) ratio of 0.45 after 77% hydration. In Fig. 4, size refers to the size of the admixture particles, while reactivity is equivalent to the pozzolanic activity discussed above. Small silica fume particles (System F) added at a 20% replacement for cement, on a mass basis,

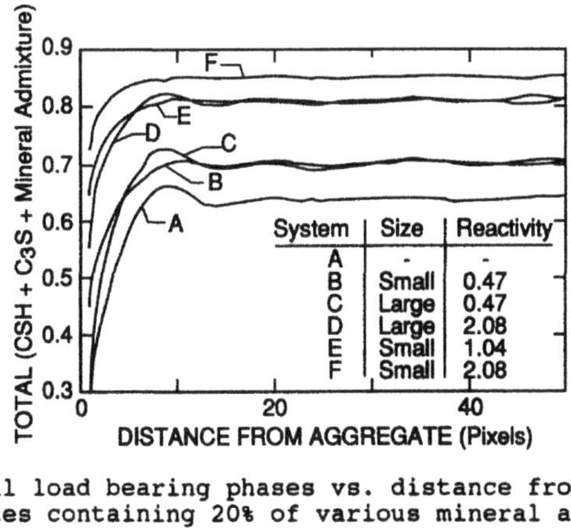

Fig. 4. Total load bearing phases vs. distance from aggregate for concretes containing 20% of various mineral admixtures.

result in a nearly uniform distribution of these load bearing phases throughout the microstructure. Fly ash is not nearly as effective as silica fume, although the results suggest that utilizing very fine fly ash should be beneficial if its ultimate pozzolanic activity of 1.04 is achieved (System E) due to the exposure of all available silica surfaces. Berry et al. (1989) have obtained experimental results in support of this hypothesis. Thus small-sized pozzolanic mineral admixtures are seen to somewhat offset the wall effect while not affecting the one-sided growth effect, but they offer the additional benefit of converting CH into a larger volume of pozzolanic C-S-H.

4.2 Use of lightweight absorptive aggregate
Recently, Zhang and Gjorv (1990) have characterized the microstructure of the interfacial zone in lightweight concrete as being more dense and homogeneous than that found in normal weight concrete. One possible reason for this observation is the rearrangement of the cement particles in the vicinity of an aggregate due to water absorption by the aggregate. As discussed earlier, this phenomena can be conveniently simulated using the microstructural model.

To perform this investigation, the coarse aggregate was assumed to have an absorption of 14% on a dry mass basis (data from Ben-Othman and Buenfeld (1990)) and to be present in the concrete at a 42% volume fraction. Assuming that dry aggregate was added to the mix, an original w/c ratio of 0.586 would be required to produce a final w/c ratio in the paste of 0.39 after aggregate saturation. The final unhydrated configuration for such a system is given in Fig. 5a. Here, cement particles were placed to achieve the starting 0.586 w/c ratio and then were moved a pixel at a time towards the flat plate aggregate surface until the final w/c was produced. Of course, mixing of concrete would be expected to disturb this arrangement of particles, but this simple approach should adequately serve as a model representation of the phenomena. This system can be contrasted against a system with normal weight aggregate and a w/c of 0.39.

Results for phase fractions vs. distance from the aggregate surface after 77% hydration are provided in Fig. 5b for both porosity and C_3S +C-S-H. The lightweight absorptive aggregate improves both of these phase distributions relative to those found for the normal aggregate system. Because the cement particles are more densely packed near the aggregate, as the aggregate is approached from a distance the porosity decreases slightly while the C_3S+C-S-H increases slightly. Within 5 to 10 pixels of the aggregate, the porosity increases and the C_3S+C-S-H

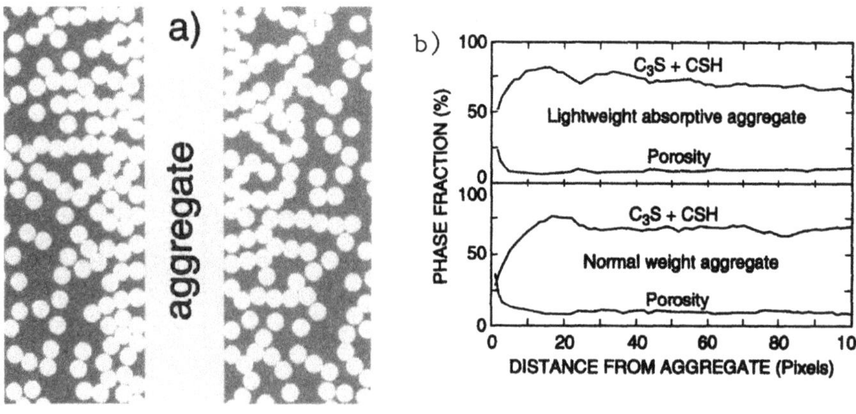

Fig. 5. Hydration of lightweight absorptive aggregate concrete a) initial microstructure and b) quantitative phase analysis.

decreases but not as drastically as for the system with normal weight aggregate. Thus the lightweight absorptive aggregate partially eliminates the wall effect although it is still present to an extent as even packing spheres against a flat wall results in a high porosity zone near the wall surface. Here, the scale of the roughness of the cement particles and the aggregate may play a role in real concrete. In addition to roughness providing mechanical interlocking, for absorptive aggregate, one might expect the cement particles to be drawn tightly against the rough aggregate surface in a "lock-and-key" fashion to substantially reduce the wall effect. In any case, the one-sided growth mechanism would still be operative.

4.3 Use of cement clinker as aggregate

Berger (1974) has investigated the use of cement clinker aggregate in concrete. He attributed the increased strengths of these systems to an improved cement paste-aggregate bond. Since the clinker aggregate will hydrate on its surface, the one-sided growth mechanism should be eliminated. Additionally, since cement clinker is generally porous, it should affect interfacial zone microstructure in a manner similar to the lightweight aggregates discussed in section 4.2. Once again, simulation provides a convenient means for studying these systems.

To model cement clinker aggregate, the aggregate is assigned a phase identifier of C_3S so that it is eligible for dissolution and reaction just like the smaller cement particles. Because of its small surface-to-volume ratio, only the exterior will undergo significant hydration as shown in Fig. 6a for an absorptive clinker aggregate (12% absorption by mass as measured on a typical dry clinker in our lab) with w/c=0.39 and α=75% for the bulk cement paste. From the phase fractions plotted in Fig. 6b, it can be seen that the distribution of porosity and of C_3S+C-S-H are now nearly homogeneous throughout the system suggesting a superior paste-aggregate bond. Due to the high cement content at the aggregate surface, porosity decreases slightly and C_3S+C-S-H increases within a few pixels of the aggregate. Not only has the one-sided growth mechanism been eliminated, but now the aggregate side of the local volume is composed entirely of cement.

Non-absorptive clinker aggregate can also be easily simulated. In this case, improvements in interfacial zone microstructure are not as drastic but are substantial as seen in Fig. 7 for a fully hydrated system with a w/c ratio of 0.47. Like the plots shown in Figure 2 for conventional concrete, the porosity increases as the aggregate is approached from a distance, but a decrease is observed in the

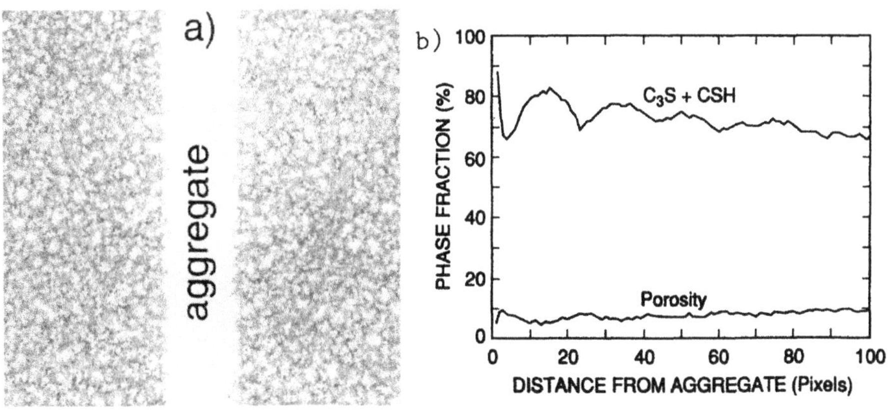

Fig. 6. Hydration of absorptive clinker aggregate concrete a) hydrated microstructure and b) quantitative phase analysis.

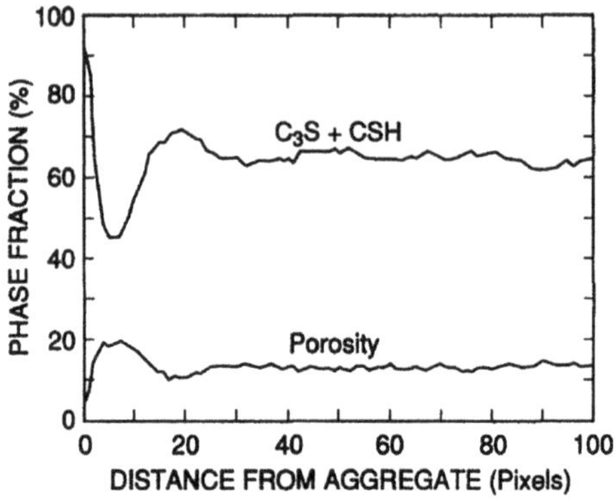

Fig. 7. Quantitative phase analysis for non-absorptive clinker aggregate concrete.

immediate vicinity of the aggregate due to the now beneficial effects of the aggregate-side growth. Similar trends are also observed for the C_3S+C-S-H phase fraction. Using cement clinker as aggregate produces a superior interfacial zone microstructure because both the one-sided growth mechanism and wall effect can be addressed, depending on the clinker's absorption characteristics.

5 Comparison between model and real microstructures

Bentz et al. (1992) have verified the cement microstructural model by comparing the interfacial zone microstructures of model and real specimens for systems in which 0, 10, and 20 mass percent of the cement has been replaced by silica fume. Specimens consisting of cement paste and coarse aggregate were prepared and phase fraction distributions were measured using scanning electron microscopy and image analysis. Results are shown in Fig. 8 for the total cement+C-S-H+silica fume and porosity phase fractions. In Fig. 8, the model results have been scaled along the x-axis so that the average cement particle size, about 40 μm, is similar in the two systems. In general, agreement between the two systems is quite good. As seen earlier, silica fume improves the concrete microstructure by decreasing the porosity and increasing the cement+C-S-H+silica fume phase fraction present in the interfacial zone as well as in the bulk paste. These phase fractions are nearly homogeneous throughout the 20% silica fume concrete.

6 Summary

A digital-image-based model has been applied to investigating a variety of concrete interfacial zone microstructures. In the absence of bleeding, the model suggests that interfacial zone microstructural features are due to both wall and one-sided growth effects and that the interfacial zone thickness will largely be controlled by the median particle size of the cement. Denser, more uniform interfacial zone microstructures can be achieved in a variety of ways. Fine pozzolanic mineral admixtures improve interfacial zone microstructure by partially offsetting the wall (packing) effect and producing extra C-S-H due to their pozzolanic reaction with CH. Both size and pozzolanic reactivity are paramount in modifying interfacial zone

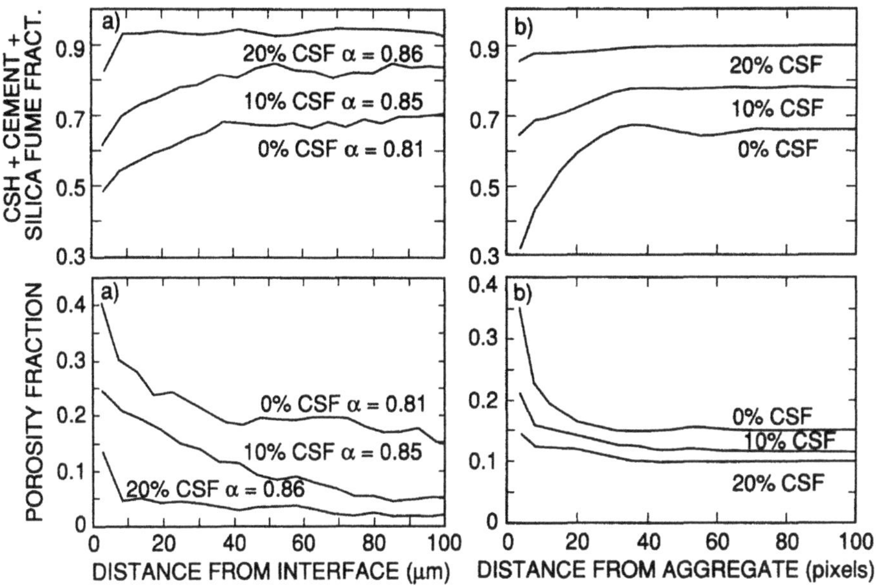

Fig. 8. Comparison between a) real and b) model specimens.

microstructure with small (1 μm) highly reactive admixtures such as silica fume providing the most improvement. Lightweight absorptive aggregates also partially offset the wall effect by rearranging the distribution of cement particles near the aggregate. Some decrease in the density of the paste is still observed in the immediate vicinity of the aggregate due to inefficient packing and the one-sided growth effects. The use of porous cement clinker as aggregate reduces both the one-sided growth and wall effects.

The microstructural model has been verified by comparing the microstructure of model and real concretes containing various amounts of silica fume. The good agreement obtained suggests that the model accurately represents microstructural development in the interfacial zones of concrete. As the demand for high performance concrete increases, the capability to engineer interfacial zone microstructure will become more important. Simulation of microstructural development appears to be a powerful tool in achieving this goal.

7 Acknowledgements

The authors would like to thank the National Science Foundation Science and Technology Center for Advanced Cement-Based Materials for partial support of this work.

8 References

Ben-Othman, B., and Buenfeld, N.R. (1990) Oxygen permeability of structural lightweight aggregate concrete, in **Protection of Concrete**, E & FN Spon, pp. 725-736.

Bentur, A. (1991) Microstructure, interfacial effects, and

micromechanics of cementitious composites, in **Advances in Cementitious Materials** (ed S. Mindess), American Ceramic Society, Westerville, OH, pp. 523-550.
Bentz, D.P., and Garboczi, E.J. (1990) Digitized direct simulation model of the microstructural development of cement paste, in **Physical Phenomena in Granular Materials** (eds G.D. Cody, T.H. Geballe, and P.Sheng), Materials Research Society, Pittsburgh, PA, pp. 523-530.
Bentz, D.P., and Garboczi, E.J. (1991a) A digitized simulation model for microstructural development, in **Advances in Cementitious Materials** (ed S. Mindess), American Ceramic Society, Westerville, OH, pp. 211-226.
Bentz, D.P., and Garboczi, E.J. (1991b) Simulation studies of the effects of mineral admixtures on the cement paste-aggregate interfacial zone, **ACI Mat. J.**, 88, 518-529.
Bentz, D.P., Stutzman, P.E., and Garboczi, E.J. (1992) Experimental and simulation studies of the interfacial zone in concrete, accepted by **Cem. and Conc. Res.**
Berger, R.L. (1974) Properties of concrete with cement clinker aggregate, **Cem. and Conc. Res.**, 4, 99-112.
Berry, E.E., Hemmings, R.T., Langley, W.S., and Carette, G.G. (1989) Beneficiated fly ash: Hydration, microstructure, and strength development in portland cement systems, in **Fly Ash, Silica Fume, and Natural Pozzolans in Concrete** (ed V.M. Malhotra), ACI, Detroit, MI, pp. 241-273.
Costa, U., Facoetti, M., and Massazza, F. (1990) Permeability of the cement-aggregate interface: Influence of the type of cement, water/ cement ratio and superplasticizer, in **Admixtures for Concrete: Improvement of Properties** (ed E. Vazquez), Chapman and Hall, London, pp. 392-401.
Fagerlund, G. (1978) Frost resistance of concrete with porous aggregate, Report of the Cement and Concrete Institute in Sweden.
Garboczi, E.J., and Bentz, D.P. (1991) Digital simulation of the aggregate-cement paste interfacial zone in concrete, **J. Mat. Res.**, 6, 196-201.
Hooton, R.D. (1986) Permeability and pore structure of cement pastes containing fly ash, slag, and silica fume, in **Blended Cements** (ed G. Frohnsdorff), ASTM, Philadelphia, PA, pp. 128-143.
Ping, X., Beaudoin, J.J., and Brousseau, R. (1991) Flat aggregate-portland cement paste interfaces, I. Electrical conductivity models, **Cem. and Conc. Res.**, 21, 515-522.
Scrivener, K.L., Bentur, A., and Pratt, P.L. (1988) Quantitative characterization of the transition zone in high strength concretes, **Adv. in Cem. Res.**, 1, 230-237.
Scrivener, K.L., and Gartner, E.M. (1988) Microstructural gradients in cement paste around aggregate particles, in **Bonding in Cementitious Composites** (eds S. Mindess and S.P. Shah), Materials Research Society, Pittsburgh, PA, pp. 77-85.
Uchikawa, R. (1988) Similarities and discrepancies of hardened cement paste, mortar, and concrete from the standpoint of composition and structure, **J. Res., Onoda Cem. Co.**, 40, 1-24.
Van Breugel, K. (1991) **Simulation of Hydration and Formation of Structure in Hardening Cement-Based Materials**, Ph.D. thesis, Dept. of Civil Engineering, Delft University of Technology, The Netherlands.
Young, J.F., and Hansen, W. (1987) Volume relationships for C-S-H formation based on hydration stoichiometries, in **Microstructural Development During Hydration of Cement** (eds L.J. Struble and P.W. Brown), Materials Research Society, Pittsburgh, PA, pp. 313-322.
Zhang, M.H., and Gjorv, O.E. (1990) Microstructure of the interfacial zone between lightweight aggregate and cement paste, **Cem. and Conc. Res.**, 20, 610-618.
Zimbelmann, R. (1985) A contribution to the problem of cement-aggregate bond, **Cem. and Conc. Res.**, 15, 801-808.

PART 1.3

INTERFACIAL MICROSTRUCTURE AND PROPERTIES: MECHANICAL PROPERTIES - TECHNICAL METHODS

(Microstructure et propriétés de l'auréole de transition: Characterisation de la microstructure de l'auréole: Propriétés mecaniques - methodes d'etude)

13 A STUDY OF PROPERTIES OF THE PASTE-AGGREGATE INTERFACE

K. MITSUI
Takenaka Corporation, Tokyo, Japan
ZONGJIN LI, D. A. LANGE and S. P. SHAH
Northwestern University, Evanston, Illinois, USA

Abstract
The properties of the interfacial zone between aggregate and cement paste was studied. A newly developed push-out method was used to evaluate the interfacial properties of transition zone including the stiffness, strength, friction and fracture energy of the interface. A closed loop controlled set-up was used to continuously measure force and slip-displacement during loading. The experimental data were analyzed by using a recently proposed theoretical model. The transition zone was further investigated by using backscattered electron imaging and energy dispersive analysis of x-rays to characterize the microstructure of the interface. The relationship between mechanical properties and microstructure of interfacial zone was studied. Type of aggregate, treatment of aggregate surface, mixing process and incorporation of silica fume were investigated. It was found that the microstructure of the interfacial zone can be significantly improved by aggregate pretreatment and mixing methods, and thus the mechanical interfacial properties can be enhanced.
Keywords: Interfacial zone, Bond-slip relationship, Fracture energy, Silica fume, Aggregate, Shear bond strength

1 Introduction

A number of studies have been conducted to characterize the chemical nature, morphology and mechanical properties of interfacial zone. The microstructure of the interface has been investigated using a scanning electron microscope(BSE)[1]. The formation of a calcium hydroxide film adjacent to the aggregate surface has been observed. By using EDAX, the presence of ettringite and the preferentially oriented calcium hydroxide crystals in the interfacial zone has been demonstrated[2]. BSE imaging has been used to quantify the transition zone[3]. It was found that the porosity at the interface decreases towards the bulk paste.

Attempts have been made to characterize the mechanical properties of the interface. Microcracks tend to propagate around the aggregate which implies that the fracture resistance of the interface is lower than that of the bulk paste[4]. To measure aggregate-paste bond strength, different types of test methods, such as three or four point bending test[5,6], direct tensile test[7,8], axial compression test and shear bond test[6], have been employed. It has been shown that bond strength varies

Interfaces in Cementitious Composites. Edited by J. C. Maso. © RILEM.
Published by E & FN Spon, 2 - 6 Boundary Row, London SE1 8HN. ISBN 0 419 18230 6.

with curing condition, type of aggregate, and type of cement. To better characterize the bond behavior, one needs to know the bond-slip response of the interface and the influence of the microstructure on the mechanical properties. However, few studies have concentrated on these subjects.

The objective of this research is to investigate the relationship between mechanical properties and microstructure of the transition zone. This may lead to better understanding of the toughening mechanism of concrete. In the study, the bond-slip relationship between aggregate and cement paste was measured. Interfacial mechanical properties such as interfacial stiffness, shear bond strength and interfacial surface energy were obtained by analyzing the bond-slip relation-ship. The investigation included evaluation of the aggregate-paste interface by BSE imaging for the corresponding specimens.

2 Experimental

2.1 Mechanical test set-up

The test set-up is shown schematically in Fig.1. The specimen was put on a flat circular plate which was connected with a servohydraulic actuator. A steel rod made the contact with the top surface of the aggregate and pushed the central cylindrical aggregate downwards. Two LVDTs were used to measure slip displacement of the top of the aggregate relative to the cement matrix annulus. The average output of the LVDTs was used as feedback signal to control the servohydraulic system.

2.2 Materials

The paste matrix was prepared by using Type I ordinary Portland cement. Silica fume(SF) used was condensed silica fume slurry. High range water reducer used was graft co-polymer with functional sulfonic group and carboxyl group. The aggregate used were Indiana limestone and granite.

2.3 Specimen preparation

Aggregate was cored out from stone block as a 12.7 mm diameter cylinder. The surface of some aggregates were pre-treated with several kinds of processing listed in Table 1. After mixing cement paste matrix, aggregate was put in the center of the 50.8mm diameter cylindrical mold and then covered with matrix. Specimens were cured in water under the temperature of 20^{0}C until testing ages of 14 and 28 days. Specimens were sliced perpendicular to the cored aggregate into appropriate thickness and ground and polished on both top and bottom surface before push-out test. Three specimens for each mixture were tested.

3 Microscopic study

3.1 Sample preparation

Samples were also cast for study in SEM by using the same procedures described earlier. These specimens were cut into thin slices on a diamond saw at 14 and 28 days. Samples were impregnated with epoxy and then polished to 0.25μm.

3.2 Evaluation procedure

The interfacial zone was examined with SEM with EDAX. An image magnification of 400x was selected based on precedence of work by Scrivener[9]

Table 1. Test program

SPECIMEN No.	MATRIX		AGGREGATE		Age at test(days)
	material	W/C	material	surface	
1(control)	cement	35%	limestone	none	14,28
2	cement	65%	limestone	none	28
3	cement	35%	granite	none	28
4	C+SF[1]	35%	limestone	none	14,28
5	C+SF[1]	35%	limestone	C+SF[2]	14,28
6	cement	35%	limestone	cement[3]	14,28
7	cement	35%	limestone	SF[4]	14,28

[1] Mixture of cement with 10% by weight of silica fume
[2] Paint C+SF slurry on aggregate and then expose to room temperature(20^0C) for 1 hour before casting;
[3] Paint cement slurry on aggregate, expose in room temperature for 1 hour then cast;
[4] Immerse aggregate in SF slurry, dry out 6 hours, then cast.

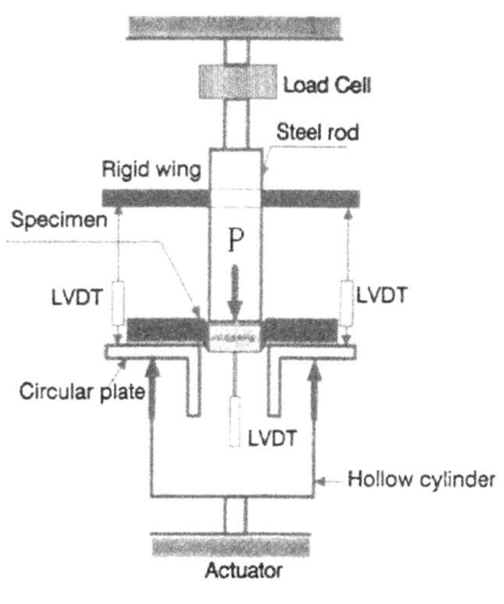

Fig.1 Push-out test set-up

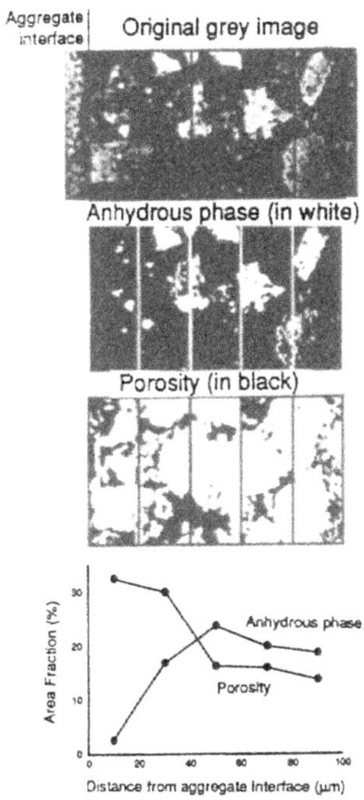

Fig.2 Typical micrograph of aggregate–paste interface (top) is segmented to create binary image of anhydrous phase (middle) and porosity (bottom). The gradients of anhydrous and porous phases are represented in the graph below the three images.

and others. Each region of different brightness in the images was analyzed with EDAX to identify the different chemical components, such as unhydrated cement, calcium hydroxide, C-S-H and pores. It was found that grey level in BSE images could be used to identify phases in the microstructure. The micrographs, captured from the SEM, were digitized by an image analysis system. The digitized images are 512 x 512 pixels with 256 grey levels. The acquisition magnification of 400x in a resolution of about 0.5 microns per pixel width.

Each image of the aggregate-paste interface was subdivided into bands with a width of 20μm and up to 200μm away from the aggregate surface. Fig.2 shows a typical image to illustrate this technique. Binary images of the unhydrated phase and the porosity phase were then created for each subdivided image according to the threshold of grey level. The area percentage, which is representative of the volume fraction of material, was measured for each band. Gradient plots for anhydrous material and pore space created by plotting the average area percentage versus the distance from the aggregate surface for each image.

4 Analysis method for mechanical property test

As shown in Figures 5-8, it should be noted that the push-out load versus slip displacement curve clearly shows a linear-elastic stage, a non-linear ascending stage and a gradual descending stage. This implies that the push-out process includes a phenomena of interfacial crack propagation. However, the conventional analysis of the behavior of interface generally only uses the peak load to calculate the average bond strength, ignoring the progressive interfacial crack propagation. To improve our knowledge about the nature of the aggregate-cement paste interface, a newly developed mathematical model[10], in which the fracture concepts have been taken into account, was employed. The summary of this new theoretical model is given below.

Fig.3 represents the schematic figure of the mathematical model. In Fig.3, L and A represent the aggregate embedded length and cross-sectional area. The aggregate is assumed to be elastic with Young's

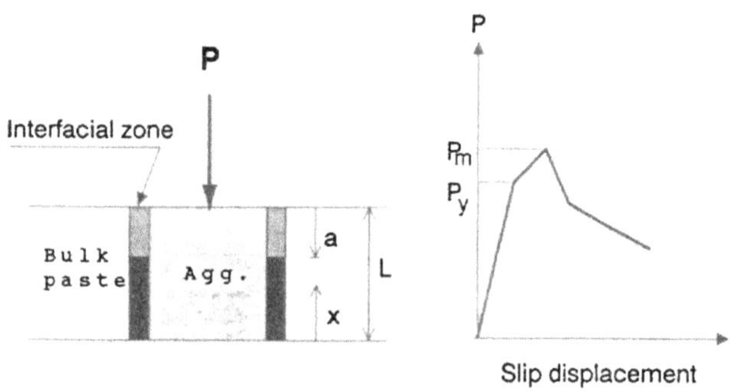

Fig.3 Mathematical model

Modulus of E_a. The bulk matrix is assumed to be rigid except for the interfacial layer which is idealized as an elastoplastic shear layer.

It is assumed that debonding has occurred over a certain length, a, starting at x=L. Treating the boundary layer as a shear lag and assuming that a constant shear stress are acting at the debonded interface, The following equations can be written,

$$q = \begin{cases} kU(x) & 0 < x < L-a \\ q_f & L-a < x < L \end{cases} \tag{1}$$

where k is the stiffness per unit length of the interfacial layer for small deformation, q is the shear force per unit length acting on the aggregate, q_f is the frictional shear force per unit length, and U is the aggregate displacement.

Introducing the equilibrium equation and the constitutive relationship for the aggregate, the following differential equation for U can be obtained,

$$\begin{aligned} U_{,xx} - \omega^2 U &= 0 \qquad 0 < x < (L-a) \\ U_{,xx} - \frac{q_f}{E_a A} &= 0 \qquad (L-a) < x < L \end{aligned} \tag{2}$$

Where, subscript "," indicates the differential operator.
The quantity ω is defined as $\omega=(k/E_aA)^{0.5}$
Equations (2), together with boundary conditions and continuity conditions, constitute a complete set of equations for the determination of U(x). Solving this set of equations the following closed form expression for the slip displacement at the loading end, U^*, is

$$U^* = U(L) = \frac{P^* - q_f a}{E_a A \omega} \coth(\omega(L-a)) + \frac{P^* - 0.5 q_f a}{E_a A} a \tag{3}$$

Furthermore, the relationship between the push-out force and debonding length, a, derived from both the shear strength criterion (material properties represented by q_y, and q_f) and fracture energy criterion (material properties represented by Γ, and q_f) take the following forms;
for the shear strength criterion:

$$P^* = q_f a + \frac{q_y \tanh \omega (L-a)}{\omega} \tag{4}$$

for the fracture energy criterion:

$$P^* = q_f a + \left[\frac{q_f}{2\omega} + \left\{ \left(\frac{q_f}{2\omega} \right)^2 + 2 E_a A p \Gamma \right\}^{0.5} \right]^{0.5} \tanh(\omega(L-a)) \tag{5}$$

Where, Γ is the specific fracture energy.

Note that three material parameters are needed for either model. They are ω, q_y and q_f or ω, Γ and q_f.

The stiffness parameter, ω can be determined from the initial slope of the experimental load-slip displacement curve. The interfacial yield parameter, q_y, the interfacial frictional forces, q_f, and the specific energy, Γ, can not be independently obtained from a load-slip experimental curve. A method which utilizes the maximum load, P^*_{max}, and the slip displacement corresponding to P^*_{max} is proposed[11]. The formula used to calculate q_f, q_y and Γ are Eq.6, 7 and 8, respectively,

$$q_f = \frac{\omega P^*_{max}}{a\omega + \sinh(\omega(L-a))\cosh(\omega(L-a))} \tag{6}$$

$$q_y = q_f \cosh^2(\omega(L-a)) \tag{7}$$

$$2E_f Ap\Gamma = \left(\frac{q_f}{\omega}\right)^2 [\cosh^4(\omega(L-a)) - \cosh^2(\omega(L-a))] \tag{8}$$

Note that an addition equation (Eq.9) is needed to determine the bonding length, a.

$$P^2_{max}\frac{0.5(\omega a)^2+\cosh^2(\omega(L-a))+\omega a\sinh(\omega(L-a))\cosh(\omega(L-a))}{\omega a + \sinh(\omega(L-a))\cosh(\omega(L-a))} - U^*_{peak}E_a A\omega = 0 \tag{9}$$

In Eq.(9), U^*_{peak} represents the slip displacement corresponding to the peak load. The shear strength, τ_y and frictional stress τ_f can be obtained by dividing q_y and q_f by perimeter of the aggregate.

5 The results of the analysis

The calculated results of these parameters for push-out specimens tested at 14 and 28-day ages are listed in Table 2. As an illustration, Fig.4 shows a comparison of the interface for three different samples. Fig.4(a) is an image of control (reference) with a w/c ratio of 0.35. A clear, porous transition zone of about 100µm is observed. When the w/c ratio is increased to 0.65, the amount of porosity increased as shown in Fig.4(b). On the other hand, the pre-coating procedure together with an incorporation of silica fume into the matrix shows a dramatic densification of the microstructure. As shown in Fig.4(c), porosity in the vicinity of the aggregate decreased and transition zone was almost eliminated. These observations may help to explain the behavior of mechanical properties of the interface.

Fig.5 indicates the influence of the w/c ratio of the paste. It can be seen from Fig.5(a) that the specimen with w/c of 0.35 has the higher push-out load and steeper initial slope. Consequently, it can be shown that all the interfacial properties decrease with the increase of the w/c ratio, especially the fracture energy which becomes less than half(Table 2). Fig.5(b) shows an obvious difference of the porosity

Table 2 Test results with 14 and 28-day curing

	ω (1/m)		τ_y (MPa)		τ_f (Mpa)		Γ (N/m)	
No.	14-day	28	14	28	14	28	14	28
1	17.87	20.30	2.87	4.07	2.83	3.98	2.10	4.74
2	-	19.09	-	3.76	-	3.73	-	2.01
3	-	13.09	-	3.33	-	3.26	-	3.67
4	15.70	23.02	2.94	5.88	2.90	5.71	2.77	9.64
5	20.48	27.36	6.28	11.44	6.12	10.34	12.11	59.02
6	23.57	23.78	5.55	7.81	5.35	7.57	9.77	16.74
7	15.41	16.18	1.80	2.63	1.78	2.59	1.43	1.92

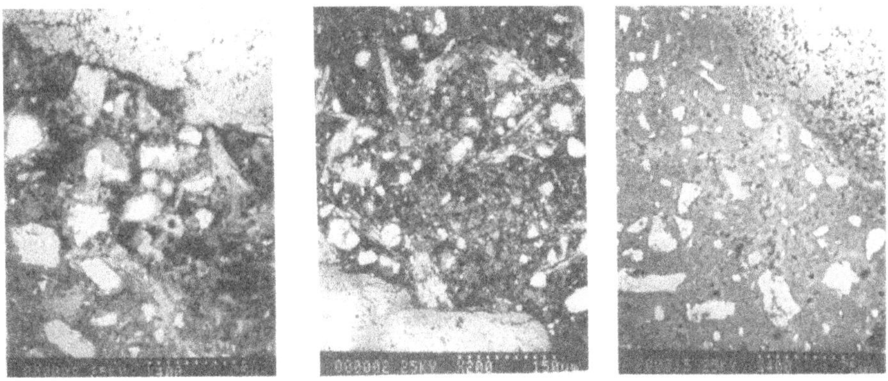

(a)control (w/c=35%) (b)W/C=65%,cement matrix (c)w/c=35%,matrix and coating:C+SF

Fig.4 Bacscattered electron micrograph of interface

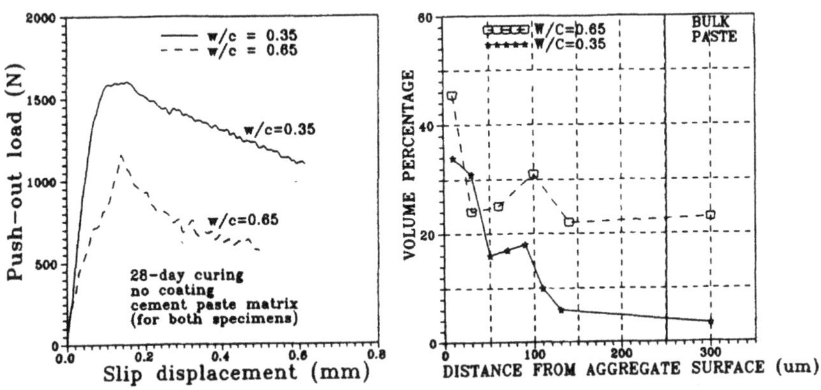

(a)Load-slip displacemant curve (b)Pore distribution

Fig.5 The influence of W/C to microstructure and mechanical properties gradient of the porosity for two different w/c ratio. Note that one of the reasons of the existence of the interfacial zone is that bleed water stays under coarse aggregate and creates a porous zone under aggregate. Therefore, the higher the w/c ratio, the greater the bleeding and the more porous will be the interfacial zone. This phenomena is helpful to explain the reduction in bond mechanical properties.

It has been pointed out that the mineralogical property of aggregate can affect the microstructure of the interface[6]. A possible reaction between calcite in limestone and calcium hydroxide in hydrated cement may cause microstructural changes in the interfacial zone as shown in earlier studies[12]. The test results of two mineralogically different aggregate, limestone and granite are shown in No.1 and 3 on Table 2. In the case of granite, all the interfacial properties are lower than those of limestone, especially the stiffness of the interface. This reflects the possible microstructural difference of the interfaces.

The effects of silica fume can be attributed to the fact that it reacts with calcium hydroxide in concrete and makes the structure dense through pozzolanic reaction and a "packing" effect due to smaller particle size. As a result, the amount of interfacial porosity is reduced. Fig.6(b) compares the response of the control specimen with the specimen with silica fume modified matrix. A dramatic reduction of the porosity within the transition zone is shown in Fig.6(a). Furthermore a significant increase in the value of interfacial surface energy is demonstrated in Table 2. The additional effect of treating the aggregate with slurry of the cement plus silica fume is demonstrated in Fig.7. It can be seen that for the specimen incorporated silica fume both in matrix and surface coating, the porosity of the interface is further reduced and the maximum push-out load is about twice as high as that of the specimen with silica fume modified matrix alone. The influence of silica fume modification is graphically illustrated in Fig.8. It can be seen that incorporating silica fume improved all the interfacial properties, especially the interfacial specific fracture energy. The surface fracture energy increased about two times for the specimens with SF modified matrix and about 10 times for the specimens

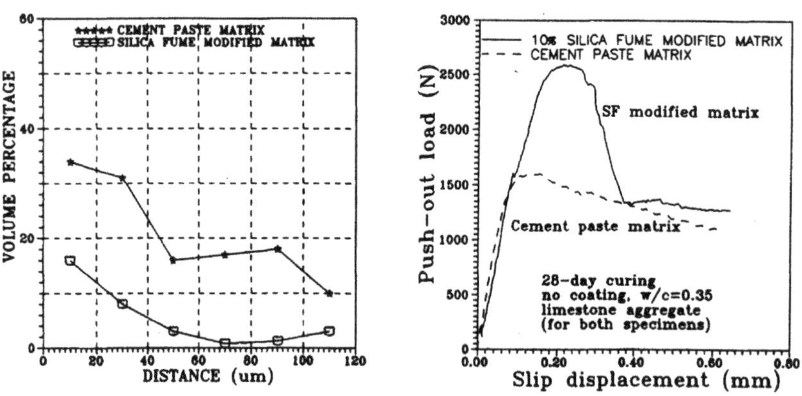

(a)comparison of porosity gradient (b)comparison of load-slip curve

Fig.6 Effect of silicafume modified matrix

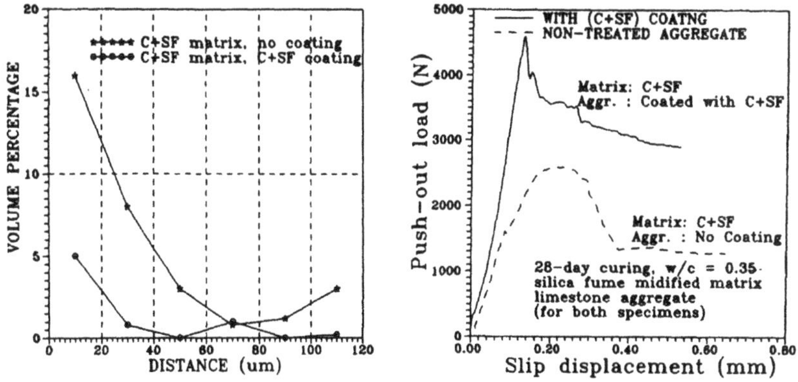

(a)comparison of porosity gradient (b)comparison of load-slip curve

Fig.7 Effect of coating on silicafume modified matrix

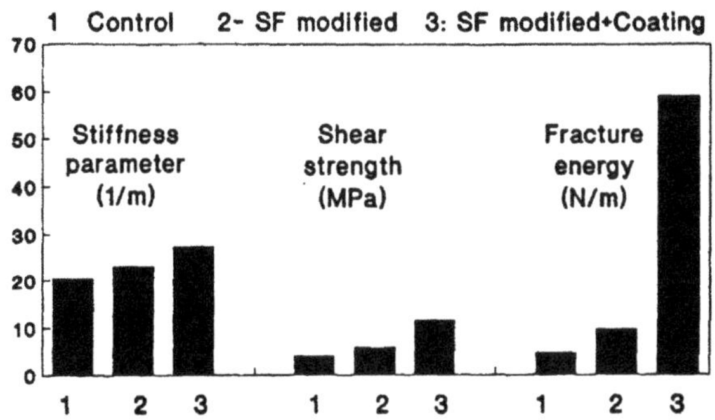

Fig.8 Effects of SF modified matrix and SF+Cement coating on the interfacial properties

with SF modified matrix together with cement and SF slurry coating when compared with the control specimen. Furthermore, the values of interfacial fracture surface energy computed in this study are lower than fracture surface energy of bulk concrete reported in previous research[13]. This confirms that the interface is the weak link in concrete. However, by changing the properties of matrix and aggregate, surface energy can be significantly enhanced by improvement of the interface. This implies that the processing methods examined in this study have potential for improving the properties of concrete.

5 Conclusions

The load-bond slip relationship between cement paste and aggregate was obtained by using a newly developed push-out test. Using a recently proposed mathematical model, the overall behavior of the load-slip curve was analyzed and the mechanical properties of the interfacial zone were obtained. These properties can be represented by the parameters, such as interfacial stiffness, shear bond strength and interfacial fracture energy. These properties were varied with curing age, type of aggregate and properties of matrix. Interfacial parameters, especially fracture energy, were improved significantly by incorporating silica fume in matrix and by coating the aggregate with a cement slurry containing silica fume.

Examination of the transition zone by using BSE shows that the microstructure of the interface can be significantly improved by lower w/c ratio and incorporation of silica fume into the cement paste matrix. The pre-coating of aggregate with cement and silica fume slurry can further reduce the porosity of the transition zone. Direct comparison of the gradients of the porosity in transition zone and the interfacial mechanical properties reveals that the microstructure of the interface is the key factor which determines the mechanical behavior.

Acknowledgements
The authors gratefully acknowledge the financial support of the National Science Foundation Center for Science and Technology of Advanced Cement Based Materials (NSF-ACBM).

6 References

1. B.D.Barnes,S.Diamond and W.L.Dolch(1978)The contact zone between Portland cement paste and grass aggregate, **Cem.Conc.Res.**,8,2,233-244
2. P.J.M.Monteiro,J.C.Maso and J.P.Oliver(1985)The aggregate-mortar interface, **Cem.Conc.Res.**,15,6,953-958
3. K.L.Scrivener,A.Bentur and P.L.Pratt(1988)Quantitative characterization of the transition zone in high strength concrete, **Advances in Cement Research**, 1,4,230-237
4. Maji and S.P.Shah(1988)Initiation and propagation of bond cracks as detected by laser holography and acoustic emission, **MRS Symp., Proc.114**
5. N.Iwasaki and Y.Tomiyama(1974)Bond strength between cement paste and aggregate, **Annual report of Cem.Tech.**, 28,193-196
6. Y.Kosaka,Y.Tanigawa and F.Ota(1975)Effect of coarse aggregate on the fracture behavior of concrete, **J.Arch. Inst.Japan.**, 228,1-12
7. S.P.Shah and F.O.Slate(1968)Internal microcracking, mortar aggregate bond and the stress strain curve of concrete, **Int.Conf.on the structure of concrete**
8. R.Zimbelmann(1987)A method for strengthening the bond between cement stone and aggregate, **Cem.Conc.Res.**, 17,4,651-660
9. K.L.Scrivener(1989)The use of backscattered electron microscopy and image analysis to study the porosity of cement paste, **MRS Proc.**, 137,129-140
10. H.Stang,Z.Li and S.P.Shah(1990)The pull-out problem the stress versus fracture mechanical approach, **J.ASCE,EM**, 116,10,2136-2150
11. Z.Li,B.Mobasher and S.P.Shah(1991)Characterization of interfacial properties in fiber reinforced cementitious composites, **J.Am.Cer.Soc.**,9
12. P.J.M.Monteiro and P.K.Mehta(1985)Interaction between carbonate rock and cement paste, **Cem.Conc.Res.**, 16,2,127-134
13. S.Mindess(1984)Fracture toughness testing of cement and concrete, **Fracture Mechanics of Conc.**, Martinus Nijhoff Publishers, 67-110

14 THE INFLUENCE OF ROCK AND CEMENT TYPES ON THE FRACTURE PROPERTIES OF THE INTERFACIAL ZONE

M. G. ALEXANDER
University of the Witwatersrand, Johannesburg, South Africa
S. MINDESS and LIE QU
University of British Columbia, Vancouver, Canada

Abstract
Fracture tests have been carried out on rocks, pastes, and paste/rock interfaces, to attempt to study the interaction of these materials and to assist in explaining their influence on properties of concrete. The test is based on ISRM Method I, and can also be used to test interfaces. Results are presented and discussed with reference to the influence of silica fume (S.F.) addition on pastes and interfaces, and of two different rock types on the interaction with pastes. It is shown that S.F. has a minor influence at the mix ratios and ages used. Paste/andesite rock interfaces have substantially higher fracture energies than paste/dolomite rock interfaces.
Keywords: Interfaces, Silica Fume, Fracture, Paste-aggregate bond.

1 Introduction

Concrete is often simplistically viewed as merely a two-phase mixture of granular aggregates and a binder based on portland cement. Such a model fails to account for the influence of the interfacial zone between the phases, an influence which on occasions is profound. While concrete strength may well be a function of interfacial bond strength, it is now being postulated that concrete stiffness (elastic modulus) may also depend on the interfacial region to an important extent (Alexander (1992a)). Thus there is a need for reliable data to test hypotheses concerning interfacial zone influence on concrete properties.

Several reviews and collected papers on this subject have been published (Mindess (1989), Mindess and Shah (1988)), resulting in often contradictory conclusions. This is not surprising considering the use of different test methods, materials and mix proportions, curing, and ages of test. This should act as a stimulus to further work, in order to resolve outstanding problems. From the materials engineering perspective, it is wise to keep our objectives in mind, namely, to understand the influence of the interfacial zone on the properties of cementitious composites, and thereby to exercise control over these properties so as to satisfy engineering criteria. The work reported in this paper has just such an overall objective in view.

Interfaces in Cementitious Composites. Edited by J. C. Maso. © RILEM.
Published by E & FN Spon, 2 - 6 Boundary Row, London SE1 8HN. ISBN 0 419 18230 6.

The results herein stem from a collaborative research programme between the Universities of the Witwatersrand (Johannesburg) and British Columbia (Vancouver). The interfacial region between several different rock types and cement pastes, some including silica fume, is being studied by mechanical test techniques and SEM observation. Details of the tests and early results are given elsewhere (Diamond and Mindess (1992), Mindess and Diamond (1992), Alexander (1992b), and Alexander (1991)). This paper concentrates on interfacial fracture properties in the light of measured properties of concrete incorporating different aggregate types.

2 Experimental details

The technique used is an adaptation of the ISRM Test Method I for determining fracture properties of rock cores (ISRM (1988)). Chevron-notched cylindrical beam specimens are used, and the fracture toughness K_{CB} and the work of fracture R_{CB} are calculated from the results. A correction to K_{CB} is possible for non-linear effects although it was not applied for these tests. Rock cores and pure cement paste cylinders of 41mm diameter were tested. The paste specimens (and paste/rock specimens) were prevented from drying during testing, by wrapping in plastic film.

To determine properties of paste/rock interfaces, paste was cast against half-cores from the rock core fracture tests. Paste and paste/rock specimens were placed under plastic in a fog room for 40 to 48 hours before being stripped very carefully, and placed in lime water at standard temperature.

Materials used were:-

Rocks	Andesite and dolomite.
Pastes	Ordinary portland cement (OPC) (ASTM Type I); Condensed Silica Fume (SF). A superplasticiser was also used.

A water/(cementitious content) of 0,3 was used. Silica fume replacement was at 15%.

Pure OPC and OPC-SF pastes were cast, as well as these pastes in combination with andesite and dolomite half-cores.

All the rock cores, and roughly half the paste and paste/rock specimens were tested at the University of the Witwatersrand, the test age for the paste-related beams being 28d. The balance of the specimens was cast and tested using identical (imported) materials at the University of British Columbia, the test ages being between 44 to 70d after casting. Specimen remnants at UBC were retained for subsequent SEM examination (ref. paper by Diamond et al (1992), these Proceedings).

3 Results and discussion

3.1 Load-deflection curves

Figure 1 gives typical envelope load-deflection curves for the various combinations of materials, and the individual constituents. The rock specimens have substantially higher peak loads than the paste-related beams. Pure OPC beams have higher peak loads than OPC-SF beams. Composite OPC/dolomite beams have similar peak loads to corresponding pure cement beams, while composite OPC/andesite beams have significantly higher peak loads compared with pure cement specimens.

The displacement at peak load for rock specimens is markedly higher than that for cement paste specimens, while the corresponding displacement of paste/rock specimens is lower than that of pure paste beams. The initial linear portions of the curves represent elastic moduli of the cement and rock materials; in the case of paste/rock specimens, an effective stiffness of the composite can be calculated, which will depend to some degree on the interface region.

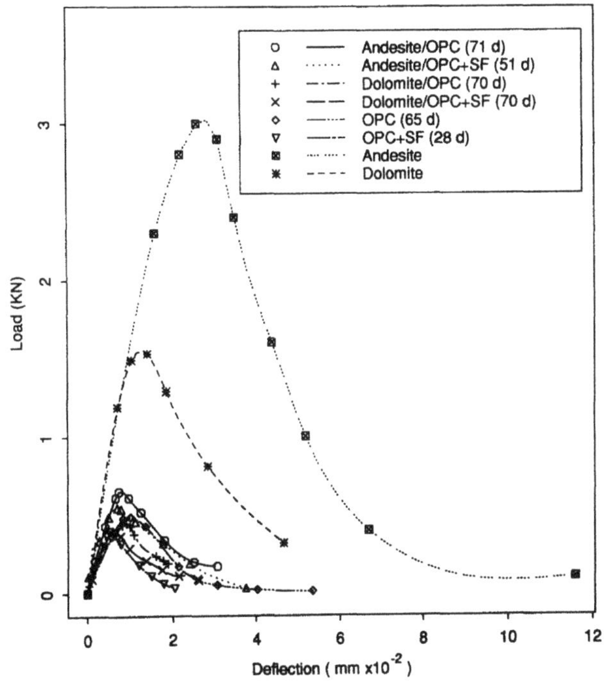

Fig. 1
Typical load-deflection curves for paste specimens, rock specimens and paste/rock composite specimens.

3.2 Age effects

Figures 2 to 4 show that, over the period 28d to 70d, there is no clearly identifiable trend for the parameters measured, with the general conclusion that no significant aging effects appear to occur over this time frame. This is not surprising in view of the low w/c ratio and relatively long ages in question.

3.3 Fracture properties

3.3.1 Fracture toughness K_{CB}

See Figures 2(a) and (b). For the pure paste specimens, OPC-SF has marginally lower values on average than OPC, although it is doubtful if this would be significant in practice. Andesite exhibits higher fracture toughness than dolomite, and both rocks are substantially tougher than the pastes.

For the interfaces, the use of silica fume appears not to significantly influence K_{CB}, for the mix proportions and ages used. This observation may well not hold for OPC-SF pastes at higher w/c ratios and shorter ages, where strengthening and densification of the interfacial zone by silica fume is a demonstrable phenomenon. It also appears that paste/andesite interfaces exhibit on balance slightly higher toughness than paste/dolomite interfaces.

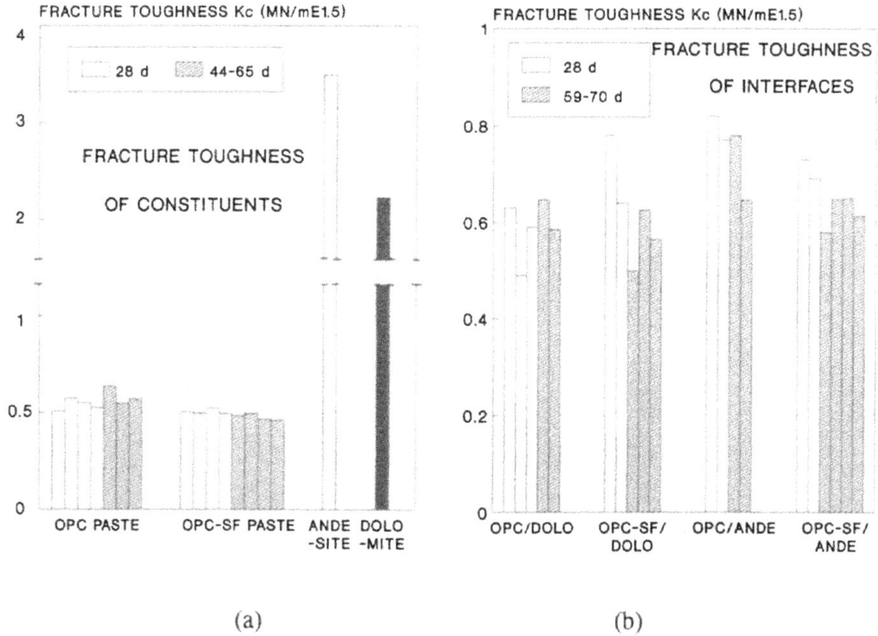

Fig.2
Fracture toughness for (a) constituents and (b) interfaces.

3.3.2 **Fracture energy R_{CB} (Total work of fracture)**

See Figures 3(a) and (b). The greater brittleness of OPC-SF is seen more clearly here. Likewise, andesite is seen to have a substantially higher fracture energy than dolomite, and both rock types have much higher fracture energy values than pure pastes. Silica fume appears to reduce the fracture energy of pastes by about 30% for these relatively mature high strength materials.

Referring to interfacial fracture energies, there is a clear differentiation between dolomite and andesite, with the paste/andesite interfaces achieving substantially higher values. This can be related to the higher ultimate strength of concretes made with aggregates derived from this andesite rock (Alexander and Ballim (1986)). The results may possibly be interpreted as showing that silica fume enhances interfacial fracture energy for paste/dolomite, but the opposite trend appears to occur for paste/andesite.

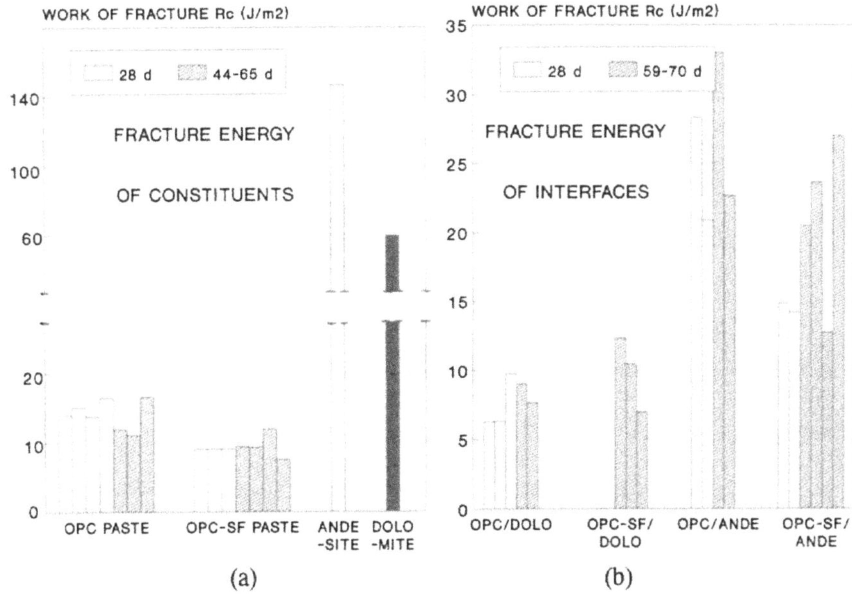

Fig. 3
Work of Fracture for (a) constituents and (b) interfaces.

3.3.3 **Comparison of fracture properties of pure pastes and of paste/rock interfaces**

For the OPC/dolomite and the OPC-SF/dolomite specimens, fracture occurred at the interface, while for the OPC/andesite specimens, the fracture could occur either at the interface or in the paste immediately adjacent to the interface. However, for the OPC-SF/andesite specimens, fracture occurred in the paste zone rather than at the interface.

Table 1 shows that andesite rock has the ability to enhance interface fracture toughness and work of fracture quite remarkably, whether for OPC or OPC-SF pastes. It also appears possible to create interfaces tougher and stronger than the paste, thus initiating paste fracture. What is extremely interesting in this case is that fracture toughnesses appear to be enhanced in the paste zone adjacent to the interface.

Table 1. Fracture properties of pastes and paste/rock interfaces

	Mean K_{CB} (MN/m^{1,5})	Mean R_{CB} (J/m^2)
OPC	0,56	14,2
OPC - SF	0,49	9,3
OPC/DOLOMITE	0,59	7,8
OPC-SF/DOLOMITE	0,62	9,9
OPC/ANDESITE	0,75	26,2
OPC-SF/ANDESITE	0,65	18,8

Dolomite rock has little influence on interfacial fracture toughness in relation to pure paste values, whether for OPC or OPC-SF. There seems to be an anomalous set of fracture energy results for OPC in relation to OPC/dolomite, which is not the case for OPC-SF specimens.

3.4 Stiffness properties

See Figures 4(a) and (b). For the pure pastes, the later age results (from UBC), are slightly higher than the 28d results (from Wits); there is no practical difference between OPC and OPC-SF pastes. For the rocks, the dolomite is marginally stiffer than the andesite.

With reference to the effective interfacial stiffness in composite specimens, silica fume seems to lower the stiffness for both rock types, although only marginally. This is despite the equivalence of the stiffness values for the corresponding pure pastes.

The later age OPC/andesite results seem out of phase. Leaving these aside, there seems to be little to choose between interfacial stiffness for the two rock types. Results seem to indicate that for very high strength concretes, the full potential stiffness of andesite concrete should be realised, and should be of the same order as that for dolomite concrete. This is in contrast to low to medium strength concretes where the full stiffness potential of andesite is not as readily realised as it is for dolomite (Alexander (1992a)).

3.5 General discussion

According to linear elastic fracture mechanics (LEFM), the fracture toughness should be independent of the crack length, or relative notch depth. As shown in Figure 5, K_{Ic} of the interface decreases with increasing crack length. Hillemeier

(1977) has also shown that as a crack grows from an initial notch, K_{Ic} of the interface decreases with increasing crack length, and approaches a constant value.

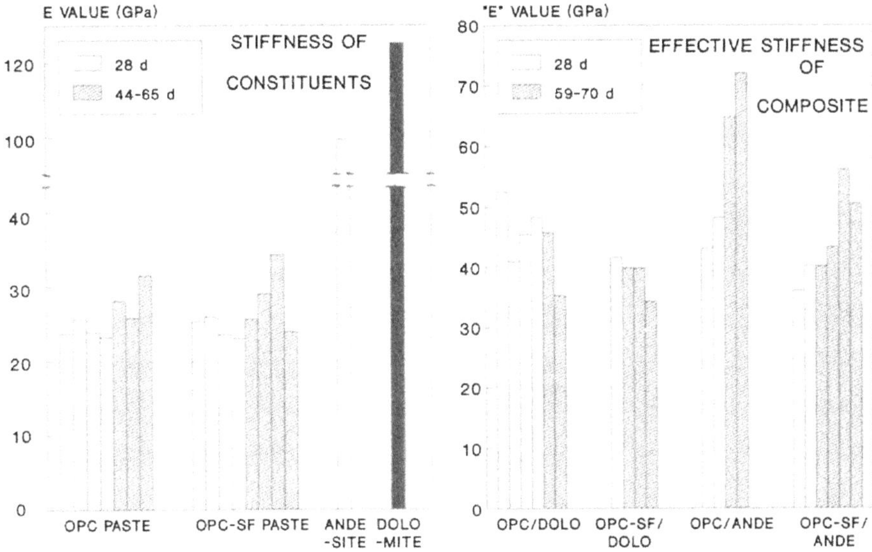

(a) (b)

Fig. 4
Stiffness of (a) constituents and (b) composites.

In Figure 5, it appears that the fracture toughness is dependent on cement type and rock type. For the high strength mature pastes tested here, it may be concluded that the interface between rock/modified paste does not necessarily have a higher value of K_{Ic} than unmodified interfaces, and that the interface between andesite/OPC has higher fracture toughness than other interfaces.

The differences between the OPC/andesite and OPC/dolomite composites may be due, in part, to the differences in microstructure of the fractured rock surfaces, as described by Diamond and Mindess (1992). However, any differences in the gross roughness of the rock surfaces were not apparent to the naked eye.

The K_{Ic} values of the chevron notch specimens obtained in our experiments are different from those of other specimens, such as those tested by Ziegeldorf (1983) and Hornain (1982), although these results were close to each other. Among different fracture methods there is a large degree of variability, and it is not yet clear which one is preferable for carrying out K_{Ic} measurements.

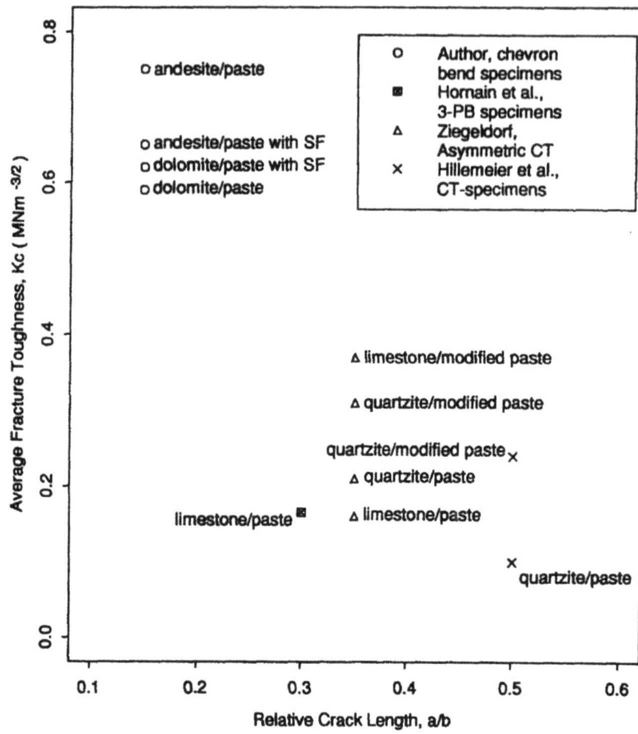

Fig. 5
Fracture toughness of interfaces measured by different investigators

4 Conclusions

Fracture tests on low w/c ratio mature pastes, rocks, and paste/rock interfaces have shown:-

(a) The rock types chosen (andesite and dolomite) have substantially higher fracture and stiffness values than pure pastes.
(b) Aging effects in the period 28d to approximately 70d appear to be minimal in paste-related specimens.
(c) The use of silica fume appears to embrittle pure pastes, but has smaller influence on interface properties.
(d) Andesite rock has the ability to enhance interface fracture toughness, often resulting in fracture occurring adjacent to, rather than at, the interface.

5 Acknowledgements

The authors acknowledge the support of Anglo Alpha Ltd and Hippo Quarries Ltd (South Africa), in particular Mr. Bill Coull, for logistic support; also a contribution of the Canadian Network of Centres of Excellence on High Performance Concrete for work done at the University of British Columbia.

6 References

Alexander, M.G. (1991) Fracture energies of interfaces between cement paste and rock, and application to the engineering behaviour of concrete, in **Proc. Int. Conf. on Fracture Processes in Brittle Disordered Materials**, E & FN Spon, London, pp. 337-346.

Alexander, M.G. (1992a) An experimental critique of the BS 8110 method of estimating concrete elastic modulus. **Mag. Conc. Res.**, 43(157), 291-304.

Alexander, M.G. (1992b) Two experimental techniques for studying the interfacial zone between cement paste and rock. Submitted to **Cem. and Conc. Res.**, 1992.

Alexander, M.G. and Ballim, Y. (1986) The concrete-making properties of andesite aggregates. **Civ. Eng. in S.A.**, 28(2), 49-57.

Diamond, S. and Mindess, S. (1992) SEM investigations of fracture surfaces using stereo pairs: I Fracture surfaces of rock and of cement paste. **Cem. and Conc. Res.**, 22(1), 67-78.

Diamond, S., Mindess, S., Lie Qu and Alexander, M.G. (1992) SEM investigations of the contact zones between rock surfaces and cement paste, in **RILEM Int. Conf. on Interfaces in Cementitious Composites**, Toulouse, E & FN Spon, London.

Hillemier, H.B. and Hilsdrof, H.K. (1977) Fracture mechanics studies on concrete compounds, **Cem. and Conc. Res.**, 7(5), 523-536.

Hornain, H., Mortureux, B. and Regourd, M. (1982) Physico-chemical and mechanical aspects of the cement paste-aggregate bond, Colloque Inter., Toulouse, C.56-C.65.

Int. Soc. for Rock Mech. (1988). Suggested methods for determining the fracture toughness of rock. **Int. J. Rock. Mech. Min. Sci.**, 25(2), 71-96.

Mindess, S. (1989) Interfaces in concrete. In **Materials Science of Concrete**, ed. J. Skalny, American Ceramic Society, 163-180.

Mindess, S. and Shah, S.P. (1988) (eds.) **Bonding in Cementitious Composites**. Materials Research Society, Pittsburgh, PA.

Mindess, S. and Diamond, S. (1992) SEM investigations of fracture surfaces using stereo pairs: II Fracture surfaces of rock-cement paste composite specimens, **Cem and Conc. Res.**, in press.

Ziegeldorf, S. (1983) Fracture mechanics parameters of hardened cement paste, aggregate, and interfaces, in **Fracture Mechanics of Concrete**, (ed. F.H. Wittmann), Elsevier, Amsterdam, pp 371-410.

15 FRACTURE OF MORTAR-AGGREGATE INTERFACES IN CONCRETE COMPOSITES

O. BUYUKOZTURK and K. M. LEE
Department of Civil Engineering, Massachusetts Institute of Technology, Cambridge, Massachusetts, USA

Abstract
Deformation and failure behavior of concrete composites is influenced by the characteristics of mortar-aggregate interfaces. In recent years, there has been a renewed interest in the study of the role of interfaces in composite behavior, particularly, due to the need for the development of high performance cementitious materials. In this paper, an interface fracture mechanics based methodology is described for the characterization of mortar-aggregate interface toughnesses using sandwich specimens. Also, a criterion based on energy release rate considerations is discussed for the study of crack propagation in interface regions. Finally, an application of the method using a two-phase concrete composite beam model is presented.
Keywords: Mortar-aggregate Interface, Concrete Composite, Interface Fracture Mechanics, Energy Release Rate, Phase Angle, Cracking

1. Introduction

Over the past twenty years, considerable research has been conducted to study the microcrack development, the nonlinear deformation behavior and the failure mechanisms of concrete. It has been generally established that in normal strength concrete the development of bond cracks at the mortar-aggregate interfaces plays a significant role in the inelastic deformation behavior and that final failure occurs through the formation of continuous cracks in mortar, bridging the bond cracks (Shah, 1966; Buyukozturk, et al., 1972; Liu, et al., 1972; Struble, et al., 1980). For high strength concrete, however, with respect to the deformation behavior and failure mechanisms, different observations have been made (Gerstle 1979; Carrasquillo, et al., 1981; Zaitsev, 1983; Chen, et al., 1985; Bentur and Mindess, 1986). In some cases cracks through aggregates were observed, indicating a less pronounced effect of crack arrest by the aggregates, while in other cases progression of bond cracks around the aggregates were reported. These different obtained trends are indicative of the importance of relative strength and stiffness properties of mortar, aggregates and the interface in the

Interfaces in Cementitious Composites. Edited by J. C. Maso.
Published by E & FN Spon, 2 - 6 Boundary Row, London SE1 8HN. ISBN 0 419 18230 6.

characterization of the overall composite behavior.

Development of advanced concrete composites with improved toughness and durability requires a fundamental understanding of the behavior of the interfaces introduced in these materials. Various scenarios of crack initiation and crack propagation in the matrix vs. at the interface, or crack penetration into the aggregate need to be studied in order to engineer the material for an optimum behavior. An essential step, for a systematic study in that respect, is the development of a quantitative methodology for the assessment of the fracture toughness (that is a property) of the interfaces between various mortar/aggregate systems. Furthermore, criteria are needed for the crack, once initiated in one of constituent materials when concrete is modelled as a two-phase composite or in the interfacial region, to propagate along the interface vs. to penetrate into the second material. Solutions for studying such cracking scenarios are rare. Buyukozturk, et al. (1972) and Yamaguchi and Chen (1991) studied the microcrack propagation in two-phase concrete models by the finite element analysis incorporating the smeared cracking model for mortar, and the interface finite elements for bond. However, the results obtained in this way are limited due to the phenomenological nature in which the interface bond property was assumed and the bond cracking criterion was established.

In this paper a methodology is described for the assessment of the mortar-aggregate interface fracture toughness using sandwich specimens. Also, results of a numerical simulation study are given for a composite beam model consisting of a mortar matrix with an aggregate inclusion to study the crack penetration vs. crack deflection scenarios in the interface regions of the composite. The predicted results from the numerical scheme are compared with those obtained from testing of a physical laboratory model.

2. Basic interface fracture concepts

Consider a semi-infinite free crack lying along an interface between two homogeneous isotropic half planes, with material 1 above and material 2 below (see Fig. 1). The moduli mismatch parameters of Dundurs are

$$\alpha = \frac{\overline{E}_1 - \overline{E}_2}{\overline{E}_1 + \overline{E}_2}\quad,\qquad \beta = \frac{1}{2}\,\frac{\mu_1(1-2\nu_2)-\mu_2(1-2\nu_1)}{\mu_1(1-\nu_2)+\mu_2(1-\nu_1)} \tag{1}$$

where $\overline{E} = E/(1-\nu^2) = 2\mu/(1-\nu)$, and E, μ, and ν are Young's modulus, shear modulus, and Poisson's ratio, respectively; the subscripts 1 and 2 refer to the two materials. The oscillation index ε is defined as

$$\varepsilon = \frac{1}{2\pi}\ln\left(\frac{1-\beta}{1+\beta}\right). \tag{2}$$

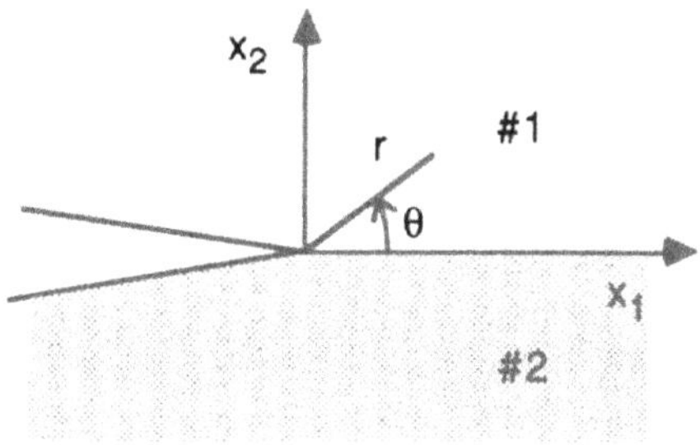

Fig. 1. Interface crack configuration

Fracture at a bimaterial interface can be expressed in terms of two parameters (Hutchinson, 1990). The first parameter is the fracture energy, Γ_i, a property that characterizes the fracture resistance of interfaces. Γ_i can be evaluated by energy release rate calibration of the test specimens. The energy release rate, *G*, per unit length of extension of the crack along an interface is

$$G = \frac{(1/\overline{E}_1 + 1/\overline{E}_2)}{2\cosh^2(\pi\varepsilon)}(K_1^{\,2} + K_2^{\,2}) \tag{3}$$

where K_1 and K_2 are the interface stress intensity factors and $\cosh^2(\pi\varepsilon) = 1/(1-\beta^2)$. The second parameter is the loading phase angle, ψ, which is a measure of the contribution of shear to opening experienced by the interface crack surface. The phase angle ψ is defined as

$$\psi = \tan^{-1}\left(\frac{\mathrm{Im}(KL^{i\varepsilon})}{\mathrm{Re}(KL^{i\varepsilon})}\right) \tag{4}$$

where $K = K_1 + iK_2$ and *L* is a reference length the choice of which is somewhat arbitrary (Rice, 1988). If Γ_i is defined as a function of ψ, one can establish a fracture energy trend of an interface.

3. Assessment of interface fracture toughness

3.1 Sandwich specimens

An interface crack model including a thin layer of material 2 sandwiched in a homogeneous body of material 1 is shown in Fig. 2. Each material is assumed to be isotropic and linearly elastic. If the thickness of the sandwich layer *h* is small compared to the crack length and to all other relevant in-plane length quantities, the following universal asymptotic relation exists between the interface intensity factors, K_1 and K_2, and the stress intensity factors, K_I and K_{II}, for the homogeneous problem, i.e., with no layer, (Suo and Hutchinson, 1989)

$$Kh^{i\varepsilon} = \sqrt{\frac{1-\alpha}{1-\beta^2}} K^{\infty} e^{i\omega(\alpha,\beta)} \tag{5}$$

where $K^{\infty} = K_I + iK_{II}$, $|h^{i\varepsilon}| = 1$ and the real function $\omega(\alpha, \beta)$ represents the shift in phase of the interface stress intensity factors relative to the applied stress intensity factors. The universal relation shown in Eq. (5) can be expressed as

$$\psi^* = \phi + \omega(\alpha,\beta) \tag{6}$$

where ψ^* is a real phase angle of $Kh^{i\varepsilon}$ and $\phi = \tan^{-1}(K_{II}/K_I)$. The energy release rate can be calculated from Eq. (3).

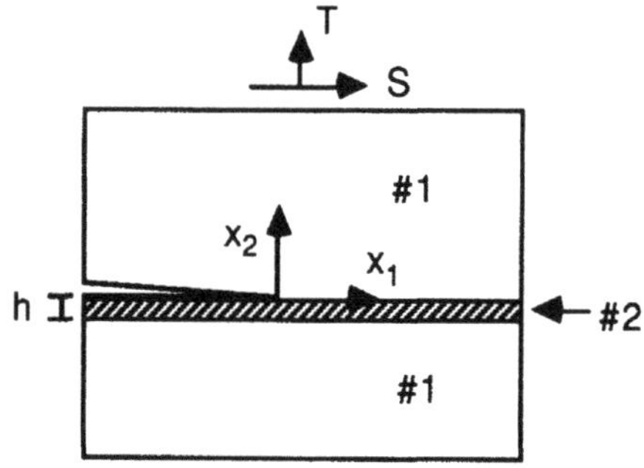

Fig. 2. Sandwich crack model

In order to apply the universal relation given in Eq. (5), the thickness *h* of the layer should be small compared with the crack length. A four-point bending specimen with a sandwiched aggregate layer is shown in Fig. 3a. In fact, since the shift, ω, is not large (Suo and Hutchinson, 1989), this specimen can be considered to be essentially in mode I. A sandwiched Brazilian disk specimen shown in Fig. 3b was proposed for mixed mode fracture testing of mortar-aggregate interfaces in concrete by Lee, et al. (1992). In this specimen mixed mode stress states ranging from pure mode I to pure mode II can be achieved by selecting, θ_i, the angle of inclination of the central through-crack.

3.2 Testing of sandwich specimens

In order to generate the fracture toughness curves of the mortar-aggregate interfaces, two types of sandwich specimens presented in Fig. 3 were tested.

The dimensions of the sandwiched beam specimen for mode I loading test were 152.0 mm (length) x 50.8 mm (height) x 38.1 mm (thickness). Radius (R) and thickness (t) of the sandwiched Brazilian disk specimen for the mixed mode testing were 38.1 mm and 25.4 mm, respectively. The thickness of the aggregate layer, h, was 2.54 mm for both specimens. The relative crack size (a/R) in the Brazilian disk specimen was fixed to be 0.25 and the relative crack size (a/d) in the beam specimen was 0.375.

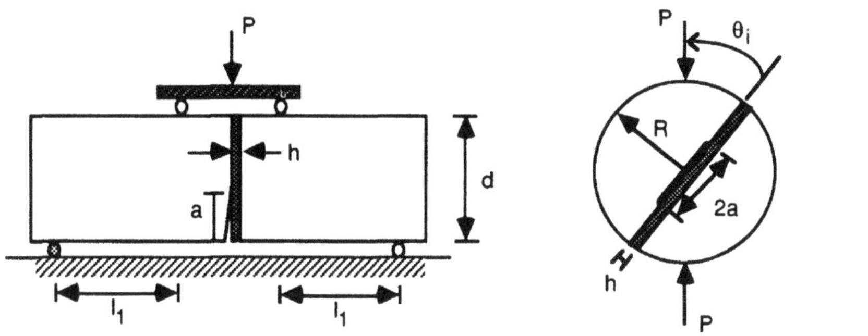

(a) Four-point bending specimen (b) Brazilian disk specimen

Fig. 3. Two types of sandwich specimens

Two mortar mixes of one normal strength and one high strength mortar were used in combination with granite aggregate layers. Thus, in the present experiment two mortar/aggregate combinations were considered. Table 1 shows the mix proportions for the two mortar mixes. Mechanical properties of the tested materials are listed in Table 2. For the two mortar/aggregate combinations the mismatch parameters, α, β, the oscillation index, ε, and the phase shift angle, ω, are presented in Table 3.

Table 1. Mix proportions for the mortar mixes (by weight)

	W/(C+SF)	Sand/(C+SF)	SF/C (%)	HRWR/C (%)
Mortar 1	0.50	2.0	0.0	0.0
Mortar 2	0.28	2.0	10.0	2.0

Table 2. Mechanical properties for the mortars and granite

	σ_c (MPa)	E (GPa)	G_{IC} (J/m^2)	ν
Mortar 1 (M1)	42.5	27.8	10.3	0.22
Mortar 2 (M2)	80.5	39.2	16.7	0.20
Granite (G)	140.1	55.3	17.5	0.16

Table 3. α, β, ε, and ω values for two bimaterial combinations

	α	β	ε	ω (°)
Mortar1/Granite	-0.320	-0.099	0.032	2.8
Mortar2/Granite	-0.163	-0.049	0.015	1.5

From the tests we observed that the interface fracture energy markedly increases as the loading phase increases, i.e the shear effect increases. This is attributed to the the shear interlocking effects at the interfaces with increased shear loading. We also found that due to the silica fume effects on the interfaces the fracture characteristics of mortar-aggregate interfaces in silica fume concrete are much different from those of the interfaces in ordinary concrete. These results are depicted in Figs. 4 and 5 showing the trends for interface fracture toughness vs. loading phase angle for both M1/G and M2/G combinations. As seen from Figs. 4 and 5, for the cases that nearly represent mode I condition the M2/G interface fracture toughness is more than twice that for the M1/G interface. However, for higher phase angles the difference between the interface fracture toughnesses of the two material systems is relatively small. This may be attributed to different failure modes observed in the tested specimens. For high phase angle values, M2/G interface exhibits a mixed failure mode of mortar and interface cracking, while M1/G interface fails by interface cracking only. Therefore, the measured fracture toughness values for M2/G interface with higher phase angles may be interpreted as lower bound values.

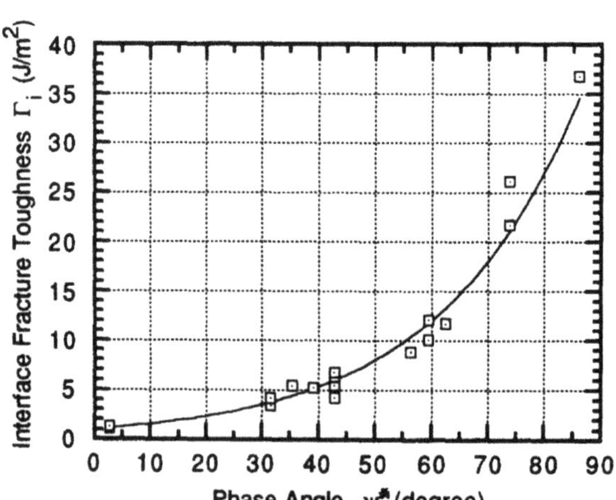

Fig. 4. Fracture toughness vs phase angle curve for M1/G interface

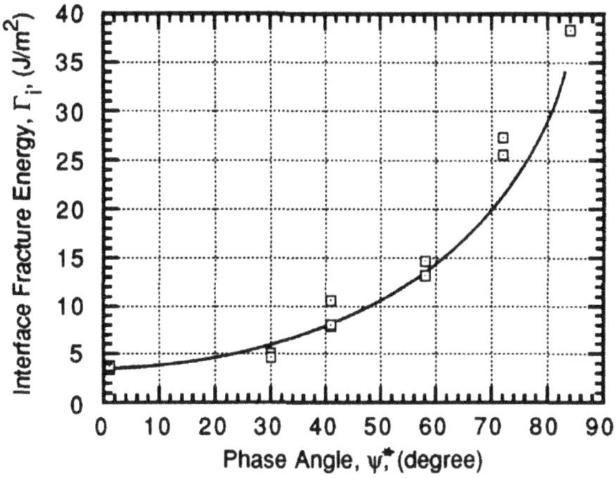

Fig. 5. Fracture toughness vs. phase angle curve for M2/G interface

4. Crack propagation in interfacial regions

4.1 Crack penetration vs. crack deflection

In elastically homogeneous brittle solids, cracks propagate in such a way that mode I conditions are maintained at the crack tip. This crack path criterion is clearly not valid when a crack advances in the region of an interface because in this case the relative magnitudes of the fracture toughnesses between the interface and the constituent materials is also involved. In a bimaterial system a crack impinging an interface may advance by either penetrating into material 1 or deflecting along the interface (see Fig. 6). The impinging crack is likely to be deflected if

$$\frac{\Gamma_i}{\Gamma_1} < \frac{G_d}{G_p^{max}} \tag{7}$$

where Γ_i is the toughness of the interface as a function of the loading phase angle, Γ_1 is the mode I toughness of material 1, $G_p{}^{max}$ is the maximum energy release rate of the penetrated crack (Fig. 6a), and G_d is the energy release rate of the deflected crack (Fig. 6b). The ratio $G_d/G_p{}^{max}$ on the right hand side of Eq. (7) can be computed by solving the crack problem for a semi-infinite body (He and Hutchinson,1989) or a numerical analysis scheme (Buyukozturk and Lee, 1992).

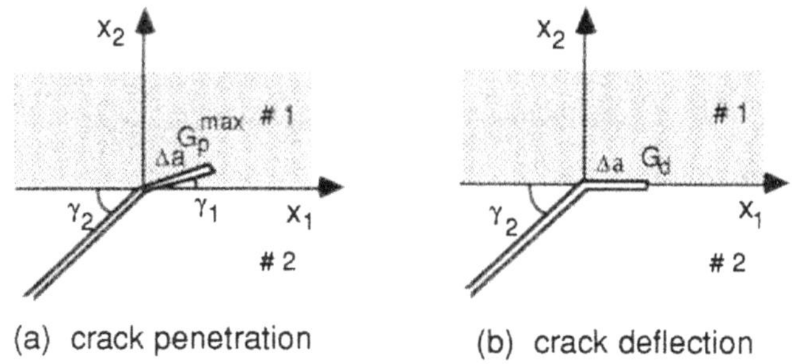

Fig. 6. Crack geometry

4.2 Application : Two-phase beam composite model

A composite model shown in Fig. 7 was tested and analyzed as an application to study the crack penetration vs. crack deflection in interfacial regions. This model consists of a beam of material 1 (mortar matrix) and a slab inclusion of material 2 (aggregate). The dimensions of the inclusion were 50.8 mm (length) x 12.7 mm (height) x 25.4 mm (width). The inclusion was oriented at a right angle with respect to the impinging crack. Mortar 1 and granite materials were used in manufacturing the beam composite models. Four-point bending tests on the beam specimens were performed. The failure load was 2.56 kN (2P). It was observed that in these beam specimens the crack went around the granite inclusion, i.e. the interface cracking occurred.

For the beam composite model tested, the energy release rate and the phase angle for the deflected crack were calculated from the finite element analysis (Buyukozturk and Lee, 1992). For this, a small crack Δa of 2.286 mm was introduced along the interface. Materials constituting the beam model were assumed to be linear and elastic. Finite element meshes were generated using 2D isoparametric eight node elements and in the crack tip region the mesh was refined using 24 element rings. The total number of elements was 1488. For the calculation of the phase angle the fixed length was selected to be 2.54 mm.

From the finite element analysis the phase angle for the deflected crack was 37°. From the tests the mode I fracture energy of granite, Γ_g, was 17.5 J/m^2 and $\Gamma_i(37^o)$, was approximately 5.0 J/m^2 (Fig. 4). The ratio G_d/G_p^{max} was calculated to be 0.45. Therefore, the ratio $\Gamma_i(37^o)/\Gamma_g$ which is 0.35 is found to be less than the ratio G_d/G_p^{max} which is 0.45. Thus, according to Eq. (7) the impinging crack can be predicted to be deflected. This agrees with the test results showing the interface cracking. With $\Gamma_i(37^o)$ = 5.0 J/m^2 the initial cracking load is predicted to be 2.25 kN from the numerical analysis. This value is 12 % lower than the measured value from the beam model testing. This difference can be attributed to the numerical and experimental errors. It is seen that the criterion given by Eq. (7) provides a reasonable prediction for this case, and may be used for studying the crack penetration vs.deflection scenarios in the interfacial regions of particulate composites.

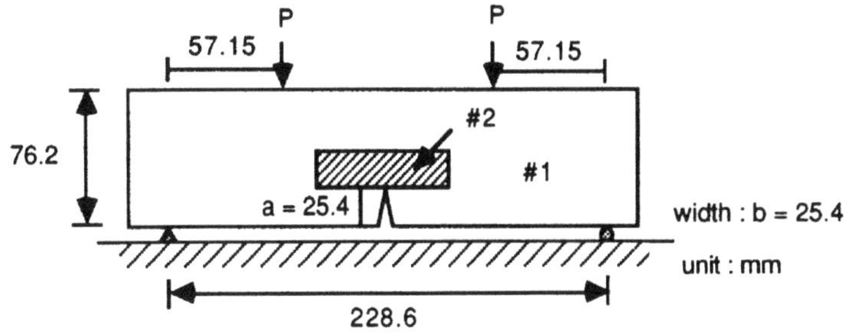

Fig. 7. Beam type composite model

5. Conclusion

In this paper, sandwich specimens are used in conjunction with the interface fracture mechanics concepts to assess the fracture toughness of mortar-aggregate interfaces in concrete composites. A criterion based on energy release rate considerations is discussed to study the crack penetration vs. crack deflection scenarios in interfacial regions. It is concluded that this combined experimental and analytical fracture mechanics based methodology can be appropriately applied to the study of the interface fracture in concrete composites. This study represents an initial step for the use of such methods in providing a fundamental understanding of the material behavior. The present approach is based on a linear elastic interface fracture mechanics concepts. This aspect together with other considerations such as size effects needs to be further studied.

6. Acknowledgement

Support of this work was provided by the National Science Foundation through Grant No. MSM-9016814. The cognizant NSF program official was Dr. K. P. Chong; his support is gratefully acknowledged. The authors would like to thank Professor J. W. Hutchinson of Harvard University, and Dr. C. K. Y. Leung of MIT for their valuable comments.

7. References

Bentur, A., and Mindess, S. (1986), "The Effect of Concrete Strength on Crack Patterns," *Cement and Concrete Research*, Vol. 16, pp. 59-66.

Buyukozturk, O., and Lee, K. M. (1992), "Finite Element Simulation of Cracking in Interfacial Regions of Particulate composites," *Proc. of International Conference on Application of Computers to Civil Engineering*, June 15-18, Istanbul, Turkey.

Buyukozturk, O., Nilson, A. H., and Slate, F. O. (1972), "Deformation and Fracture of Particulate Composite," *Journal of Engineering Mechanics*, ASCE, Vol. 98, June, pp. 581-593.

Carrasquillo, R. L., Slate, F. O., and Nilson, A. H. (1981), "Microcracking and Behavior of High Strength Concrete Subject to Short-Term Loading," *J. of ACI*, May-June, pp. 179-186.

Chen, R. C., Carrasquillo, R. L., and Fowler, D. W. (1985), "Behavior of High Strength Concrete under Uniaxial and Biaxial Compression," *High Strength Concrete*, ACI SP-87, pp. 251-273.

Gerstle, K. (1979), "Material Behavior under Various Types of Loading," *Proc. of a Workshop, High Strength Concrete,* Univ. of Illinois at Chicago Circle, pp. 43-78.

He, M.-Y., and Hutchinson, J. W. (1989), "Crack Deflection at an Interface between Dissimilar Elastic Materials," *Int. J. Solids Structures*, Vol. 25, pp. 1053-1067.

Hutchinson, J. W. (1990), "Mixed Mode Fracture Mechanics of Interfaces," *In Metal-Ceramic Interfaces*, Eds., M. Ruhle et al., Pergamon Press, New York, pp. 295-306.

Lee, K. M., Buyukozturk, O., and Oumera, A. (1992), "Fracture Analysis of Mortar-aggregate Interfaces in Concrete," Accepted for publication by ASCE in Journal of Engineering Mechanics.

Liu, T. C. Y. , Nilson, A. H., and Slate, F. O. (1972), "Stress-Strain Response and Fracture of Concrete under Uniaxial and Biaxial Compression," *ACI Journal*, V. 69, May, pp. 291-295.

Rice, J. R. (1988), "Elastic Fracture Concepts for Interfacial Cracks," *J. Appl. Mech.*, Vol. 55, pp. 98-103.

Shah, S. P., and Winter, G. (1966), "Inelastic Behavior and Fracture of Concrete," *ACI SP-20*, pp. 5-28.

Struble, L., Skalny, J., and Mindess, S. (1980), "A Review of the Cement-Aggregate Bond," *Cement and Concrete Research*, Vol. 10, pp. 277-286.

Suo, Z., and Hutchinson, J. W. (1989), "Sandwich Test Specimens for Measuring Interface Crack Toughness," *Materials Sci. and Engrg.*, A107, pp. 135-143.

Yamaguchi, E., and Chen, W.-F. (1991), "Microcrack Propagation Study of Concrete under Compression," *Journal of Engineering Mechanics*, ASCE, Vol. 117, No. 3, pp. 653-673.

Zaitsev, Y. (1983), "Crack Propagation in a Composites Material," *Fracture Mechanics of Concrete*, F.H. Wittmann, ed., Elsevier Science Publ., Netherlands.

16 THE INTERFACE ZONE AROUND EXPANDED SHALE GRAIN IN HARDENED CEMENT PASTE

U. SCHNEIDER and S. CHEN
Technische Universität Wien, Vienna, Austria

Abstract
The morphology of the interface between expanded shale grains and the hardened cement paste was observed by means of scanning electron microscopy (SEM). The microhardness of the interfacial zone around the expanded shale grains in the cement paste was measured. The results show, contrary to normal concrete, that there exists a transition layer of about 100-200 μm in thickness which is denser and stronger irrespectively whether the layer develops around dry grains or wet grains. The results also demonstrate that the lightweight aggregate is of benefit to structural formation and mechanical properties of the lightweight aggregate concrete as a result of the rough surface and the so-called "micro-waterpump-effect" of the porous aggregate in the lightweight aggregate concrete.
Keywords: Interface, Transition layer, Micromorphology, Microhardness, Lightweight aggregate, Lightweight aggregate concrete, Lightweight concrete.

1 Introduction

As is known, cement concrete is a multiphase composite construction material in which different phases work together through the medium of the interfaces amongst the phases. It can be assumed that during the whole life of concrete, from structural formation to damage and breakdown, there is a permanent process of interface-disappearance, interface-appearance, and interface-transformation in the concrete. In short, the whole life of concrete is an interfacial process.

Concrete is simply considered as a two phase composite construction material in which the disperse phase is the aggregate grains and the matrix is the cement paste. In normal concrete the interface between aggregate and cement paste is usually considered to be a weak link and to have a dominant influence on mechanical behaviour of the concrete. Therefore more and more attention is paid to the interface study throughout the world.

Interfaces in Cementitious Composites. Edited by J. C. Maso. © RILEM.
Published by E & FN Spon, 2 - 6 Boundary Row, London SE1 8HN. ISBN 0 419 18230 6.

J.Farran has initiated the interfacial research in 1950 and advanced the concept of "transition ring" in 1972.

In 1980J.C.Maso reported the hypothesis of the interfacial formation which has laid the theoretical foundation of the interface study and has promoted the rapid development of the interface study in concrete all over the world.

It is known that there are usually the following three types of combinations between aggregate grains and the cement paste in concrete:

—physical bond (adsorption etc.)

—mechanical bond (friction, meshing etc.)

—chemical bond (chemisorption, chemical reaction etc.)

In fact, the so-called interface between aggregate and cement paste in the concrete is not a real mathematical plane but an transition layer (transition ring) which structure and property are not quite the same neither as compared to the aggregate nor as to the cement paste. Many experts have done a large number of experiments in order to determine the thickness of the transition layer between different aggregate grains and the cement paste. According to J.Grandet and J.P.Ollivier the thickness of the transition layers between the cement paste and limestone, quartz and polyethylene respectively are all about 40 μm.

The publications in regard to the transition layer between cement paste and lightweight aggregate are not so many as far as information goes. Jitendra K.Bhargava has determined the interface between the hardened cement paste and the expanded clay as well as the pumice by means of radiography. The results indicated that the porosity of the hardened cement paste in the interface is ca. 13±2% lower than that of the adjacent region. The range of the dense transition layer between the cement paste and the expanded

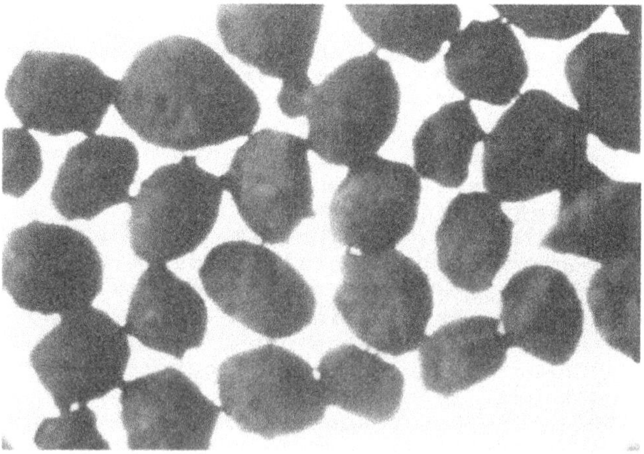

Fig.1. View of the expanded shale grains(ESG)

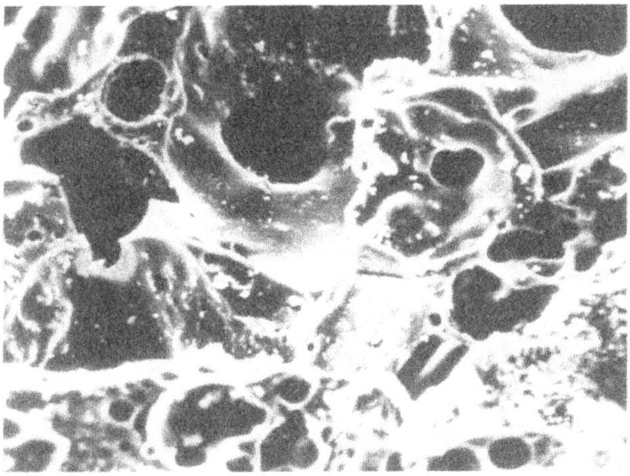

Fig.2. The inner pore structure of the ESG

clay grains amounts to 200-300 µm. But Jitendra K.Bhargava indicated that the value of 200-300 µm may be too large.

2 Morphology of the interface between expanded shale grain and the hardened cement paste

The expanded shale grain (following for short ESG) which was calcined at high temperature has a rough surface(Fig.1) and contains a lot of open pores and closed pore spaces (Fig.2).

The reflecting microscopy and the scanning electron microscopy were used to observe the interface between the

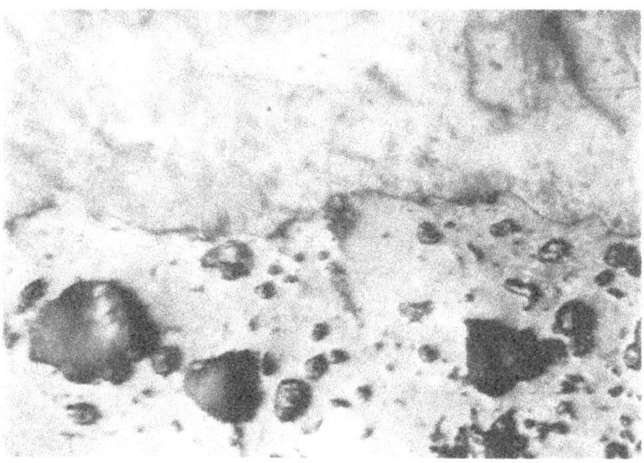

Fig.3. The interface of the HCP and the ESG
(above:HCP below:ESG)

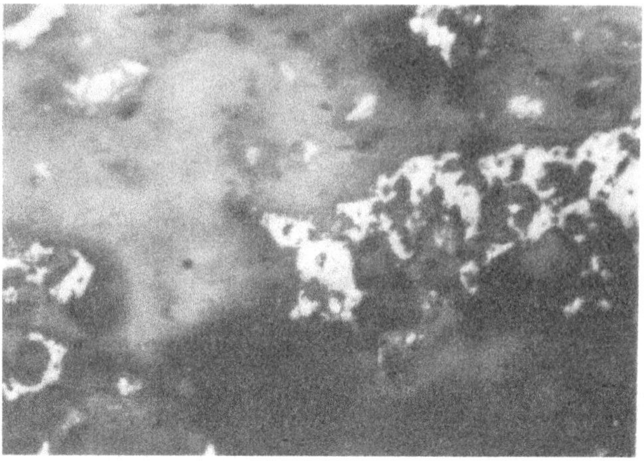

Fig.4. The interface of the HCP and the ESG
(above: HCP below: ESG)

expanded shale grain and the hardened cement paste (following for short HCP).

The results were quite clear. In spite of the existence of air bubbles and flaws in the interface the cement paste was strongly adhered to the aggregate in the main cases under observation(Fig.3-4,Fig.6).

The Fig.4 shows that the fresh cement paste was sucked into the open capillaries of the aggregate grain resulting in a significant strengthening of the mechanical bonds.

These photographs show also that the hydration products in the hardened cement paste close to the aggregate surface are very fine, dense and uniformly distributed(Fig.3-6).

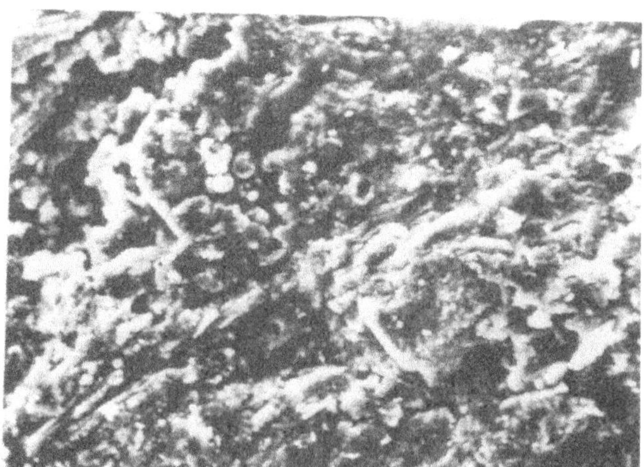

Fig.5.View of the HCP close to the ESG

3 The microhardness determination in the transition zone

The microhardness of the hardened cement paste close to the expanded shale grains was measured by means of automatic microhardness tester which is part of the polarization microscope "ORTHPLAN".

The dry expanded shale grains which were dried out under 105 °C and the wet grains which were soaked in water for 24 hours were used as samples. At the age of 110 days (dry aggregate grains) and of 100 days (wet aggregate grains) under standard condition (20±3 °C, over 90% r.h.) the microhardness was determined.

Fig.6 shows the actual indentation distribution in the transition area. It is clear that the farther the distance from the surface of the expanded shale grain, the bigger the indentation. This means that the hardened cement paste near to the surface of the expanded shale grain is relatively denser and its microhardness is relatively higher. The obtained results of the microhardness determination are shown in Fig.7.

The results indicated that the microhardness of the hardened cement paste ca. 20 μm near to the dry aggregate grain surface is about two times as high as that of ca. 2000-4000 μm distant from the aggregate surface. The mean microhardness of the hardened cement paste in the range up to 200 μm is 22.5 % higher than that of the distant one.

The contact area between the hardened cement paste and the saturated expanded shale grain is weaker as compared with the neighbouring region but about the same as that of the distant one (2000-5000 μm). The mean microhardness of the hardened cement paste in the range up to 200 μm is about 15 % higher than that of the distant one (2000-5000 μm).

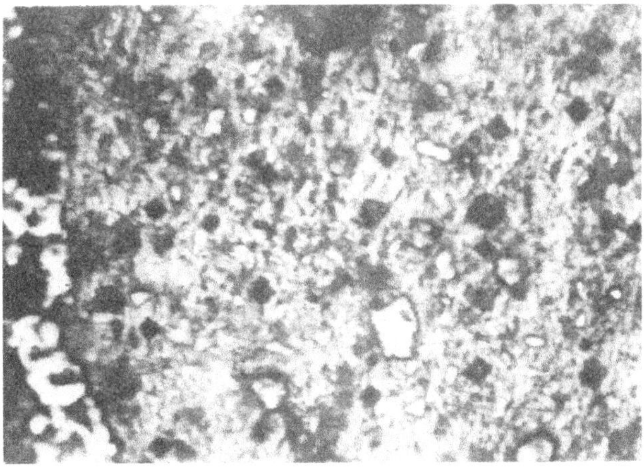

Fig.6. The indentation distribution (left: dry ESG right: HCP)

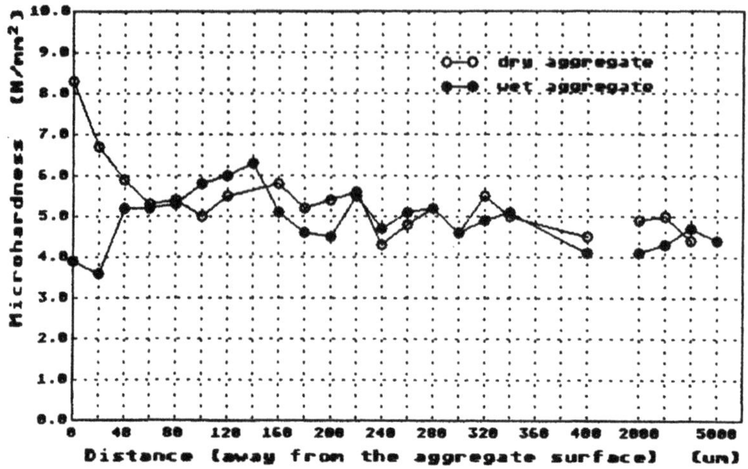

Fig.7. The microhardness of the HCP around the ESG

4 Analysis and discussion

Expanded shale is an artificial building material which is calcined under high temperatures ranging from 1100-1250 °C. The active silica and active alumina in the surface of the expanded shale grains is able to combine chemically with the liquid hydrated lime in the cement paste and to form the contact points between aggregate and cement paste.

According to physicochemical knowledge, the finer the grains, the higher the electronstatic attraction and the higher the surface energy. The smaller the interspaces, the denser the structure and the higher the critical fracture stress. On these grounds the transition layer between the expanded shale grain and the hardened cement paste has certainly high microhardness as well as high strength because there exist fine particles of hydrated cement and dense microstructure in the transition layer.

The characteristic of the expanded shale grain is the rough surface and the open capillaries which not only enhance the mechanical bonds but also increase the actual contact area between expanded shale grain and cement paste so that the physical adsorption and chemical bonding between the two phases is increased. Ozol (1979) indicated that the bond strength between hardened cement paste and aggregate depends basically on the real contact area irrespectively whether the bond strength depends on mechanical bonds or chemical bonds. On the basis of experimental results from Alford and Poole the critical strain energy release rate (G_{IC}) increases with the increase of the veins and edges on the aggregate grain surface. The above mentioned results confirm that the porous and rough surface of the expanded shale grain is able to contribute to the interfacial bond between the aggregate grain and the cement paste in concrete.

Since expanded shale is a porous material, the open capillaries are able to suck the mixing water from the fresh concrete before the setting and the hardening of the concrete. This means that the real W/C-ratio around the aggregate grains would be reduced and its structure would be closer and denser. But after the concrete hardening the sucked water within the aggregate grains would be sucked back into the hardened cement paste due to the temperature, humidity and capillary pressure difference between the aggregate grain and the hardened cement paste. This means that the unhydrated cement kerne in the hardened cement paste could further hydrate and the structure of the hardened cement paste would become much denser. The effect is usually called "micro-waterpump-effect" of lightweight aggregate in the cement paste. In other words, lightweight aggregate concrete is able to activate so-c lled " self vacuumizing", " self densifing", " self moistening" and " self curing".

These effects are dominant reasons for the development of a dense transition layer between expanded shale and cement paste irrespectively whether the aggregate grains were dry or wet, whereby the dense transition layer is able to be ca. 100-200 μm thick.

At the same time confirmation was found in our experiments that the growth rate of the compressive strength, splitting strength and the E-modulus of the concrete with wet expanded shale was much higher than that with the dry one. It can be considered that it was attributed to the effects of the " self moistening" and " self curing" as well.

5 concluding remarks

As previously stated, contrary to the normal aggregate, lightweight aggregate (for example expanded shale, expanded clay) which has a rough surface and porous structure is advantageous to combine with the cement paste. Especially the so-called "micro-waterpump-effect" of the porous aggregate in concrete is the basis of the dense transition layer between the lightweight aggregate grains and the cement paste.

The contact zone between the dry expanded shale grain and the hardened cement paste is rather strong. The microhardness of the hardened cement paste ca. 20 μm near to the dry grain surface is about two times higher than that of the distances of about 2000-4000 μm.

The contact zone between the wet expanded shale grain and the hardened cement paste is comparatively weak, but much the same as that of the far distant microhardness at an age of 100 days.

There exists a dense and hard transition layer between the hardened cement paste and the expanded shale grain, which is ca. 100-200 μm in thickness, irrespectively whether the aggregate grain was dry or wet. It has been proved that a hard cement paste shell forms around the aggregate grains in the lightweight aggregate concrete, which significantly improves the structure and the mechanical property of the concrete, so that it is possible to make high strength lightweight concrete by the use of porous aggregate.

6 References

Bhargava,Jitendra K.(1970) Radiographic Study of Paste-Aggregate Interface in lightweight Concrete. Royal Institute of Technology Stockholm, Sweden.

Bombery,Makk.Moisture Flow Through Porous Building Materials.

Diamond,S.(1986) in **8th ICCC**,vol.1, p122.

Farran, J.(1956)**Rev.Mater.Constr.**(490-491)155;(492)191.

Farran,J.,Javelas,R., Maso,J.C. and Perrin,B.(1972)**CR Acad. Sci. Paris** D275,1467.

Glucklich,J.(1971) The Strength of Concrete as a Composite Material. in **1th Int.Conf.on Mechanical Behavior of Materials** vol.4,pp.104-112.

Hanson,J.A.(1968) American Practice in Proportioning lightweight-aggregate Concrete. in **1th Int. Cong. on Lightweight Concrete,** London.

Harmathy,T.Z.(1967) Moisture Sorption of Building Materials, Ottawa.

Lea,F.M.(1970) **The Chemistry of Cement and Concrete** (3rd Ed.), Arnold, London.

Maso, J.C.(1980) in **7th ICCC,** vol.1, p.VII-1/3.

Massaza,F. and Costa,U.(1986) in **8th ICCC,** vol.1,p.158.

Monteiro,P.J.M., Maso,J.C. and Ollivier,J.P.(1985) **Cem. Concr.Res. 15,** 953.

Struble,L. and Mindess,S. (1983) **J.Cem.Comp.Lightwt Concr. 5,** 79.

Struble,L., Skalny,J. and Mindess,S.(1980) **Cem.Concr.Res. 10,** 277.

Taylor, H.F. (1990) **Cement Chemistry.** ACADEMIC PRESS LIMITED, London.

WU,Z.-W., Liu,B.-Y.and Xie,S.-S. (1982) in **Liaisons Pâtes de Ciment Matériaux Associés** (Proc.RILEM Collog.),p.A28, Laboratoire de Génie Civil, Toulouse.

PART 1.4

INTERFACIAL MICROSTRUCTURE AND PROPERTIES: MECHANICAL MODELLING

(Microstructure et propriétés de l'auréole de transition: Modelisation)

17 NUMERICAL SIMULATION OF ALTERING MECHANICAL PROPERTIES OF CONCRETE DUE TO MATERIAL BEHAVIOUR OF THE INTERFACES

H. SCHORN
University of Bochum, Germany

Abstract
The Bochum Method for numerical simulation of damage process in material structure has been developed since 1984 to that state of practicability which is needed for studying interface problems sufficiently. The investigations were made by two types of mechanical models: A simple plain model of a body with one single particle of aggregate in a binder and a 3-D-model of concrete. Comparing the results of tensile and compression tests it can be shown how sensitive a material as concrete alters its properties due to interface behaviour.
Keywords: Numerical Simulation, Modelling of material structure, Interface dependent properties

1 Introduction

In experimental investgation particular material properties depending on material structure - e.g. stress strain curve, strength, elongation, modulus of elasticity etc. - can be measured. The damage process itself consisting of microcrack formation and microcrack accumulation due to loading up to material failure by macrocrack opening can not be observed in experiments easily. The Bochum Method of Numerical Simulation allows to study damage processes on a mechanical model of material structure. The stress strain curves or other proporties calculated by numerical simulation can be compared with properties measured in experimental investigations.

The influence of interface behaviour was studied by a three-dimensional simulation on two types of material models:

* A single round shaped gravel is situated in a homogeneous binder (matrix)
* An aggregate as usual for concrete bonded by the same binder as above

Interfaces in Cementitious Composites. Edited by J. C. Maso. © RILEM.
Published by E & FN Spon, 2 - 6 Boundary Row, London SE1 8HN. ISBN 0 419 18230 6.

The crack mechanisms are studied according to different bond behaviour of the interface aggregate/binder under tensile and under compressive loading.

2 Simulation Method

The Bochum Method of Numerical Simulation has been developed since 1984. It is described in detail in the doctor thesis of Diekkämper [1] and the doctor thesis of Rode [2] in German . A short description in English language is given in [3]. Following, a short descriptive introduction of the method is given:

A simulation method showing fracture mechanisms requires a mechanical model, which degenerates due to loading path and crack opening until the system collapses absolutely. Complex numerical problems have to be solved and the use of very powerful computers of present generation is unavoidable. As a consequence of modelling the real material structure has to be abstracted for working out a mechanical system; the problem is: to find out that mechanical model which, on the one hand, is complex enough for delineating a real structure onto the model, and, on the other hand, simple enough for giving a chance to solve the numerical problems of the degenerating system without using tremendous time for operating in computer. The idea of delineating material structure to a mechanical model used for Bochum Method is explained by fig. 1.

The circles are round-shaped aggregate particles. Size and position of particles come out of a computer program, the picture shown is a cross section of a three-dimensional example. It is not necessary to delineate this typical concrete structure in its geometrically exact state. The idea is: Not the particles themselves, but the effects of the particles in stress flow between equidistant points in space, must be delineated to a mechanical model. There are distances between the equidistant points in fig. 1 which are determined by aggregate only, distances determined by binder only, and those determined by binder and aggregate as well as the interface behaviour. Following this idea, the mechanical model must be a framework with horizontal, vertical and diagonal struts, arranged symmetrically to all directions, see fig. 2. All struts of a macro-element are shown in the right part of the figure.

This model seems to be very similar to the well-known Hrennikoff models, published already in 1941. But there are important differences: Hrennikoff tried to delineate a continuum, not a real material structure. In Hrennikoff-models no crack opening occurs; the simulation of a fracture process is impossible.

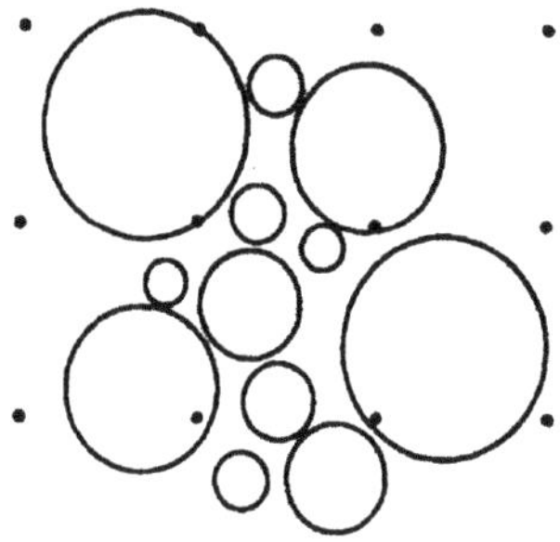

Fig. 1. Concrete model for numerical simulation

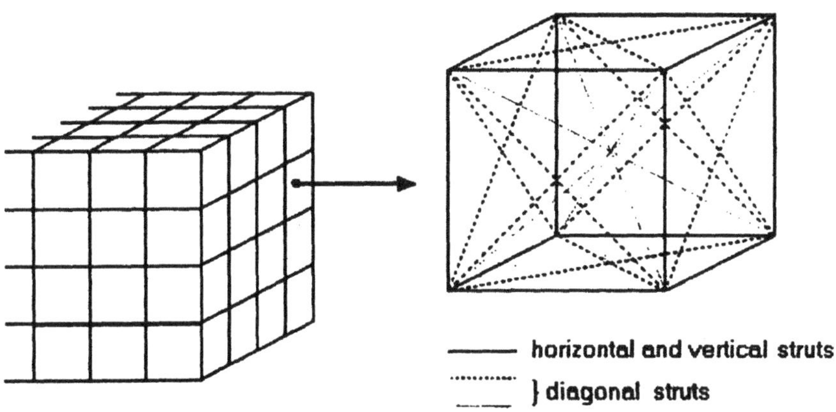

Fig. 2. 3-D-Framework model

3 Damage Process shown on a Simple Model

The mesh for Finite Element calculation as a plain model is shown in fig. 3. It represents a very simple model consisting of one single round shaped aggregate particle surrounded by a binder.

As input only four terms are required:

1 Element failure due to compression
2 Element failure due to tension
3 Modulus of elasticity
4 Element fracture energy

All terms are related to aggregate, binder and - concerning all struts cut by the circle in fig. 3 - bond behaviour.

Paramters used are:

Tensile strength of binder 6 N/mm^2
Tensile strength of bond 3 N/mm^2
Compressive strength 48 N/mm^2
Fracture energy of binder G_F = 0,05 N/mm
Modulus of elasticity of aggregate 40 000 N/mm^2
Modulus of elasticity of binder 26 500 N/mm^2

To simulate inhomogenouity of material the stiffness of the struts becomes distributed stochastically.

The tensile stress strain curve as calculated is shown in fig. 4. At all points marked with numbers, the state of damage of material structure is given in fig 5. Fig. 5 (1) belongs to number 1 in fig. 4; the direction of uniaxial tensile load is vertical in fig. 5. On the left side the loaded model is shown. It contains cracked struts round the upper part of the round shaped gravel. These cracks are drawn in the figure with an enlarged scale. On the right hand side only the crack pattern is shown. Small points or lines characterize microcracks, longer lines accumulations of microcracks or macrocracks.

Beginning from point 2 in fig. 4 the binder besides the aggregate particle is interspersed with microcracks symmetrically, see fig. 5 (2, 3, 4). The aggregate becomes nearly separated from the binder not only by cracks normally to load direction but round the entire surface. Even at point 3 in fig. 4, there do not exist any macrocracks in the binder besides the gravel, see fig. 5 (3). This crack suddenly opens and cuts the cross section in that way shown in fig. 5 (4).

4 Damage Process shown on a Concrete Model

A mesh for Finite Element calculation was made as a three-dimensional model. A cube will be generated and filled with balls of diameters according to an aggregate. Fig. 6 shows a cross section. Along the dotted line the material is cut similar to cutting a specimen by sawing.

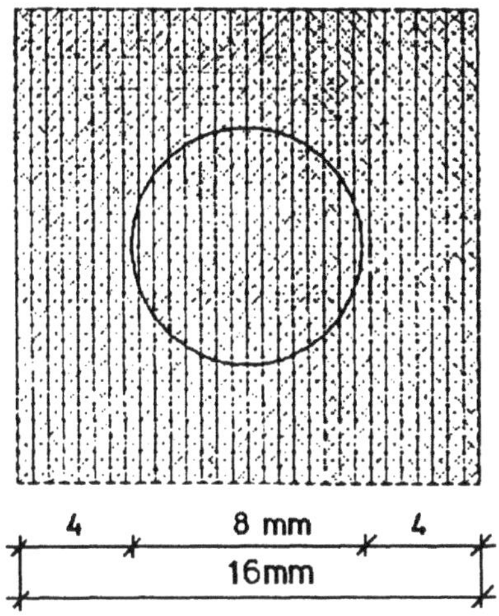

Fig. 3. Mesh for a simple plain concrete model

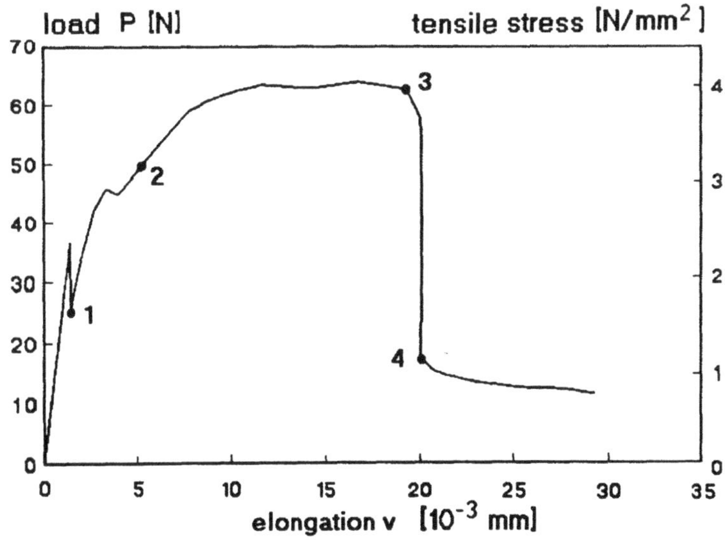

Fig. 4. Tensile stress strain curve calculated for the model fig.3

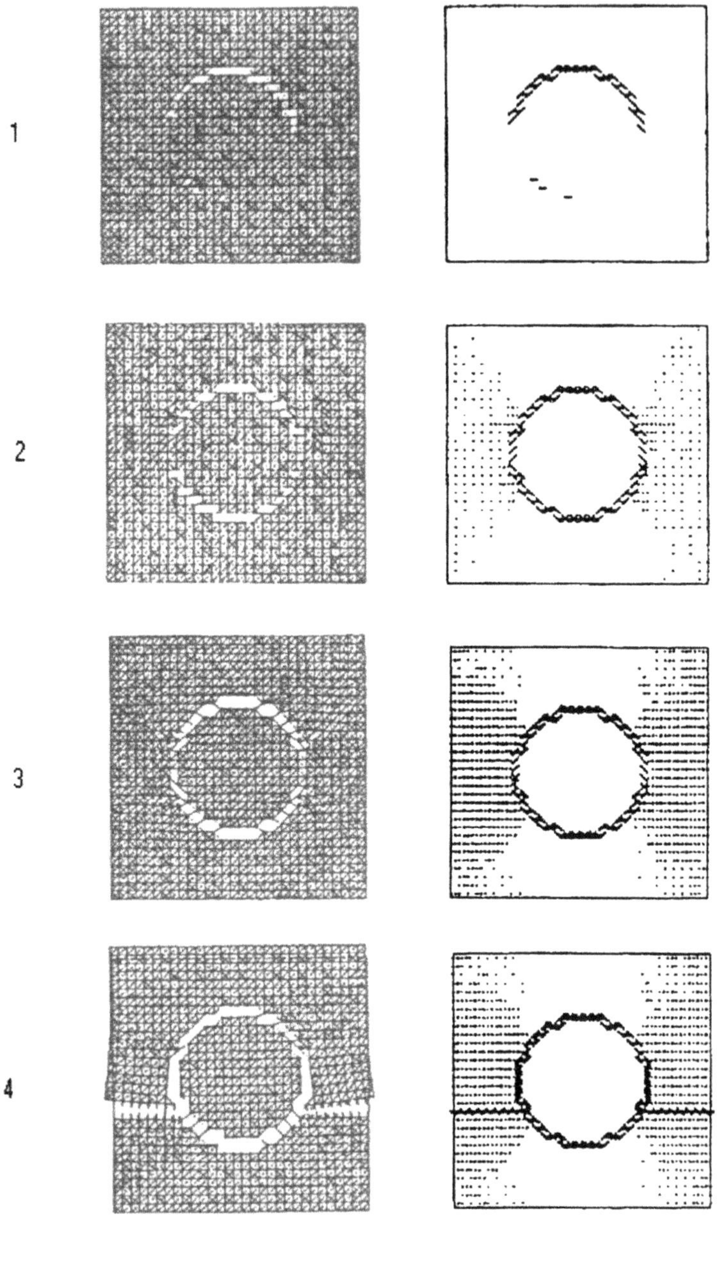

Fig. 5. Deformation state (left) and crack pattern (right) calculated for the fig. 3

The stress strain curve calculated for the concrete model (see fig. 7) is quite different from that curve taken from the simple model, see fig. 4. The reason is not the use of a 3-D-model instead of a plain model, but the very different damage process of a concrete model and of a simple one gravel-model. In a tensile loaded body a microcrack pattern occurs mainly orientated normally to load direction, see fig. 8(1) showing the crack state at point 1 in fig. 7.; that is the point of strength. In the descending branch of the stress strain curve only accumulations of microcracks occur which are forming the beginning of a macrocrack, see figure 8 (2). Deep in the descending branch, point 4 in fig. 7, the macrocrack propagates and leads to the final failure of the body.

If the behaviour is poorer or better as calculated in the example shown in fig. 7, the stress strain curve of the tensile loaded body will alter significantly. Fig. 9 contains three stress strain curves:

1 Tensile strength of bond = tensile strength of binder
2 Tensile strength of bond = 1/2 tensile strength of binder
3 Tensile strength of bond = 1/6 tensile strength of binder

Decreasing bond strength from 1 to 1/2 and 1/6 of the tensile strength (and the bending strength) also decreases but in lower ranges: 1 to 2/3 to 1/4.

If the same mechanical model is loaded not in tensile but in compression, the stress strain curves 1,2,3 only differ in a very small range; they are approximately identical.

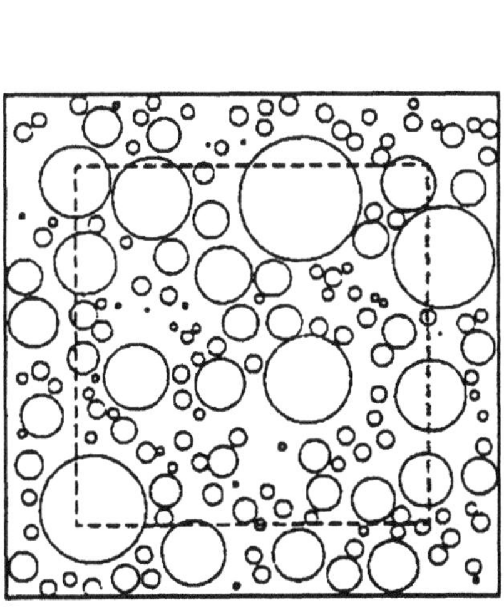

Fig. 6 Cross section of a 3-D Concrete model for numerical simulation

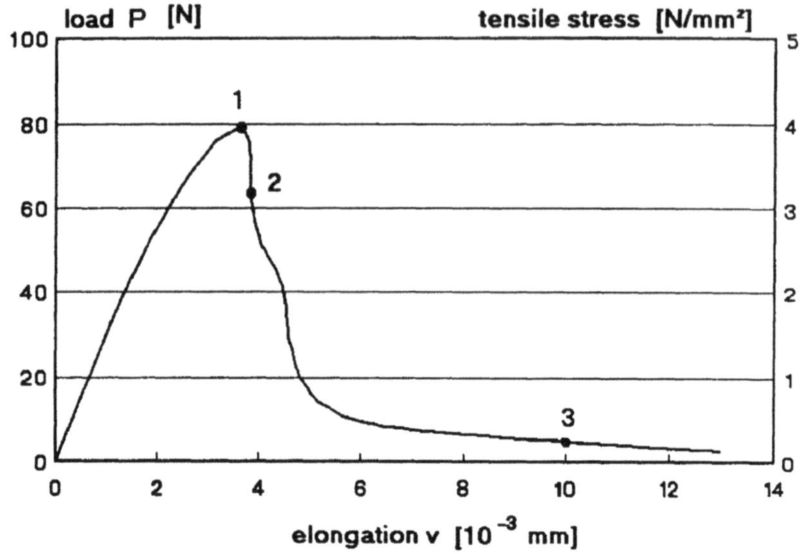

Fig. 7. Tensile stress strain curve calculated for the concrete model

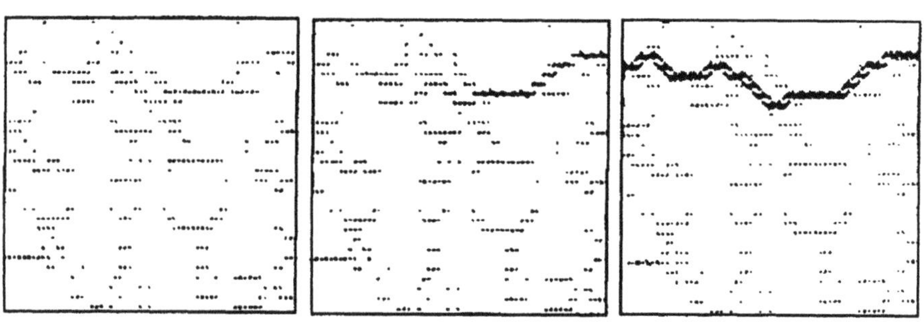

1 2 3

Fig. 8. Crack pattern of the concrete model at stress strain states as shown in fig. 7

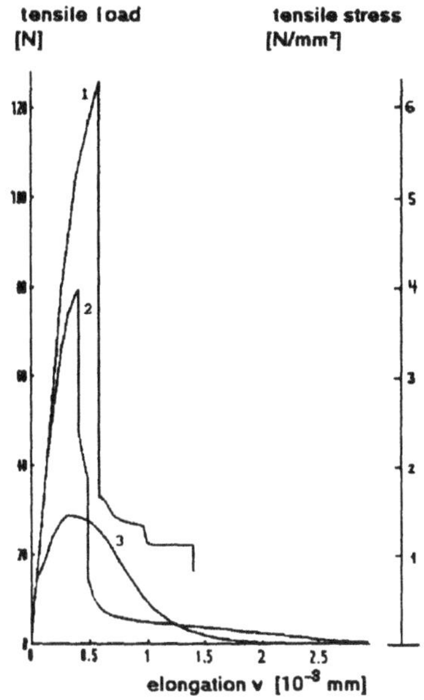

Fig. 9. Stress strain curves caused by different bend behaviour

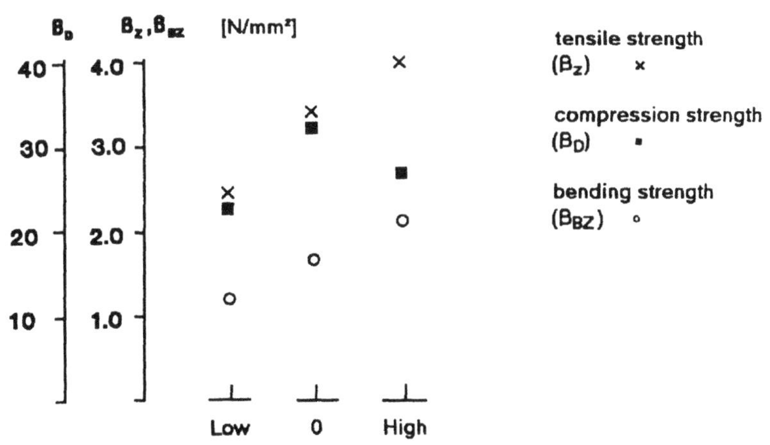

Fig. 10. Comparison between test results of concrete with lower and higher bond quality as usual

5 Experimental Control

Specimens shaped 40 mm x 40 mm x 160 mm were made from concrete with aggregate of maximum diameter of 8 mm. Reference (0 in fig. 10) was a concrete with its natural bond behaviour. Using polymer modified mortar as an envelope of all aggregate particles their surface will be prepared for a better bond behaviour (high in fig. 10). With an oil contacted surface the opposite effect has been realized (low in fig. 10).

Test results from measured bending strength, tensile strength and compressive strength are compared with each other in fig. 10. As expected tensile as well as bending strength follows the altering bond quality. The compressive strength shows no tendency according to bond behaviour - as expected - but contrary to the result of numerical simulation the differences in strength are relatively great. More tests seem to be necessary.

6 Conclusions

- The Bochum Method for numerical simulation can be used as a helpful tool in investigating damage process.
- Tests -experimental as well as in numerical simulation- must be made on concrete structure; simplifying leads to very different damage processes under load.
- Tensile strength and bending strength are correlated to bond strength but not linear. Decreasing bond strength leads to decreasing tensile strength of lower ranges.
- Compressive strength is not strongly correlated to altering bond strength (if bond strength is different from zero).

7 References

[1] Diekkämper, R. (1984) Ein Verfahren zur numerischen Simulation des Bruch- und Verformungsverhaltens spröder Werkstoffe, Technisch-wissenschaftliche Mitteilungen der Instituts für Konstruktiven Ingenieurbau Nr. 84-7, Ruhr-Universität Bochum.

[2] Rode, U. (1991) Ein Verfahren zur numerischen Simulation lastbedingter Gefügeänderungen im Baustoff Beton, Technisch-wissenschaftliche Mitteilungen der Instituts für Konstruktiven Ingenieurbau Nr. 91-5, Ruhr-Universität Bochum.

[3] Schorn, H. and Rode, U. (1987) Three Dimensional Numerical Simulation of Mechanical Behaviour of Concrete, International RILEM Congress From material Science to Construction Materials Engineering, Paris, Sept. 1987.

18 MECHANICAL MODELIZATION OF INTERFACES IN CEMENTITIOUS FIBER COMPOSITES

P. HAMELIN and R. M. COURTADE
Materials and Mechanical Engineering Laboratory,
Université Claude Bernard - IUTI, Lyon, France

Abstract

This paper proposes two approaches to analysing the mechanical behavior of cementitious fiber composites : experimental characterization and numerical modelling. The first part of this work is devoted to the description of the experimentation procedure : the laser granularity technique is used and gives the displacement field of in plane loaded specimens. It takes into account the different mechanical features of the constitutive materials. Numerical modelization is presented in the second part. A mixed hybrid finite element is derived to provide an average static continuity of the stress vector along the interfaces between materials. It can be used for homogeneous materials or for areas containing inclusions of other materials.

Keywords: Laser Granularity, Displacement Field Measurements, Hybrid Finite Elements, Inclusion.

1 Introduction

The analysis of the mechanical behaviour of composite materials, taking into account either the notion of stiffness or the notion of failure criteria, shows that the physical continuities at the interface between each elementary phasis are very important.

If we examine the configuration state of the micro structure in this transition area, we remark that the mechanical continuity cannot be approached by classical models as displacement models or force models.

Interfaces in Cementitious Composites. Edited by J. C. Maso. © RILEM.
Published by E & FN Spon, 2 - 6 Boundary Row, London SE1 8HN. ISBN 0 419 18230 6.

We present in our paper the methodology followed, on the experimental point of view, for an actual description of the interface between fiber and matrix. On the other hand, we present the formulation of specific hybrid finite element able to evaluate the stress distribution along the interfacial area.

2 Experimental procedure

According to the nature of the composite material, the displacement field under uniaxial and multiaxial forces can vary from a few microns to, at the upper limit, several milimeters. In order to use the same experimental and theoretical methodology to evaluate the fields of displacements, we have developed an experimental procedure : measurement by means of laser granularity.

2.1 Measuring the field of displacement by laser granularity

Optical tracking techniques are particularly effective as far as anisotropic materials are concerned. Recent developments in signal processing (digitalization of the image) allow data to be used more easily. Among recent publications on the subject we can cite the work of FRANCON [1978], BERTHAUD [1989] and ARCHBOLD [1970].

Experimental set-up

The experimental equipment developed by the Engineering Mechanics and Materials Laboratory at Université Claude Bernard, Lyon I, consists of :

- A new neon helium lazer ;
- A CCD camera ;
- Operating software.

Operational principals

Generally, the surfaces of materials are not macroscopically smooth, they can be assimilated to rough surfaces. Therefore, when we illuminate an object by a coherent light, each point of its surface acts as a light source. The elementary waves produced by any given point interfere with each other to form a granular field called speckle.

The laser granularity technique is divided into two phases :
Firstly, the granularity (figure 1) is recorded. We record on the same negative two consecutive images of a surface which is mechanically unloaded then loaded while being lighted by a coherent laser beam.
Then the film is exposed to a laser beam to produce interference fringes which are captured by a CCD camera and exploited with a frame analysis (figure 2).
The displacements and thus the slopes are simply related to the number of fringes.

FIGURE 1

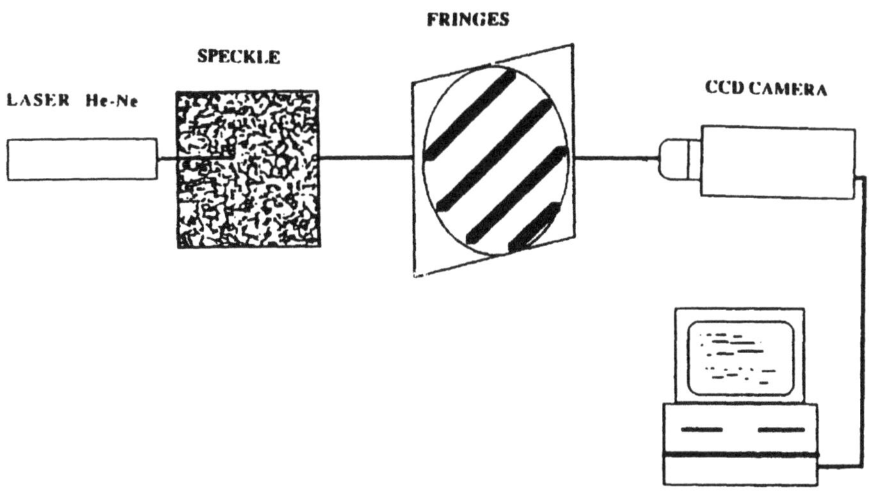

FIGURE 2

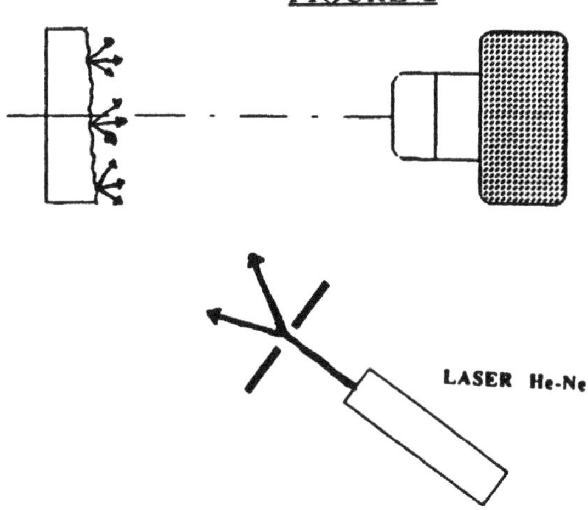

Limitation of the method :

The main factors which influence the measurements are, according to BERTHAUD [2] and ARCHBOLD [3] :

- Minimum mesurable displacement :

We have to see more than two black fringes

$Umin = \lambda D/Rh$ $Umin = p/2, 44$

Where p is diameter of the grain on the negative.

- Maximum mesurable displacement :

It is limited by the visibility of the fringes
in general : 5 Umax = diameter of the lighted surface

$1 \mu m < Umax > 150 \mu m$.

Operational error in measurements is 3%.

3 Pian hybrid variational principle

3.1 General scheme

One of the main features of PIAN's principle [1964] was primarily to take into account a dualized continuity of the stress vector at interfaces between two different areas. This static continuity of the stress vector is not introduced simply in structural modelizations : either it is very poor in primal approaches or too severe in mixed models using Reissner's principle. Partial solutions have been proposed for example by VERCHERY [1973] and can be used in the numerical analysis of multilayered and sandwich structures. The approach suggested by Pian is also partial, but it has an advantage in that internal static unknowns can be solved in terms of displacements inside each hybrid element because they are independent from one element to the other; consequently, an equivalent stiffness matrix is available and the hybrid element(s) can be assembled with other kinematically compatible finite elements. This remarkable feature allows us to insert only one hybrid element around a singularity in a finite element mesh: the stress field of this element is the analytical stress field of singularity.

3.2 Analytical details

Consider a volume V composed up of N elements $V^{(K)}$ (figure 3):

$$V = \prod_{K=1}^{N} V^{(K)} \tag{1}$$

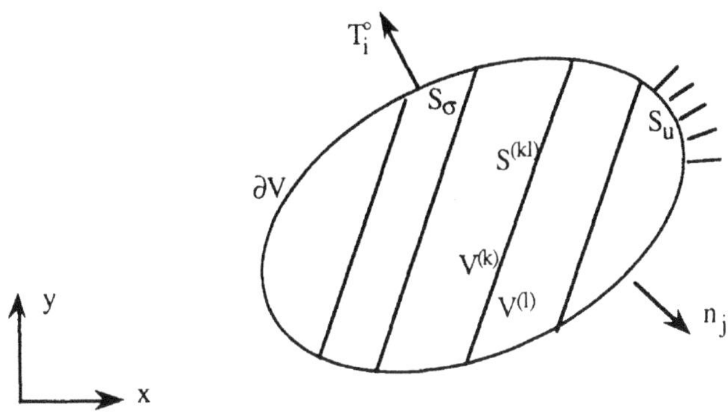

FIGURE 3 : Geometry, loading and boundary conditions

On the part S_u of the surface of V, the displacements are known and on the complementary part S_σ, the stress vector T_i is prescribed and equal to T_i°.

The following admissible spaces Ψ for stresses and M for displacements are defined :

$$\Psi(V) = \{\tau \,/\, \tau = (\tau_{ij}),\ \tau_{ij} = \tau_{ji}\ ,\ \tau_{ij} \in L^2(V^{(K)}),\ \tau_{ij,j} \in L^2(V^{(K)}), \quad \tau_{ij,j} = 0 \,/\, V^{(K)},\ K = 1, N\ \} \tag{2}$$

$$M(V) = \{\vec{v} \,/\, \vec{v} = (v_i),\ \exists\ \vec{w}^{(K)} \in H^1(V^{(K)}),\ \vec{w}^{(K)} = \vec{v} \,/\, \partial V^{(K)}, \quad K = 1, N;\ \vec{v} = 0 \,/\, S_u\ \} \tag{3}$$

Introducing the stress tensor σ_{ij}, the stress vector takes the following form

$$T_i = \sigma_{ij}\, n_j \tag{4}$$

Pian's principle can be written as

$$\mathcal{L}(\tau,\vec{v}) = \sum_{K=1}^{N} \Big[-\frac{1}{2} \int_{V^{(K)}} S_{ijkl}\, \tau_{ij}\, \tau_{kl}\, dV + \int_{\partial V^{(K)}} \tau_{ij}\, n_j\, v_i\, dS - \int_{\partial V^{(K)} \in S_\sigma} T_i^{\circ}\, v_i\, dS \Big] \tag{5}$$

In (4) and (5), n_j are the components of the outward normal to the surface considered for the evaluation of T_i (here, it is S_σ associated with prescribed T_i°). As plane elasticity is considered, latin indices i, j, k, and l vary from one (or x) to two (or y) only. The stationary conditions of this functional are :

$$\exists\, (\sigma, \vec{u}) \in \Psi \times M,$$

$$\forall\, \delta\tau_{ij} \in \Psi,\ \sum_{K=1}^{N} \Big[\int_{V^{(K)}} S_{ijkl}\, \sigma_{kl}\, \delta\tau_{ij}\, dV - \int_{\partial V^{(K)}} u_i\, \delta\tau_{ij}\, n_j\, dS \Big] = 0 \tag{6}$$

$$\forall\, \delta v_i \in M,\ \sum_{K=1}^{N} \Big[\int_{\partial V^{(K)}} \sigma_{ij}\, n_j\, \delta v_i\, dS \Big] = \sum_{K=1}^{N} \Big[\int_{\partial V^{(K)} \in S_\sigma} T_i^{\circ}\, \delta v_i\, dS \Big] \tag{7}$$

Relation (7) is equivalent to the virtual work principle. The first condition (6) appears as a behaviour law written for the whole structure. As previously mentioned, it allows us to solve the independent σ_{ij} of the $V^{(K)}$ th element in terms of the boundary displacements $u_i \in \partial V^{(K)}$.

3.3 Numerical details

An internal equilibrium stress field σ is defined (8) inside the volume of the hybrid element, in terms of static parameters β; it can have a polynomial form or it can be chosen as an analytical stress field; the solution for an inclusion has been used for example in a previous work by COURTADE & SURRY [1987].

$$\sigma = P\,\beta \tag{8}$$

A boundary displacement field u along the sides of the element is interpolated (9) with the nodal displacements U_{nod}; either linear or quadratic shape functions are used; in isoparametric finite elements, particular care has to be taken with the treatment of the components of the outward normal to the sides.

$$u = L\,U_{nod} \tag{9}$$

As soon as the mechanical characteristics of the materials are known, an elementary compliance matrix H is calculated with the aid of a Gaussian quadrature for integration over the surface of the hybrid element; a boundary matrix T is calculated by integration along the sides of the element. The system of equations derived from relations (6) and (7) is written in terms of H, T, β and U_{nod}

$$\begin{Bmatrix} 0 \\ F_{nod} \end{Bmatrix} = \begin{bmatrix} H & T \\ T^t & 0 \end{bmatrix} \begin{Bmatrix} \beta \\ U_{nod} \end{Bmatrix} \tag{10}$$

F_{nod} are the equivalent nodal forces which resemble closely to those of the classical displacement method. The stiffness matrix K of the hybrid element is :

$$K = T^t\,H^{-1}\,T \tag{11}$$

4 Conclusion

We pursue our research by analying different elementary cases described in the following figures.

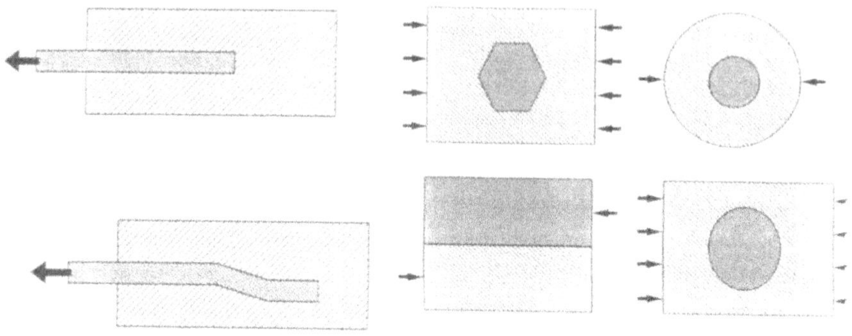

For each geometry of grain, each geometry of fibers, we compare the stress and strain distributions measured with the result of calculations (classical numerical modelization and mixed hybrid finite element).

5 References

[1] Archbold E, Burch J.M. and Enwos A.E. (1970), **Optica acta** n° 12883-898, Vol.17.

[2] Courtade R.M. and Surry C. (1987) Hybrid elements with defects in **Mechanical behaviour of adhesive joints**, (ed Pluralis), pp 389-400.

[3] Fond C., Berthaud Y. (1989) Automatisation d'un banc de mesure de champs de déplacements par granularité laser. **DEA Mécanique et Matériaux** L.M.T. E.N.S. Cachan.

[4] Francon M. (1978) **La granularité laser et ses applications en optiques**, *Paris, Edition Masson.*

[5] Pian T.H.H. (1964) Derivation of element stiffness matrices by assumed stress distribution, **A.I.A.A. Journal**,.2, 1333-1336.

[6] Verchery G. (1973) Extremal theorems in terms of mixed variables. Applications to beams and plates subjected to transverse shear, **15th Polish Solid Mechanics Conference**, Zakopane, Pologne, 258-259.

PART 2.1

RELATIONSHIPS BETWEEN MICROSTRUCTURE OF THE INTERFACIAL ZONE AND PROPERTIES OF THE COMPOSITE: MECHANICAL PROPERTIES

(Relations entre la microstructure de l'auréole et les propriétés du composite: Les propriétés mecaniques)

19 EFFECT OF AGGREGATE TYPE ON THE MECHANICAL BEHAVIOUR OF CONCRETE

H. KAWAKAMI
Fukui University, Fukui, Japan

Abstract
The effect of the mechanical properties of aggregate rock and their bond strength to mortar on the mechanical behavior of concrete are studied. Five types of rock were tested. The tests results were obtained on specific gravity, water absorption, compressive strength, elastic modulus of rock and their bond strength to mortar, as well as the mechanical behavior of concrete containing the each aggregate. The discussion revealed that the type of aggragate rock influence the failure mechanism, compressive strength, strain at the maximum stress and elastic modulus of concrete.
Keywords: Aggregate, Rock type, Bond strength, Compressive strength, Strain, Concrete failure.

1 Introduction

Studies on the bond strength between cement mortar and aggregate rock has been reported. However the effects of them on the mechanical behavior of concrete has not been clearified yet. The present paper reports the results of three groups of experiment, that is, (1) the mechanical properties of five types of rock obtained from Kuzuryuu river located at the central part of Japan, (2) their bond strength to the cement mortar and (3) the mechanical behavior of concretes containing each type of aggregate rock.

2 Mechanical properties of aggregate rock

Three drilled cores (5 cm ϕ x 10 cm) of five types of rock were tested in compression under the surface dry condition. The longitudinal strain were measured by wire strain gages at the tests. The mean value of the test results of specific gravity in surface dry condition, water absorption, compressive strength and elastic modulus are listed in Table 1.

Interfaces in Cementitious Composites. Edited by J. C. Maso. © RILEM.
Published by E & FN Spon, 2 - 6 Boundary Row, London SE1 8HN. ISBN 0 419 18230 6.

Table 1. Properties of aggregate rock

Rock type	Specific gravity ($10^{-2}N/cm^3$)	Water absorption (%)	Elastic modulus ($10^4 N/mm^2$)	Compressive strength (N/mm^2)
Granite	2.66	0.54	5.76	192
Porphyrite	2.71	0.33	6.70	161
Rhyolite	2.50	1.66	2.08	104
Andesite	2.83	0.90	5.86	219
Sandstone	2.79	0.32	8.91	342

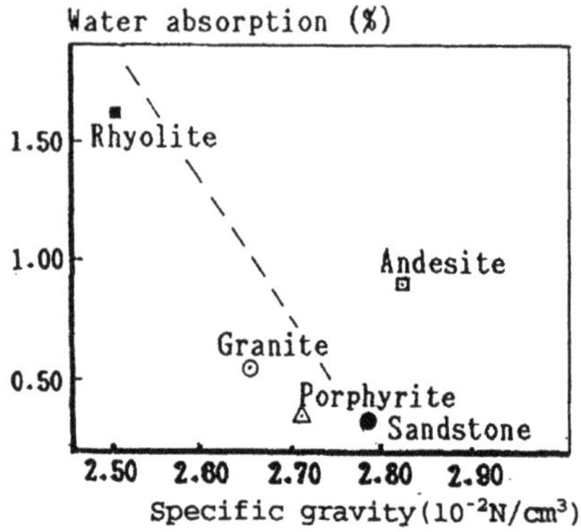

Fig.1. Water absorption and specific gravity of rocks

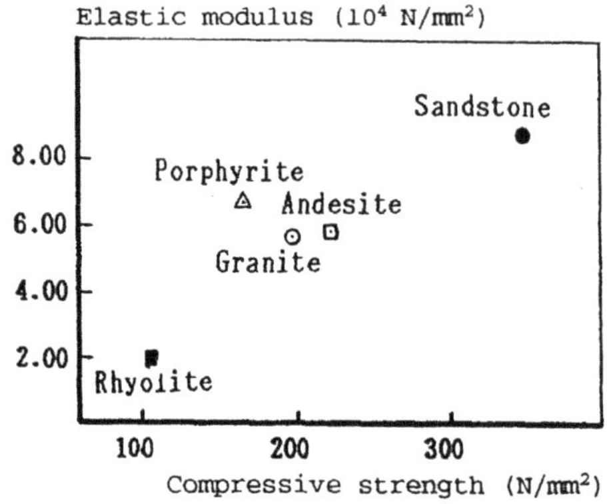

Fig.2. Elastic modulus and compressive strength of rocks

Figure 1 shows the specific gravity (surface dry) and water absorption. Figure 2 indicated that the rock with higher value of elastic modulus showed the greater compressive strength. Porphyrite showed the higher ratio of elastic modulus to compressive strength than the others.

3 Bond strength between rock and mortar

3.1 Materials and test specimens

Properties of Portland cement and sand used in tests are shown in Table 2. Table 3 shows mix proportion of mortar and concrete. The interface of the rock specimen to the mortar was polished with #120 sandpaper. Mortar were placed into the mould 50 minutes after the first mixing. The test specimens were cured in 20 °C water.

Table 2. Properties of cement and sand

Cement: Normal Portland cement					
(Bond test series)	(1)	(2)	(3)	(4)	(5)
Bending strength (N/mm^2)	6.46	6.57	6.54	7.55	7.28
Compr. strength (N/mm^2)	37.5	38.9	38.8	40.5	39.0

Sand : Kuzuryuu river 2.5 mm, Fineness modulus : 2.71
Specific gravity: 2.53 ($10^{-2}N/cm^3$), Water absorption: 2.08 %

Table 3. Mix Proportion

	W/C (%)	Slump (cm)	s/A (%)	Water (l/m^3)	Cement (l/m^3)	Sand (l/m^3)	Gravel (l/m^3)
Concrete	45	8	35.9	168	118	253	451
Mortar	45	-	----	308	217	465	---

3.2 Test procedure

The bending tests of the specimens were carried out at the age of four weeks. The size of specimens and the bending apparatus are shown in Fig.3 and Fig.4. The bond strength (f_b) was obtained by the expression (1).

$$f_b = M / Z \quad (N/mm^2) \tag{1}$$

M: Bending moment at the interfece section ($N \cdot mm$)
Z: Modulus of the interface section (mm^3)

Compressive strength of the mortars were tested in 2 x 2 cm area after the bending test.

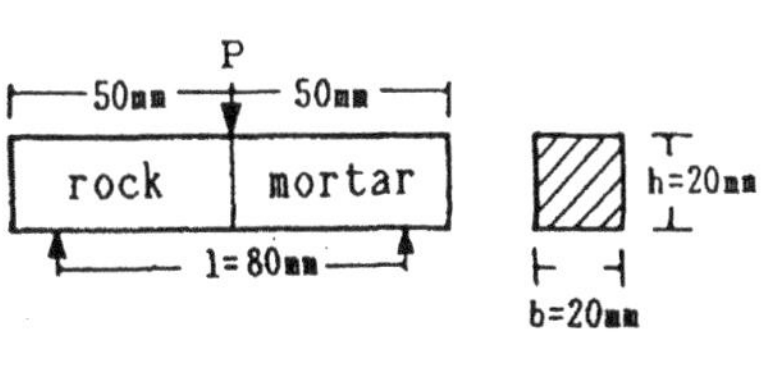

Fig.3. Bending test specimen

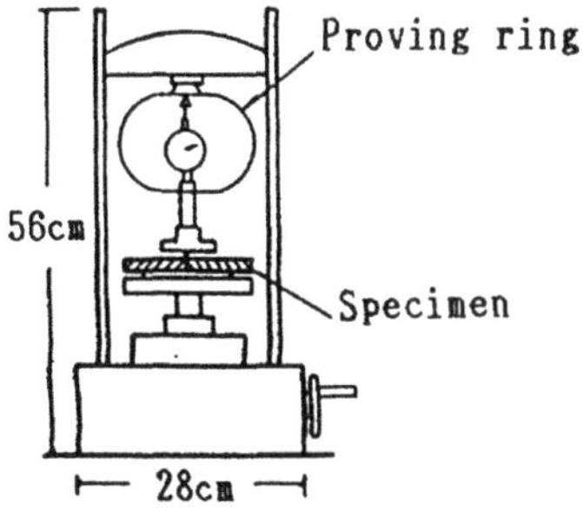

Fig.4. Bending test apparatus

Table 4. Bond strength (N/mm^2)

Rock type	Granite	Porphyrite	Rhyolite	Andesite	Sandstone
Mean Value	4.22	3.14	4.59	2.99	2.48
Standard deviation	0.39	0.34	0.44	0.33	0.26

(Compressive strength of mortar: 65.8 N/mm^2)

3.3 Test results

The results of five series of tests (16 specimens for each rock type) are summerized in Table 4. Figure 5 and Figure 6 show the bond strength in relation to the compressive strength and elastic modulus of rocks.

Rhyolite showed the highest bond strength, (1.85 times of the value of sandstone), while the compressive strength and the elastic modulus of it remained at the lowest value.

The compressive strength and the elastic modulus of sandstone showed the highest value, but the bond strength was the lowest. The bond strength of granite, andesite and porphyrite stayed within the range of the values of rhyolite and sandstone. And their test data plotted in the figures were not necessarily located on a straight broken line drawn through the data of rhyolite and sandstone.

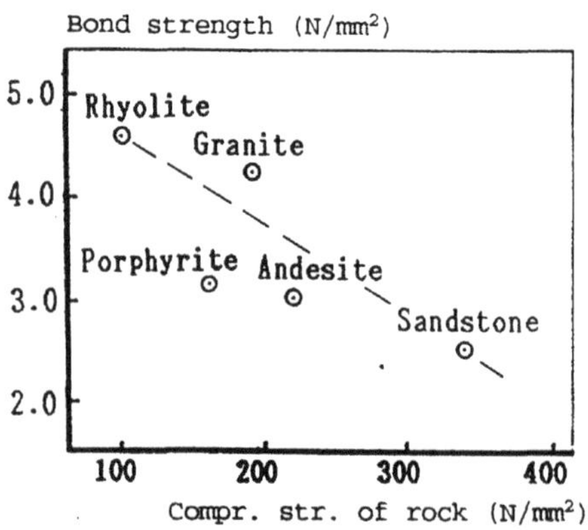

Fig.5. Bond strength and compressive strength of rocks

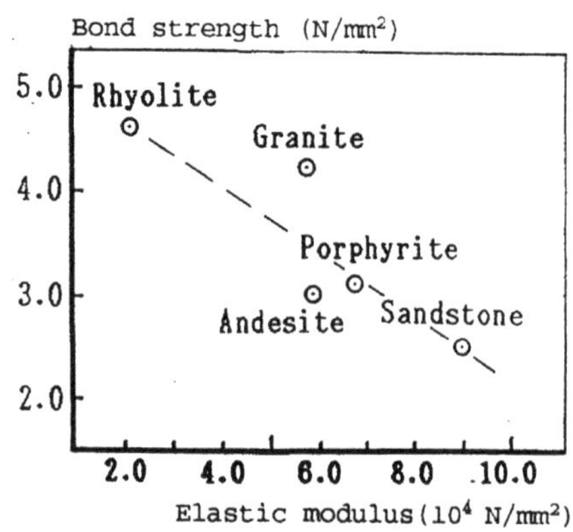

Fig.6. Bond strength and elastic modulus of rocks

4 Mechanical behavior of concrete

4.1 Materials and specimens

Properties of cement and sand are shown in Table 2. The river gravel was classified according the rock type. And they were sieved into three groups of size, 5-10, 12-15, 15-20 mm. The same amount of these pebbles were used for each concrete. And "mixed" aggregate was also prepared, containing the same amount of every rock type. The mixed mortar was covered with wet cloth for 50 minutes and mixed again for the preparation of the concrete.

The surface dry aggregate were mixed with the mortar by hand for each cylinder specimen (10 cm ϕ x 20 cm). Mortar specimens were also prepared with the same size as the concrete. Three concrete specimens for each rock type and four specimens for mortar were prepared. The compression tests were carried out at the age of 4 weeks after 20 °C water curing.

4.2 Compression tests

In compression tests the longitudinal strains at both sides of the specimen were recorded. Secant elastic modulus ($E_{1/3}$) for each specimen was determined at the stress of the one third of the compressive strength. Initial tangent modulus (E_o) was obtained by the expression (2) using 36 stress-strain data for a specimen.

$$\sigma = a + b\varepsilon + c\varepsilon^2 + d\varepsilon^3 \tag{2}$$

σ : stress (N/mm^2), ε : strain ,
a,b,c,d, : coefficients

The coefficients of correlation between the test data and the expression were larger than 0.9992.

Table 5. Mechanical properties of concrete and mortar

	F_c (N/mm^2)	me (10^{-4})	$E_{1/3}$ (10^4N/mm^2)	E_o (10^4N/mm^2)
Granite	39.7	19.5	3.00	3.53
Porphyrite	35.2	13.8	3.31	3.80
Rhyolite	34.7	21.5	2.44	2.80
Andesite	37.7	17.0	3.24	4.14
Sandstone	36.7	15.2	3.38	4.15
Mix	36.1	17.5	3.06	3.45
Mortar	52.6	35.1	2.39	2.77

(F_c:Compressive strength, me:Strain at the maximum stress, $E_{1/3}$:Secant elastic modulus, E_o:Initial tangent modulus)

4.3 Test results

The mean value of the test results are shown in Table 5 . The compressive strength of concretes were lower than that of the mortar. Granite concrete showed the highest strength, while rhyolite concrete the lowest.

5 Discussion

5.1 Compressive strength and type of failure

Rhyolite pebbles were found to be broken at the failure section of concrete cylinder. In the other kind of concretes, pebbles were not broken but separated from the mortar at their surface. The fracture of rhyolite concrete was considered to be originated from the pebble fracture, while the fracture of the other concretes were caused by the bond failure of the aggregate interface. The lowest strength of rhyolite concrete might be resulted from the lower strength of the aggregate rock.

5.2 Compressive strength of concrete and bond strength of rock

Figure 7 shows the relationship of the compressive strength of concrete and the bond strength of their aggregate rock. The bond strength of the aggregate rock was proved to govern the compressive strength of concretes containing sandstone, andesite and granite each. The lower strength of rhyolite concrete was considered to be resulted from the low strength of aggregate rock. The compressive strength of porphyrite concrete might be lowered by the higher stress concentration at the interface, resulted from the higher value of elastic modulus of aggragate rock, which is 14 % larger than the value of andesite. It is shown that the compressive strength of concrete is affected by the bond strength of aggregate rock and the stress concentration at the interface caused by the higher modulus of elasticity as well.

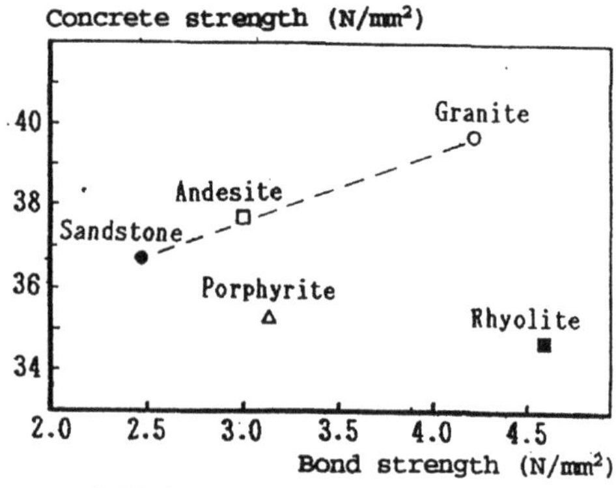

Fig.7. Bond strength(f_b) and compres. strength of concrete (F_c)

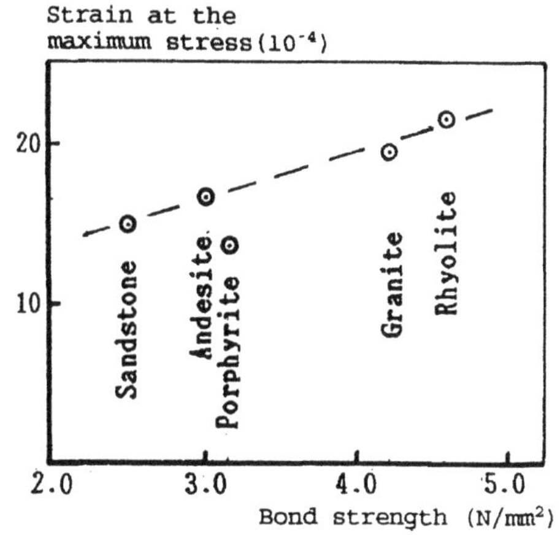

Fig.8. Bond strength and the strain at the maximum stress

The compressive strength of mixed-aggregate concrete showed 36.1(N/mm^2) and the value was 98 % of the mean value of each concrete containing single type aggregate.

5.3 Strain at the maximum compressive stress

Figure 8 shows the mean value of the strain at the maximum stress of each three specimens in relation to the bond strength of the aggregate rock to the mortar. It was observed that the concrete containing coarse aggregate with higher value of bond strength showed larger value of the strain and phophyrite concrete showed relatively smaller ratio of the strain to the bond strength. However,the effect of the bond strength and the elastic property of the aggregate rock on the strain at the maximum stress of concrete is still to be studied.

5.4 Elasticity of rock and concrete

Figure 9 shows the secant elastic modulus and the initial tangent modulus of each concrete in relation to the elastic modulus of aggregate rock. The both elastic moduli of concretes are dependent on the elastic modulus of aggregate rock.

6 Conclusion

The summary of the present investigation are as follows:

1) The rock with higher value of elastic modulus showed the greater compressive strength. Porphyrite showed the higher ratio of the elastic modulus to compressive strength than the others.

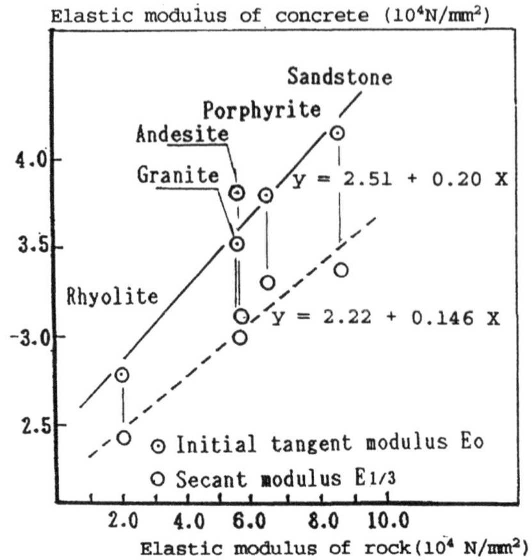

Fig.9. Elastic moduli of rocks and concretes

2) Rhyolite showed the highest bond strength, which is 1.85 times larger than the value of sandstone.

3) It was proved that the bond strength of the aggregate rock govern the compressive strength of concretes containing sandstone, andesite and granite each. The compressive strength of porphyrite concrete might be decreased by the higher stress concentration at the interface, resulted from the higher value of elastic modulus of aggregate rock. The bond strength of rhyolite was the highest among them, but the strength of the concrete stayed at the lowest level, as the aggregate pebbles were broken in the concrete.

4) The secant modulus and the initial tangent modulus of concrete were dependent on the elastic modulus of aggregate rock.

Acknowledgement

The author expresses his thanks to Prof. Shizuka MIURA in Fukui University for his advice on the rock and also to Mr. Keiichi WAKI(technician) and Mr.Yoshizumi UENO,Mr.Shigeshi HISANO and Yuhko TATUTA (students in Fukui University) for their corporation in experiments and analysis.

20 SOME MECHANICAL EFFECTS OF INTERFACE DEBONDING IN PLAIN CONCRETE

P. STROEVEN
Delft University of Technology, Delft,
The Netherlands

Abstract
A survey is presented of global characteristics of damage evolution in connection with stress-strain behaviour and engineering mechanical properties of concrete. Aspects of experimental assessment of such characteristics are discussed. The paramount influence exerted by the particle-matrix interface is outlined. Effects due to imposed modifications of interphase morphology, such as by mineral admixtures, are approached within the framework outlined.
Keywords:Interface, Bond, Damage, Stereology, Cracking.

1 Introduction

Concretes are macroscopically heterogeneous quasi-brittle materials. Hydration- and compaction-induced stresses will lead to premature cracking and will otherwise result in high residual stresses in the virgin body. The latter easily leads to further crack development under relatively low stresses. Such structural defects constitute the damage structure. The material body will gradually degradate under increasing loading due to damage evolution. Global aspects will be emphasized in relation to particular mechanical features.

On microscopical level degradation mostly starts in the particle-matrix ''interface''. Interface bonding therefore governs particular aspects of global mechanical behaviour. Models for damage evolution are designed on an intimate relationship between particle and crack features. Micro- and macromechanics are at least qualitatively connected in this way. Studies of particle-matrix interphase morphology offer insight into the structural background of ''bond''. This would allow estimating the effects of induced interphase modifications, such as by fine particle additions, on engineering mechanical properties. Some relevant aspects of the matter will be discussed.

2 Global mechanical behaviour

On engineering level the material behaviour is considered being uniquely reflected by stress-strain curves associated with certain loading regimes. Mechanical properties dealing with stiffness, strength and toughness are derived from particular characteristics of

Interfaces in Cementitious Composites. Edited by J. C. Maso. © RILEM.
Published by E & FN Spon, 2 - 6 Boundary Row, London SE1 8HN. ISBN 0 419 18230 6.

these stress-strain curves. Young's modulus is obtained from the slope of the ascending branch, which is generally over the first three-quarters, or so, quite straight. As a consequence, behaviour is interpreted in terms of linear elastic behaviour. Some evolution in the damage structure is however recognized to take place.

For good quality concretes it is found that over its quasi-elastic range of material behaviour the volume of the specimen is almost linearly declining in compression and increasing in tension (Stroeven, 1979). Since the secant value of Poisson's ratio is proportional to the slope of the volumetric versus axial strain curve, it will be roughly constant over this range. In compression this is well-established. Though the amount of experimental evidence in the tensile field is more limited, data also conform to the afore-mentioned characteristics (eg, Stroeven, 1970).

The end of quasi-elastic behaviour is indicated by a discontinuity in the respective curves of axial strain versus stress, volumetric strain and Poissoin's ratio. The sensitivity to detecting such deviations increases in this order. The two common experimental approaches for determination of this ''point'', ie pulse velocity and acoustic emission techniques, have also a different sensitivity. The discontinuity point is generally associated with the ''onset of (major) cracking'' and is thus also referred to as the ''cracking strength''. As a result it can be conceived as a lower bound to strength (roughly corresponding to the fatigue limit).

''Ultimate strength'' is defined as the stress transferred at the BOP of the stress-strain curve. In a low-stiffness test set up the specimen will rupture earlier. Hence, engineering strength depends on material properties as well as on loading conditions. In compression the minimum of material volume is arrived before BOP. This ''point'' is sometimes associated with the long term strength.

3 Global features of damage evolution

The damage evolution process can be considered uniquely reflecting certain loading regimes operative under specified circumstances. Characteristic global features of damage evolution due to monotonically increasing direct compressive loadings were already discussed by Glücklich about 30 years ago. Vile (1965) was probably the first to propose a model for meso-cracking underlying global damage evolution in direct compression. The topic also received attention by Stroeven (1973), Lusche (1972) and Perry and Gillott (1977). They also provided experimental evidence in support of the model. It recognized the interface between particles and matrix as the weakest link in the load-bearing system. This holds for most concretes. For a global approach it suffices to model the material as a continuum. Particles are sufficiently far appart to exclude interaction effects. With exclusion of the post-ultimate range, this approach has been shown to yield reliable estimates for mechanical properties.

Arguments in favour of the damage evolution model are amply supplied in the literature. The proper ones should also be in terms of the continuum approach. Elastic solutions of single inclusions in an infinite matrix predict in the direct compression case relatively unfavourable stress situations at the ''equator''. The matrix material at the pole positions is in biaxial (or for 3-D solutions: triaxial) compression. Such solutions are readily available and date back some sixty years (eg, Goodier, 1933). Goodier's 3-D solution, and the 2-D one's given by Muskhelishvili (1953) and Sezawa and Nishimura (1931) have been elaborated by Stroeven (1973) for concrete, whereupon solutions for the various stress components were presented in graphical form. The 3-D case is shown to yield the more unfavourable results. Of course, in direct tension relatively high triaxial tensile stresses will arise at the poles.

The conditions of the elastic continuum are fulfilled for photoelastic models. Such experiments should be conceived in fact as analogous computation. Thibodeau and Wood (1938) used rubber as matrix material and as a consequence even detected tensile stresses in the equatorial regions. This is in conformity with the elastic solutions. Creep deformations can cause Poisson's ratio in concrete also to rise, however, above the conventionally applied value of 1/6. Other arguments for early interface debonding are coming from micro-hardness investigations, the oldest ones being due to Lyubimova and Pinus (1962). More recent investigations however also show evidence of a ''valley''-shaped decline in local bond (Yuji, 1988). Damage evolution stages can be described in the following way:

1. Particles will debond under relatively low stresses from the matrix, in compression in the equatorial zones and in tension in the polar ones. Such isolated bond cracks exert marginal and roughly similar effects on global deformations in orthogonal directions. The bond crack population will develop a preferred orientation (Stroeven, 1973).

2. Bond cracks will further extend along the interface; in compression untill they reach the zone of triaxial compression. These zones are conically-shaped. The larger ones can contain smaller particles, which will not debond from the matrix. The smaller grains just outside these cones however do, stimulated also by secondary shear stresses. They hereby form the so called ''en echelon'' crack arrays, displayed in figure 1. This was experimentally confirmed by Stroeven (1973).

3. The ''en echelon'' crack arrays coalesce due to high local shear stresses. The number of cracks thereby starts declining, whereas average size inceases. Due to the oblique orientation of the so called ''shear bands'', the global orientation distribution looses its predominant direction. This is supported by experimental evidence obtained in the afore-mentioned study.

4. The interaction between neighbouring cracks will be of rising importance. Neighbouring cracks, partly enrobing smaller particles and unfavourably situated as to the stress field, will coalesce. This leads to a discontinuity in strain development. Cracking has two opposite effects. In a positive sence, it eliminates weak zones by releasing high residual stresses. Under compression the material continuously redistributes the internal stresses to compensate for the detrimental effect of cracking, i.e. a reduction in load-bearing capacity. From BOP on, the detrimental

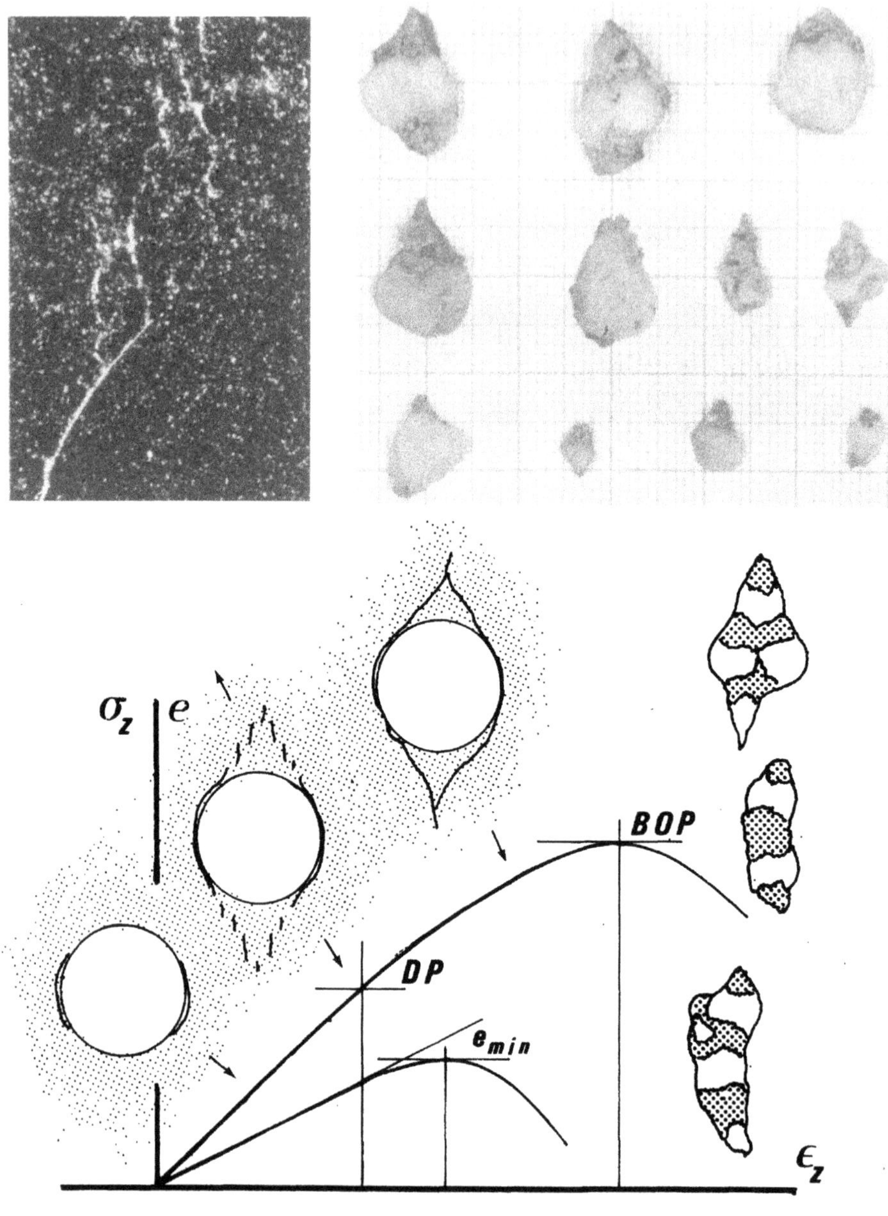

Figure 1. Curves of stress (σ_z) and volumetric strain (e) versus axial strain (ϵ_z) in compression. Damage evolution is shown from left to right: at the top left a micrograph of the transition of interface crack to en echelon crack array, at the right rupture elements of single grains (top) and clusters (bottom)

effect will be of an ever increasing dominancy. The process is of a dynamic character: a premature structural collapse can occur, unless the driving force of the system can be reduced fast enough.
The material body's possibilities for compensating the reduction in bearing capacity by continuous redistribution of its internal stress field during crack coalescence are much more restricted in case of tensile loading. The process of crack coalescence will therefore in the latter case be less advanced at BOP. By comparing simple model estimates based on the afore-mentioned charac- teristics and experimental data, this was confirmed by Stroeven (1990b). Crack coalescence between all nearest neighbour cracks already accounted in the compression case for the experimentally found specific crack surface area at ultimate.

5. During yielding, further crack development is concentrated in so called fracture process zones. Boundaries of such zones can not be defined clearly in concrete composites (Buresch, 1978). To fully exhaust the material body's cracking resistance- to separate it in two parts- major part of crack coalescence has therefore still to take place, both in tension as well as in compression, provided deformations will be controlled, of course! Section images of two-sided notched specimens, subjected to direct tensile loadings considerably beyond BOP, indeed revealed a significant portion of the fracture plane still ''in statu nascendi'' (Stroeven, 1990a). Long tails of complete stress-strain curves also prove crack coalescence far from advanced at ''ultimate''.

In agreement with observations in the virgin state, randomly distributed bond cracks can be incorporated in the model. Such cracks are resulting from shrinkage, thermal stresses and internal bleeding. Solutions for elastic stress and strain fields in a homogeneous infinite matrix containing a partly debonded cylindrical inclusion are available for about a quarter of a century (Perlman and Sih, 1967). A tendency of the smaller bond cracks is revealed to further extend along the interface under uniaxial compression until a situation is arrived at similar to the damage evolution stage described under 2. of the continuous model (Stroeven, 1973). This equally holds for the tensile loading case.

The cracked composite can be modelled for fracture mechanical purposes as an uniform material containing a regular crack array for which LEFM solution are readily available. Average size and spacing of cracks will be governed by average size and number of particles (Stroeven, 1990b and 1991). This renders possible estimating quantitatively engineering mechanical properties on the basis of micromechanical modelling.

4 Experimental assessment of global features of structural loosening

Reflective photoelasticity allows experimentally assessing the characteristic details of the global damage structure in concrete. This method has been used in numerous investigations in concrete technology during the past half of a century (e.g. Frocht, 1941). The sensitivity is governed by the thickness of the photoelastic coating (in general: a couple of millimeters). Hence, (bond) crack development below the sensitivity level will not be reflected. The larger cracks are displayed in a smeared out way. The formation of structuring elements consisting of either a single (larger) particle, or of small clusters of nearby particles, provided with the conically-shaped ''hats'', can be seen. Strings of such elements constitute the skeletton of the material body (Stroeven, 1973). After rupture in direct compression the elements can still be found among the debris, as shown by various authors. Some details are shown in figure 1.

Direct access to the damage structure of the material body, either on micro- or on macroscopical level, is provided by sections. The quantitative evaluation of this 2-D information should be based on stereological notions. This involves problems of contrast improvement, of sampling (size, location and orientation representativity), of probing (data acquisition), and of (spatial) interpretation. The experimental assessment of damage parameters is, moreover, liable to subjective biases resulting from its fundamentally resolution-dependent character (Stroeven, 1991). The latter implies the traditional classification of microcracks (in bond, mortar and aggregate cracks; see, eg, Shah and Slate, 1968) to be indeed a subjective one, which depends on sensitivity of the experimental approach! So far this aspect is still neglected even in more fundamental approaches in concrete technology (Larbi, 1991).

The information on damage presented by a field image is basically in the form of lineal features. The number of such features (crack sections!), N_c, per unit of area, A ($N_{Ac} = N_c/A$) is readily obtained. The total length of such traces, L_c, per unit of area, A (yielding $L_{Ac} = L_c/A$), can be derived from intersection counts with a superimposed line grid. In case of an isotropic uniform random (IUR) crack system, the so called method of directed secants can be employed. The number of intersections, P_c, per unit of the test line system, L, ($P_{Lc} = P_c/L$) is related to L_{Ac} by $L_{Ac} = \frac{\pi}{2} P_{Lc}$.

For arbitrary crack distributions the method of random secants should be used. Alternatively, directed secants can be successively applied in randomly selected positions, whereupon the results have to be averaged. In both cases this leads to $L_{Ac} = \overline{P_{Lc}}$. Average trace length, $\overline{c}$, equals the ratio of the two damage parameters. Hence $\overline{c} = L_{Ac}/N_{Ac}$. The averaging procedure has to be extended into the third dimension! When restricted to the image plane, the given approach leads to 2-D measures of damage. Without increasing the effort unbiased 3-D measures can be obtained, however. To that end the type of crack orientation distribution has to be specified. In the case of a direct tensile test a crack structure is developed revealing symmetry with respect to the loading direction. It is appropriate assuming the actual structure to be composed of two portions. One portion contains cracks parallel to a plane perpendicular to the loading axis, while the second is an IUR one. This is referred to as a partially-planar system. In a section parallel to the loading direction both crack portions will leave their traces. Probing with directed secants in directions perpendicular and parallel to that of the loading leads respectively to

$$P_{L\perp} = \frac{2}{\pi} L_{A3} \quad \text{and} \quad P_{L\|} = \frac{2}{\pi} L_{A3} + L_{A2} \tag{1}$$

In eq (1) the subscript 'c' has been eliminated for convenience reasons. Subscripts '2' and '3' refer to the 2-D and 3-D portions, respectively. Both parts of eq (1) can be combined to yield estimates for the 2-D damage parameters, where the degree of crack orientation is defined by $\omega_2 = L_{A2}/L_A$. Simple algebraic manipulations lead to

$$L_A = P_{L\|} + (\frac{\pi}{2} - 1)P_{L\perp} \quad \text{and} \quad \omega_2 = \frac{P_{L\|} - P_{L\perp}}{P_{L\|} + (\frac{\pi}{2} - 1)P_{L\perp}} \tag{2}$$

The assessment of 3-D damage parameters can be achieved without additional effort! The alternative for eqs (1) and (2) is $S_V = 2P_L$, in which S_V is the total crack surface area, S_c, per unit of volume, V (ie $S_c/V = S_V$). P_L should be averaged over all possible orientations for arbitrary damage structures. N_{Vc} (number of cracks per unit of volume) cannot be determined without rigorous assumptions as to the spatial shape of the cracks. Average crack size in 3-D cannot be de- termined as easily as in 2-D!

Using the concept of a partially-planar crack system, the two orthogonal intersection counts in an image plane parallel to the loading direction can be related to the composing portions of the crack system, S_{V2} and S_{V3}, by

$$P_{L\perp} = \frac{1}{2} S_{V3} \quad \text{and} \quad P_{L\|} = \frac{1}{2} S_{V3} + \frac{2}{\pi} S_{V2} \tag{3}$$

Defining a 3-D alternative for ω_2 by $\omega_3 = S_{V2}/S_V$ allows to assess the 3-D estimates of the damage structure similarly as before. Simple mathematical manipulation yields

$$S_V = \frac{\pi}{2} P_{L\|} + (2 - \frac{\pi}{2})P_{L\perp} \quad \text{and} \quad \omega_3 = \frac{P_{L\|} - P_{L\perp}}{P_{L\|} + (\frac{4}{\pi} - 1)P_{L\perp}} \tag{4}$$

For direct compression tests the concept of a partially-linear crack system can be adopted. This implies that an IUR system is combined with a portion of which the cracks are parallel to the loading axis. In a methodological sence the experimental approach is similar. The theoretical background can be established in complete analogy with the foregoing. Quantitative image analysis results obtained so far (Stroeven, 1973, 1979, 1990a) confirm the aforementioned global characteristics of damage evolution.

When interface cracks have extended to maximum size, their size distribution will be identical to that of the particles. Total crack surface area per unit of volume, S_V, can be shown to equal in that case (Stroeven, 1991): $3V_p/[2\pi d_0 \ln(d_m/d_0)]$, which is valid for a logarithmic particle size distribution function. Further, d_m and d_0 are the largest and smallest grain sizes, respectively. V_p is the volume fraction of particles encompassing all grains exceeding d_0 in size. It is obvious that estimates are resolution dependent and that, as a consequence, quantitative image analysis results dealing with damage in concrete will be resolution dependent, too.

In Stroeven (1991) it is additionally show, that when cracks would be coalescing in a single plane, a macrocrack is formed of which total surface area per unit of cross sectional area, R_S, will equal $1+2V_p(1-\frac{3}{4}\omega_3)$. Herein, ω_3 is the spatial degree of crack orientation, defined in eq (4). R_S is attributed as the (crack) surface roughness. Since V_p increases with magnification, this will also be true for R_S. The resolution-dependence of the ''roughness'', R_L, of the corresponding macrocrack profile in a section follows from the approximate expression $R_L = \frac{\pi}{4}(R_S - 1)$. Substituted into the fractal equation leads to the following expression (Stroeven, 1991)

$$\log[1 + \frac{3\pi}{8} V_p(M)] = (D-1)\log M + C \tag{5}$$

in which M is the magnification level, D the fractal dimension of the fracture surface and C a constant. Fractile dimension D follows from the slope of a $\log-\log$-plot of eq (5). Values between 1.10 and 1.15 were found, which are very close to available experimental data.

5 Interphase and micromechanics

Considerable research is nowadays devoted to the morphology of the roughly $50\mu m$ thick interphase layer between bulk matrix and aggregate particles, and on assessing its effects on ''stress transfer capability'', or simply on ''bond''. As an example, reference can be given for state-of-the-art information to 1987 MRS Symposium Proceedings on Bonding in Cementitious Composites. For the present purpose it suffices here to recall, that between the bulk matrix and a thin and (for older concrete) strong duplex layer of $Ca(OH)_2$ enrobing the aggregate, a transition zone is formed, which has a deviating structure and composition (Diamond, 1987). Among other things, large plate-like $Ca(OH)_2$-crystals perpendicular to the aggregate surface were detected (Grandet and Ollivier, 1980), reflecting the large initial porosity in the hardening state of the concrete.

Improvements to the micromechanical properties of the transition layer can be found in the literature. They can also be logically derived from the foregoing. Since strength on this level was demonstrated to be governed for an important part by van der Waals forces, packing characteristics would be of paramount importance. Obviously, strength could be expected favourably influenced by thoroughly dispersing even a very fine inert filler in the matrix. Stereological estimates would demonstrate the thickness and the porosity of the interphase layer to be (strongly) reduced, so that the growth of large $Ca(OH)_2$-crystals could be prevented. Test results of Detwiler and Mehta (1989), using among other things carbon black as filler, confirm these expectations.

Better interface bond will not lead inevitably to improved global mechanical properties. Neiter will an enhanced compressive strength due to interphase modifications automatically imply other mechanical properties to be upgraded. Once bond cracks, upon arriving at full extension in the interface layer, have initiated into the matrix, ultimate strength will be governed solely by the (bulk) matrix' properties! Better bond will nevertheless give rise to higher values of cracking strength and stiffness, thereby however reducing deformation capacity at ultimate! Ultimate strength will only profit fully from better interface bonding, once matrix properties are also and to the same degree improved as the interfacial bond. Densification by a fine filler like silica fume will probably be easier to achieve for the otherwise porous transition zone than in bulk, leading to dissimilar micro-mechanical effects. Ultra-fine fillers can generally not be dispersed properly in bulk cement pastes (as admitted in the literature). The absence of a significant improvement of the compressive strength does therefore not form conclusive evidence for the conclusion drawn by Larbi (1991), that compressive strength improvements in concretes due to fine fillers should be solely the effect of improved bonding in the interfacial zone.

6 Conclusions

The interpretation of engineering mechanical behaviour in terms of cracking mechanisms and features of damage evolution allows a structural interpretation of research into the mechanical properties, and renders possible at least qualitatively estimating mechanical characteristics on the basis of structural modifications, such as imposed by adding mineral admixtures.

7 References

Buresch, F.E. (1978) A structure sensitive K_{1c}-value and its dependence on grain size distribution density and microcrack interaction, in **Fracture Mechanics of Ceramics** (eds R.C. Bradt, D.P.H. Hasselman and F.F. Lange), Plenum Press, New York, pp 835-847.

Detwiler, R.J., Mehta, P.K. (1989) Chemical and physical effects of silica fume on the mechanical behaviour of concrete. **ACI Mat. J.**, Nov/Dec., 609-614.

Diamond, S. (1987) Cement paste microstructure in concrete, in **Microstructural Development during Hydration of Cement** (eds L.J. Struble and P.W. Brown), MRS Symp. Proc., 85, pp 21-31.

Frocht, M.M. (1941) **Photoelasticity**. Wiley, New York.

Glücklich, J. (1963) Fracture of plain concrete. **J. ASCE**, 89, Em6, 127-138.

Goodier, J.N. (1933) Concentration of stress around spherical and cylindrical inclusions and flaws. **J. Appl. Mech.**, ASME, 55, 7, 39-44

Grandet, J., Ollivier, J.P. (1980) Nouvelle méthode d'éthude des interfaces ciment-granulats, in **Proc. 7th Int. Congress on Chemistry of Cements**, Vol3, Paris, pp 85-89.

Larbi, J.A. (1991) **The cement paste aggregate interfacial zone in cement**, Ph.D. Thesis, Delft University of Technology, Delft.

Lyubimova, T.Y., Pinus, E.R. (1962) Crystallization structure in the contact zone between aggregate and cement in concrete. **Colloid J. USSR**, 24, 5, 491-498.
Muskhelishvili, N.F. (1953) **Some basic problems of the mathematical theory of elasticity**. Noordhoff, Groningen.
Perlman, A.B., Sih, G.C. (1967) Elastostatic problems of curvilinear cracks in bonded dissimilar materials. **Int. J. Engr. Sci.**, 5, II, 845-867.
Perry, C., Gillott, J.E. (1977) The influence of mortar aggregate bond strength on the behaviour of concrete in uniaxial compression. **Cem. Concr. Res.**, 7, 553-564.
Sezawa, K., Nishimura, G. (1931) **Stresses under tension in a plate with a heterogeneous insertion**. Rep. Aeron. Res. Inst., Tokyo Imp. Univ., 6, 25-43.
Shah, S.P., Slate, F.O. (1968) Internal microcracking, mortar-aggregate bond, and stress-strain curve of concrete, in **Structure of Concrete** (eds A.E. Brooks and K. Newman), C&CA, London, pp 82-92.
Stroeven, P. (1970) Poisson's ratio in uniaxial tension and anticlastic bending of concrete and perspex. **Heron**, 17, 3, 1-22.
Stroeven, P. (1973) **Some aspects of the micromechanics of concrete**. Ph.D. Thesis, Delft University of Technology, Delft.
Stroeven, P. (1979) Geometric probability approach to the examination of microcracking in plain concrete. **J. Mat. Sci.**, 14, 1141-1151.
Stroeven, P. (1990a) Some observations on mirocracking in concrete subjected to various loading regimes. **Engr. Fract. Mech.**, 35, 4/5, 775-782.
Stroeven, P. (1990b) Some stereological notions of mesoscopic structural concepts underlying mechanical behaviour of fibrous, particulate cementitious systems, in **Micromechanics of Failure of Quasibrittle Materials** (eds S.P. Shah, S.E. Swartz and M.L. Wang), Elsevier, London, pp 295-304.
Stroeven, P. (1991) Fractals and fractography in concrete technology, in **Brittle Matric Composites 3** (eds A.M. Brandt and I.H. Marshall), Elsevier, London, pp 1-10.
Thibodeau, W.E., Wood, L.A. (1938) Photoelastic determination of stresses around a circular inclusion in rubber. **J. Res. Nat. Bur. Stand.**, 9, March, 393-409.
Vile, G.W.D. (1965) **Behaviour of concrete under simple and combined stresses**. Ph.D. Thesis, Univ. London, London.
Yuji, W. (1988) The effect of bond characteristics between steel slag fine aggregate and cement paste on mechanical properties of concrete and mortar, in **Bonding in Cementitious Composites** (eds S. Mindess and S.P. Shah), MRS Symp. Proc., 114, pp 49-54.

21 MODELE EXPERIMENTAL D'ETUDE MECANIQUE ET PHYSICO CHIMIQUE DE L'INTERFACE ACIER BETON

(Experimental model for mechanical, physical and chemical study of steel-concrete interface)

R. FRANCOISE, M. ARNAUD, G. ARLIGUIE and A. AYED
Laboratoire Matériaux et Durabilité des Constructions,
INSA - UPS, Toulouse, France

Résumé *(English abstract, page 307)*

Cette communication a pour objet la présentation d'un modèle expérimental mis au point dans le but de relier l'adhérence entre acier et béton aux caractéristiques physico-chimiques de la liaison. L'étude théorique du modèle constitué de deux matrices de mortier séparées par une plaque d'acier, fait apparaître un champ de contrainte relativement complexe à l'interface. L'étude expérimentale sur une éprouvette instrumentée confirme la complexité du champ de déformation au voisinage de l'interface et permet la comparaison avec l'approche théorique. Les résultats montrent que la recherche d'un couplage résistance mécanique - caractérisation physico-chimique de l'interface ne saurait être établie, même en première approximation, par la prise en compte d'un critère global fondé sur une valeur moyenne du champ de contrainte pour toute l'interface.

Mots clés: Interface Acier-Béton, Adhérence, Cisaillement, Modélisation, Méthodologie.

1 Introduction

La liaison acier-béton est un facteur déterminant de la durabilité du béton armé. Or, pour étudier cette liaison, de nombreux essais ont déjà été réalisés permettant soit une quantification mécanique, soit des caractérisations physico-chimiques, mais aucune technique n'a, jusqu'à présent, couplé les deux types d'investigation.

Afin d'apporter une contribution dans cette voie, une recherche a été entreprise dans le but de relier l'adhérence (au sens du collage) entre acier et béton aux observations physico-chimiques de la liaison. Ainsi, il devient possible de suivre l'évolution de l'adhérence en fonction d'actions endogènes ou exogènes de type mécanique, physique et chimique.

Un modèle expérimental, permettant de satisfaire aux exigences de ce programme, a été étudié. Il est constitué par deux matrices de mortier séparées par une plaque d'acier adhérente.

La présente communication est consacrée au modèle expérimental et plus précisément à l'étude des champs de déformation et de contrainte définis au voisinage de l'interface par une approche théorique. Trois inclinaisons de la plaque d'acier sont étudiées (30°, 45°, 60°). L'étude expérimentale sur une éprouvette instrumentée permet la comparaison avec l'étude théorique.

2 Modèle expérimental

2.1 Définition du modèle

Interfaces in Cementitious Composites. Edited by J. C. Maso. © RILEM.
Published by E & FN Spon, 2 - 6 Boundary Row, London SE1 8HN. ISBN 0 419 18230 6.

Pour permettre une caractérisation physico-chimique faisant suite à une quantification mécanique de la liaison acier-béton, le modèle étudié doit satisfaire à un certain nombre d'exigences.

La première condition est bien sûr l'existence d'une liaison acier-béton. Les expériences seront donc conduites sur une éprouvette mixte. L'étude de l'interface par diffractométrie des rayons X nécessite une surface plane, ce qui exclut tous les essais classiques d'adhérence à base de barres cylindriques (pull test, beam test). L'observation de la micro-structure de la liaison acier-béton par microscopie électronique à balayage (M.E.B.) nécessite, en plus d'une interface plane, un échantillon de volume réduit afin de le mettre en place dans le M.E.B. sans préparation ou avec une préparation légère ne risquant pas d'introduire des défauts artificiels. L'analyse de l'interface par microsonde conduit aux mêmes exigences que celle du M.E.B.. En ce qui concerne la quantification mécanique, l'éprouvette doit permettre de qualifier l'adhérence par l'étude d'un paramètre global simple à mesurer (charge de rupture par exemple) et ce pour différents champs de contraintes à l'interface.

2.2 Choix du modèle

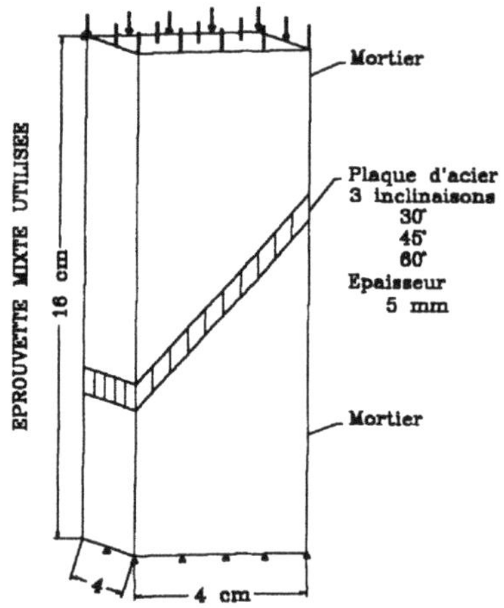

Figure 1: modèle utilisé

Pour répondre aux exigences présentées précédemment, le modèle retenu est constitué d'une plaque d'acier d'épaisseur 5 mm enrobée dans une éprouvette de mortier de dimensions 4 x 4 x 16 cm. Cette éprouvette mixte contient ainsi 2 liaisons acier-béton et 3 phases. Le centre de gravité de la plaque étant le point de symétrie de l'échantillon, les 2 interfaces sont équivalentes.

L'adhérence impliquée dans la liaison acier-béton est mise en jeu en sollicitant l'échantillon en compression monoaxiale. Le champ de contraintes et en particulier celui de cisaillement à l'interface est modifié à volonté suivant l'inclinaison de la plaque.

Un modèle similaire a déjà été utilisé par Taylor et Broms pour l'étude de la liaison pâte-granulat [1]. Alexander et al ont également étudié l'interface pâte-granulat par l'intermédiaire d'une éprouvette mixte à 2 phases seulement [2]. Dans ces 2 derniers cas, la contrainte de cisaillement τ (supposée uniforme à l'interface) était évaluée en terme d'angle de friction Φ et de cohésion C, en fonction de la contrainte normale à la plaque σ conformément à la théorie de Coulomb-Mohr : $\tau = C + \sigma.\tan\Phi$.

3 Etude théorique

3.1 Présentation du modèle étudié par éléments finis

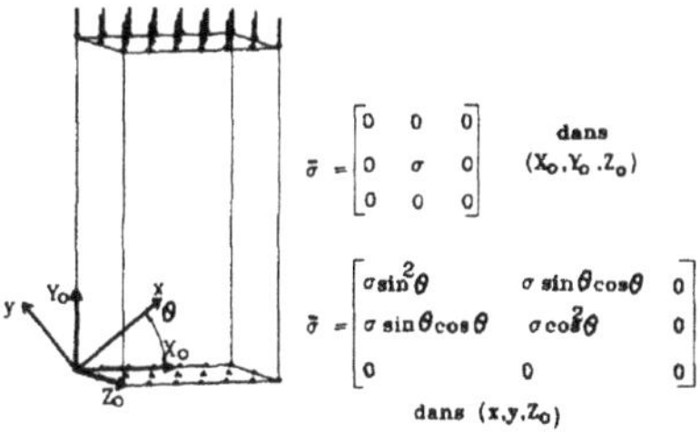

Figure 2 : Eprouvette homogène

Dans le cas d'un matériau homogène, la mécanique des milieux continus indique que le tenseur des contraintes dans le repère (Xo,Yo,Zo) possède une seule composante non nulle qui est la contrainte de compression dans le plan perpendiculaire à l'axe du chargement.

Si on s'intéresse aux contraintes sur un plan incliné de θ, on voit naturellement apparaître des cisaillements dans la direction x (direction de la pente). La figure 2 rappelle ces résultats.

Ces cisaillements sont uniformes sur le plan incliné et maximums pour une inclinaison de 45°. La valeur maximale représente alors 50% de la contrainte de compression sur un plan horizontal.

Dans le cas de l'éprouvette mixte, une plaque plane d'acier inclinée de θ sépare deux volumes de mortier. Les modules de déformation de ces deux matériaux sont très différents (rapport de 1 à 10) mais les déformations doivent être compatibles au niveau de l'interface.

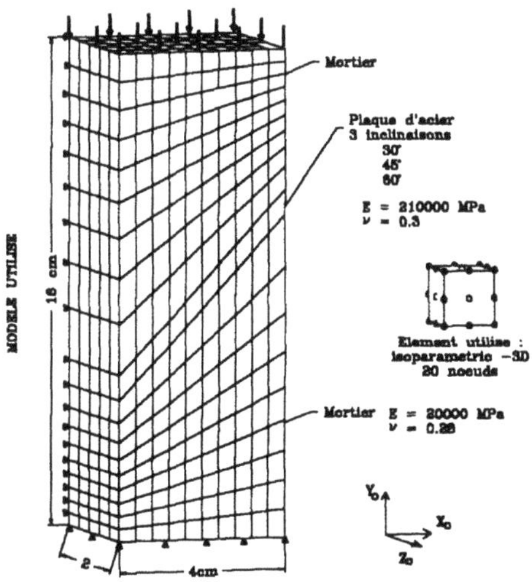

Figure 3 : maillage utilisé pour le calcul par éléments finis

Les déformations gênées des 2 matériaux vont donc conduire à la création de cisaillements dans la plaque et pas seulement dans la direction du pan incliné mais aussi suivant Zo. En effet, la dimension transversale de l'éprouvette (suivant Zo) n'est pas suffisamment petite pour faire l'hypothèse de contraintes planes. Une tentative de calcul théorique du champ de contrainte à l'interface devra donc être conduite sur un modèle 3D pour ne pas oublier les cisaillements suivant Zo.

Le maillage utilisé pour le calcul par éléments finis est présenté figure 3. Compte tenu de la symétrie de l'éprouvette par rapport à un plan vertical médian, nous pouvons limiter l'étude à une demi-éprouvette de profondeur 2 cm.

Pour tenir compte de la symétrie, les déplacements des noeuds appartenant au plan de symétrie sont bloqués dans la direction Zo. Le chargement est modélisé par une pression de 10 MPa sur la surface supérieure de l'éprouvette.

Toutes les faces des éléments appartenant à cette surface subissent cette pression de 10 MPa. Tous les déplacements des noeuds appartenant à la surface inférieure de l'éprouvette sont bloqués dans la direction du chargement Yo. Pour que le problème soit statiquement licite, un noeud situé à l'intersection du plan de symétrie et de la médiane de la surface inférieure est bloqué dans la direction Xo.

Les caractéristiques élastiques de l'acier sont le module d'Young E = 210 000 MPa et le coefficient de poisson ν = 0.3. Pour le mortier, considéré pour ce modèle comme un matériau élastique, les caractéristiques retenues sont E = 20 000 MPa et ν = 0.28 qui sont les valeurs à court terme.

Trois inclinaisons ont été retenues: θ= 30° , 45° , 60°. En deçà de 30°, les contraintes de compression obtenues sont trop importantes par rapport aux contraintes de cisaillement et l'éprouvette périt par rupture du mortier en compression. Au delà de 60° l'effet inverse est observé et conduit expérimentalement à un arc-boutement empêchant toute reproductivité des résultats.

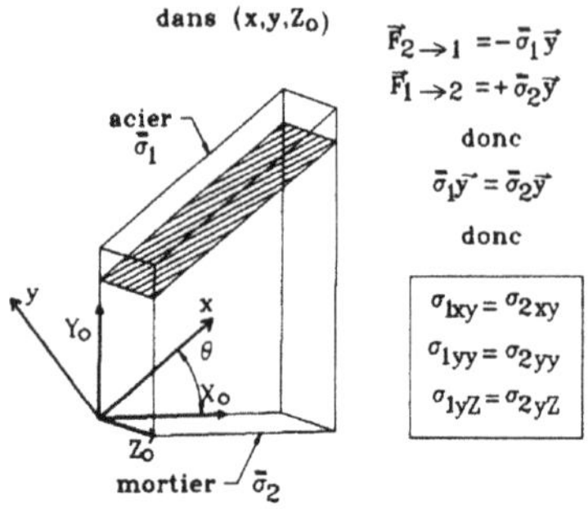

Figure 4 : Compatibilité des contraintes à l'interface

Les contraintes engendrées par la compression uniaxiale sont différentes dans les deux matériaux, mais au niveau de l'interface durant toute la phase de chargement jusqu'à la rupture, le principe de l'action-réaction permet de montrer qu'il y a continuité des composantes du tenseur définies dans le plan de l'interface.

L'évaluation des contraintes aux noeuds de l'interface, donc communs aux éléments mortier et aux éléments acier, est par conséquent correcte si on exprime les contraintes dans le repère (x,y,Zo) en se limitant aux trois composantes relatives au plan xy (σ_{xy}, σ_{yy}, σ_{yZo})

3.2 Résultats théoriques

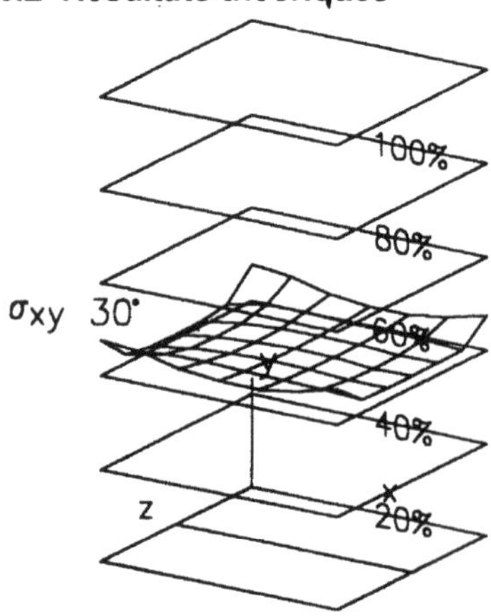

Figure 5a : Contraintes de cisaillement à l'interface pour une inclinaison de 30°

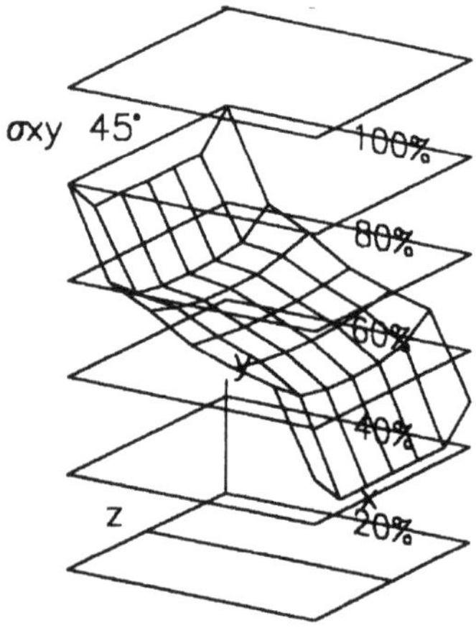

Figure 5b : Contraintes de cisaillement à l'interface pour une inclinaison de 45°

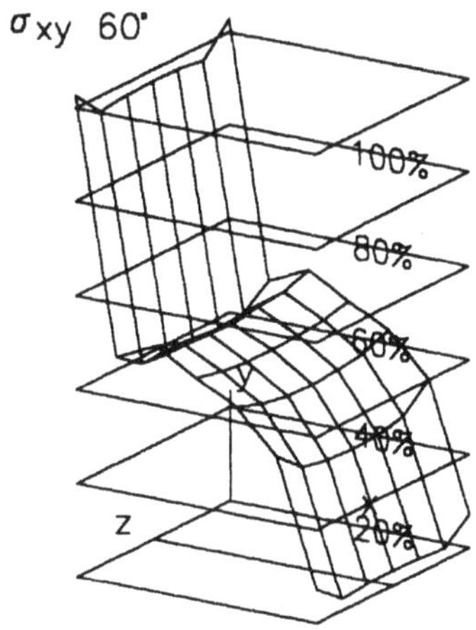

Figure 5c : Contraintes de cisaillement à l'interface pour une inclinaison de 60°

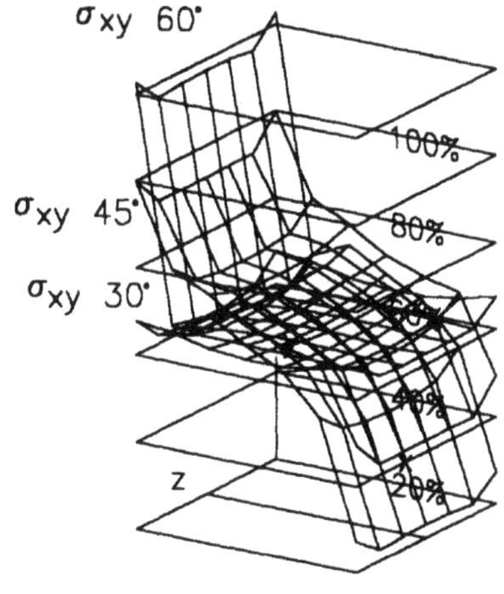

Figure 5d : Contraintes de cisaillement à l'interface : superposition des inclinaisons de 30°, 45°, 60°

Les figures 5 montrent les variations des contraintes de cisaillement exprimées dans le plan de la plaque pour les différentes inclinaisons retenues. Les contraintes sont exprimées en pourcentage de la pression exercée sur les faces de l'éprouvette, que l'on peut considérer comme la contrainte uniforme de compression dans une éprouvette homogène.

Le résultat le plus remarquable est que pour une inclinaison de 30°, la contrainte de cisaillement est quasi-uniforme sur toute l'interface (figure 5a). Les extremums obtenus pour σ_{xy30} sont 41% et 48%, la plupart des valeurs étant comprises entre 42% et 44%. Dans le cas d'une éprouvette homogène (figure 2), la contrainte de cisaillement sur une facette orientée de 30° est 43.3% de la contrainte de compression. Donc, dans le cas de l'éprouvette mixte avec une plaque d'acier orientée de 30°, le champ de cisaillement est quasi-identique au cas d'une éprouvette homogène.

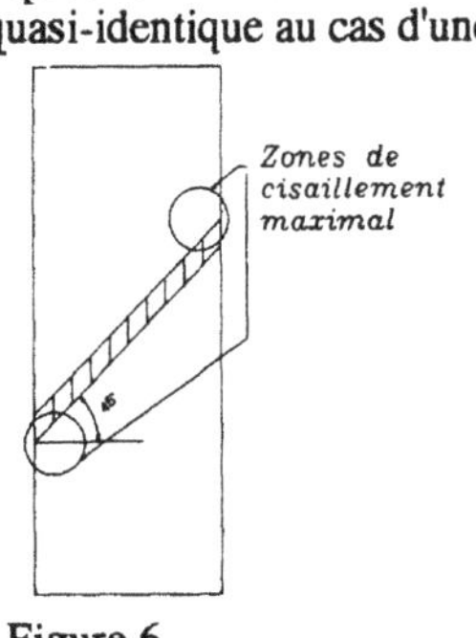

Figure 6

Lorsque la plaque est inclinée de 45° (figure 5b), des différences importantes de cisaillement apparaissent le long de la plaque (axe x). En revanche, suivant la largeur (axe Zo) les contraintes restent quasi-uniforme pour une abscisse x donnée. La zone la plus cisaillée de la plaque est la partie la plus haute pour la face supérieure et par symétrie la partie la plus basse pour la face inférieure (figure 6).

Les extremums de cisaillement sont 80% et 23% et les valeurs obtenues au milieu de la plaque sont d'environ 50% ce qui correspond au cas homogène.

Pour l'inclinaison de 60°, le phénomène décrit précédemment s'amplifie (figure 5c). Les extremums de cisaillement qui grandissent avec l'inclinaison atteignent les valeurs limites 0% et 100%. C'est pourquoi, des inclinaisons supérieures à 60° n'ont pas été étudiées.

Les contraintes de cisaillement suivant la largeur de la plaque σyz sont nulles le long de l'axe longitudinal et augmentent progressivement vers les bords de la plaque. Plus l'inclinaison de la plaque est faible et plus ces cisaillements sont importants. Pour 30°, on atteint 20% de la pression exercée

En ce qui concerne, les contraintes normales, plus l'inclinaison est faible, plus la compression est importante sur la plaque. C'est pourquoi en deçà de 30° l'éprouvette ne peut périr que par excès de compression. L'extrémité de la plaque où les cisaillements sont minimums correspond également à la compression minimale. Au delà de 60°,on obtient des tractions à l'interface avec destruction quasi-immédiate de la liaison.

4 Etude Expérimentale

Lorsqu'il s'agit de qualifier une étude théorique par une expérimentation, il est nécessaire de faire en sorte que les conditions de l'essai soient aussi proche que possible des hypothèses de l'étude.

4.1 Conditions expérimentales

Grâce à une expérience acquise au laboratoire [3], le modèle expérimental retenu est une éprouvette mixte plaque d'acier - mortier de résine (figure 7). Ce dernier a été préféré au mortier de ciment car, aux effets de fluage près, ses propriétés mécaniques sont comparables et n'évoluent pratiquement pas au cours du temps nécessaire aux essais. De plus, l'excellente adhésivité de la résine permet la mise en oeuvre de sollicitations plus importantes.

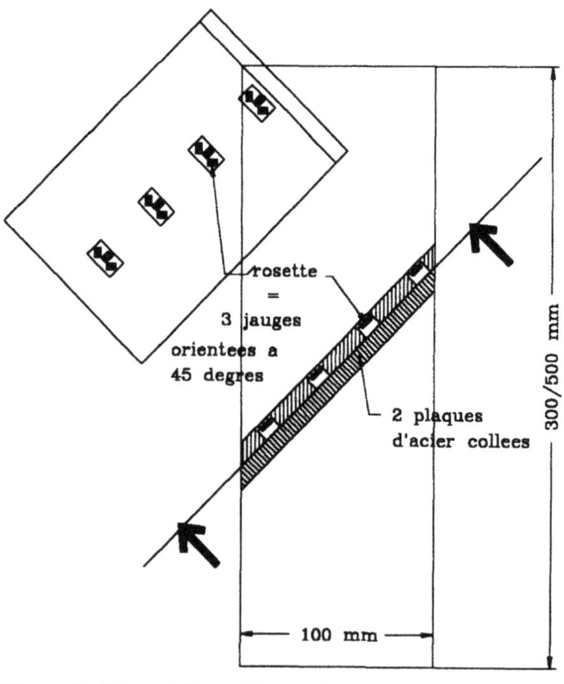

Figure 7 : Dispositif expérimental

Pour avoir accès aux déformations de cisaillement, il est nécessaire de disposer de 3 allongements relatifs dans 3 directions différentes en un "point" de mesure.

Pour étudier les déformations à l'interface acier-mortier sans perturber le champ de contraintes, nous avons collé les jauges de déformations (rosettes à 45°) au fond d'un évidement de la plaque d'acier reconstituée ensuite par collage à partir de deux éléments (figure 7), selon une technique qui a déjà fait ses preuves.[4]

Les jauges sont ainsi situées à 1 mm de l'interface. Une étude complémentaire par la méthode des éléments finis montre que la distribution des contraintes est peu différente de celle de l'interface.

Un des autres intérêts de ce dispositif expérimental est que les surfaces où sont collées les rosettes sont exemptes de contraintes normales, ce qui fait que l'on connaît une des directions principales à chaque point de mesure. Le dépouillement des relevés expérimentaux est ainsi facilité.

Nous avons réalisé des essais avec différentes vitesses de chargement variant entre 1 MPa/s à 0.01 MPa /s . Aucune différence significative de comportement n'a été relevée pour ces différentes vitesses et nous avons retenu les résultats pour une vitesse classique de chargement de 0.1 MPa/s.

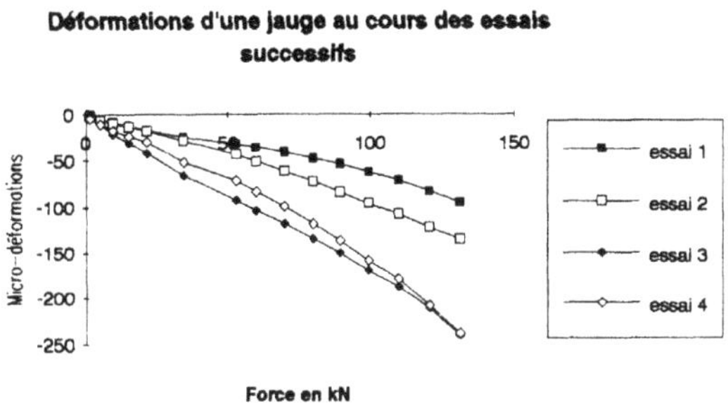

Figure 8 : Evolution des déformations au cours des essais successifs

Différentes hauteurs d'éprouvette ont été testées, de 200 à 500 mm, pour estimer l'influence de la flexion parasite lors de l'essai de compression monoaxiale. Les deux faces de l'éprouvettes en contact avec les plateaux de la presse étaient rectifiées et parfaitement parallèles.

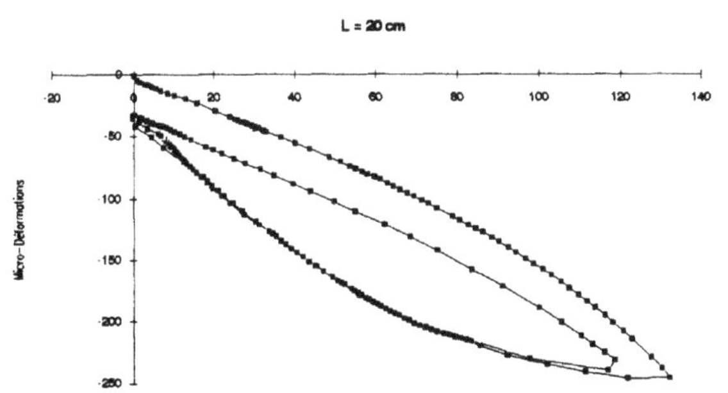

Figure 9: Evolution des déformations au cours d'un cycle

La même éprouvette étant réutilisée après sciage, on constate que les déformations des différentes jauges augmentent mais de façon non uniforme avec la réduction de longueur (figure 8). En réalisant des cycles de chargement-déchargement, on constate (figure 9) également que les déformations augmentent avec le nombre de cycles.

Bien que les rosettes soient disposées sur l'axe de symétrie du modèle, on ne constate pas une symétrie du comportement entre les jauges Gauches (jauge G) et les jauges Droites (jauge D) des rosettes (figure 10). Néanmoins le comportement des 4 rosettes disposées le long de la plaque est similaire.

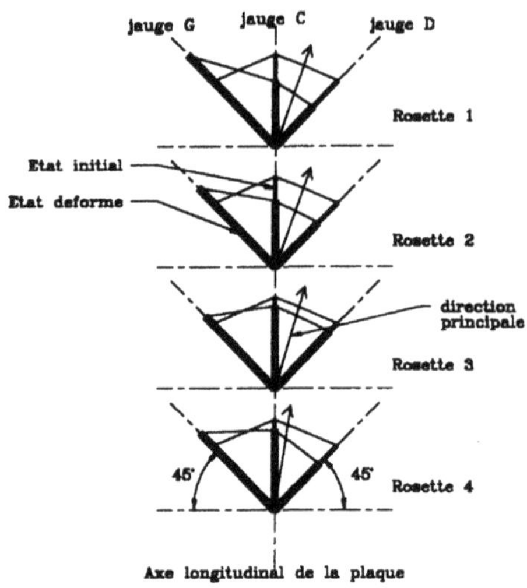

Figure 10 : Déformations des différentes rosettes

Une des directions principales appartenant au plan de la plaque est inclinée d'un angle variant de -20° à -10 ° pour les rosettes 1 à 4. En fait, seule la rosette 4 mesure l'inclinaison de -10° et les rosettes 1, 2, 3 mesurent une inclinaison de -20°.

Tout ceci montre la difficulté rencontrée pour valider le champ théorique des contraintes, les conditions expérimentales apportant une complexité supplémentaire.

La figure 11 montre les mesures brutes relevées sur les 12 jauges lors d'une mise en charge monotone de l'éprouvette jusqu'à une pression de 14 MPa.

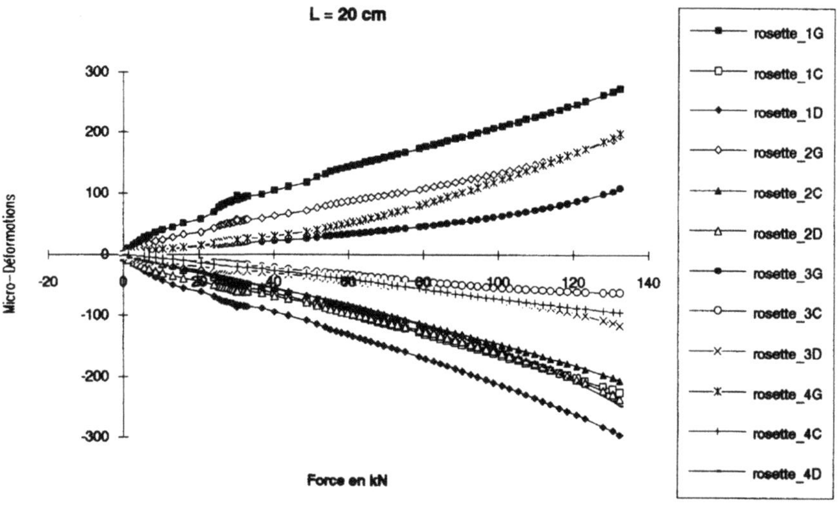

Figure 11 : Déformations mesurées sur les 4 rosettes (12 jauges) pour une longueur d'éprouvette de 20 cm

Les déformations des jauges Gauches de chaque rosette sont toujours positives, les déformations des jauges Centrales et Droites sont toujours négatives et la jauge D se raccourcit toujours plus que la jauge C pour une même rosette.

Localement pour chaque rosette, une des directions principales est perpendiculaire au plan de la plaque.

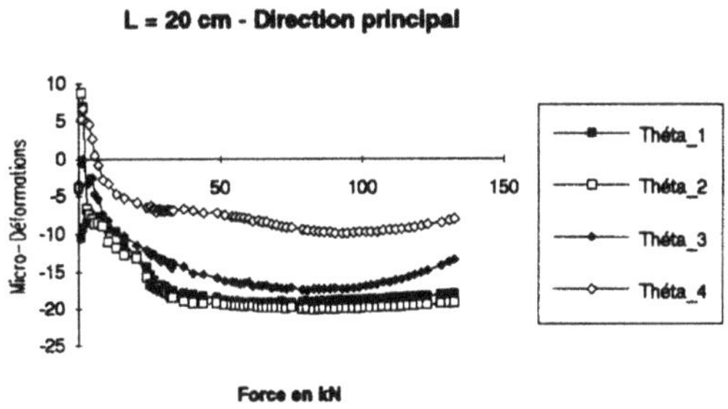

Figure 12 : Inclinaison de la direction principale par rapport à l'axe longitudinal de la plaque (axe de symétrie du modèle)

On peut donc calculer l'orientation des autres directions principales par rapport à l'axe de la plaque (figure 12). On constate alors que les rosettes 1 et 2 ont un comportement tout à fait similaire. Le comportement de la rosette 3 est peu différent de celui des 2 précédentes, mais par contre la rosette 4 se distingue par une orientation 2 fois plus faible.

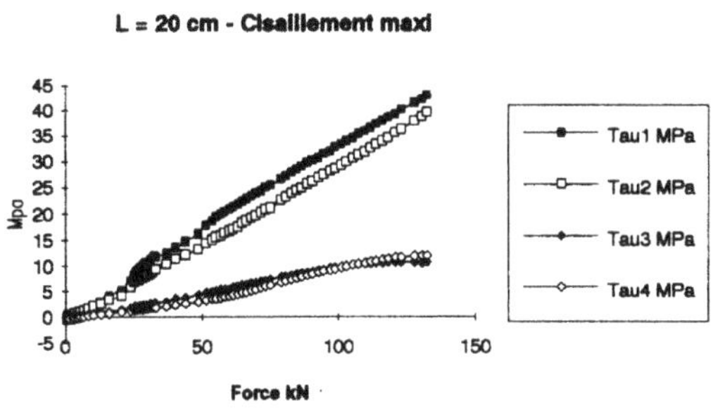

Figure 13: Contraintes de cisaillement calculées à partir des déformations de la plaque d'acier mesurées par les 4 rosettes

La figure 13 montre l'évolution de la contrainte de cisaillement maximale dans le plan de plaque calculée à partir de la différence des 2 contraintes principales non nulles. On constate que les cisaillements sont maximums au point 1 et décroissent jusqu'au point 4. Néanmoins, les différences entre t4 et t3 d'une part, t2 et t1 d'autre part sont faibles.

La figure 14 montre l'évolution du glissement suivant la direction de l'axe de la plaque calculé à partir de la différence des déformations mesurées par les jauges G et D de chaque rosette. Les différences des glissements mesurées par chaque rosette sont plus marquées mais le comportement du point 4 est différent puisqu'il est intermédiaire entre 2 et 3.

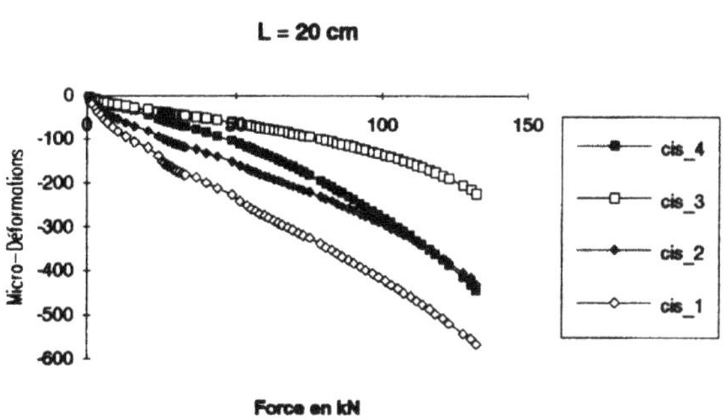

Figure 14 : Glissement dans la direction de l'axe de la plaque

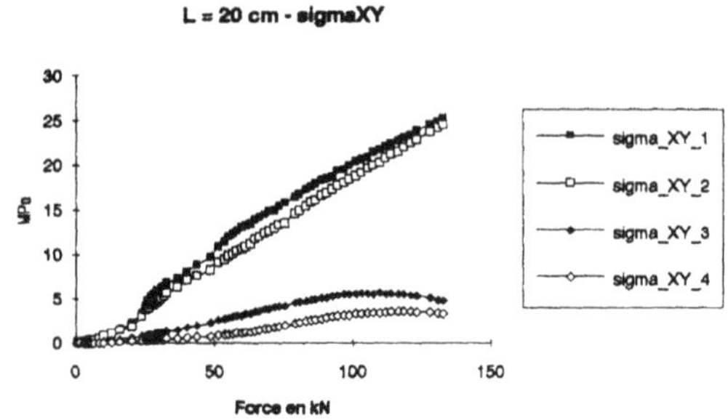

Figure 15 : Contraintes de cisaillement, dans la plaque, dans la direction de l'axe longitudinal

La figure 15 montre l'évolution des contraintes de cisaillement exprimées dans le repère de la plaque (xyZo). La différence par rapport aux cisaillements exprimés figure 13 est due au fait que les directions principales ne sont pas les mêmes pour les 4 rosettes.

On constate que les cisaillements décroissent du point 1 au point 4:

$\sigma xy(1) > \sigma xy(2) > \sigma xy(3) > \sigma xy(4)$.

4 Conclusion

La distribution des contraintes de cisaillement à l'interface plaque acier-mortier apparait au moins aussi complexe expérimentalement que théoriquement. Néanmoins, les résultats expérimentaux confirment l'étude théorique en montrant que les cisaillements ne sont pas uniformes sur toute l'étendue de la plaque. La zone de contact acier-mortier où l'angle entre la plaque et le bord de l'éprouvette est aigu est 3 à 4 fois moins cisaillée que la zone opposée. La figure 15 montre que σxy au point 1 est 4 fois plus important que σxy au point 4, ce résultat corrobore ceux obtenus théoriquement et présentés figure 5.

L'étude précédente montre que la recherche d'un couplage résistance mécanique - caractérisation physico-chimique de l'interface ne saurait être établie, même en première approximation, par la prise en compte d'un critère global fondée sur une valeur moyenne du champ de contrainte pour toute l'interface.

Ce modèle paraît bien adapté à la qualification de la liaison acier-béton par l'étude de différents facteurs tels que l'état de surface ou la réactivité des différents constituants en présence ainsi que d'autres paramètres comme la composition du mortier, les effets de cure, l'exposition à des ambiances agressives.

Bibliographie

1. M. A. Taylor and B. B. Broms (1964) Shear bond strength between coarse aggregate and cement paste or mortar. **Journal of the american concrete institute**, proceedings Vol. 61, No. 8, August 64, 939-956.
2. K. M. Alexander, J. Wardlaw and D. J. Gilbert (1965) Aggregate- cement bond, cement paste strength and the strength of concrete. **The structure of concrete**, proceedings of an international conference. September 65, 59-81.
3. M. Aggour, M. Arnaud, M. Lorrain (1983) Propriétés et mécanismes de l'adhérence acier-mortier de résines. **Colloque A.I.P.C.** Venise septembre 1983.
4. A. Bichara (1951) Etude du problème de l'adhérence dans le béton armé. **Cahiers du C.S.T.B.** N°117 et 127a, 1951.

22 MECHANICAL PROPERTIES AND MICROSTRUCTURE OF POLYESTER-CEMENTITIOUS COMPOSITES

F. SANDROLINI, M. ROSAS ROSAS, V. BONORA
and A. SACCANI
Dipartimento di Chimica Applicata e Scienza dei
Materiali, University of Bologna, Bologna, Italy

Abstract
Preliminary results of an experimental study on mechanical properties and microstructure of different mixtures of cementitius composite materials are presented.

Polymer cement mortar (PCM) and resin mortar (REM) were prepared by using several types of unsaturated polyester resins with styrene monomer(UP-St). A peroxide compound was utilized as initiator for the crosslinking reaction that occurs at room temperature.

Materials samples were then characterized by mechanical (flexural and compressive tests) and microstructures (scanning electron microscopy and mercury porosimeter).

The results so far obtained point out that resin mortar products have a considerably strength in comparison with conventional cement mortar and polymer cement mortar, chiefly due to microstructural differences arising from materials properties and mixing technology.
Keywords: Composite Materials, Resin Concrete, Mechanical Properties, Cement Mortar.

1 Introduction

In the last few years, according to the needs of new low cost products of better quality and performance for the construction field, the interest has been focused on blends of different materials in order to obtain products suitable to offer desirable combinations of properties.

Polymer concrete materials (polymer concrete cement, PCC and resin concrete cement, REC) seem to embody advanced features, such as high tensile and flexural strengths, excellent adhesion, waterproofness and good chemical resistance; they are therefore being increasingly utilized in construction and repair of structures, bridge decks waste water pipes, etc.

Ye Zhiyong found that polymer cement mortar (PCM), prepared from polymer, initiator, cement paste and aggregate, provides all the advantages of the latex modified cement ma-

Interfaces in Cementitious Composites. Edited by J. C. Maso. © RILEM.
Published by E & FN Spon, 2 - 6 Boundary Row, London SE1 8HN. ISBN 0 419 18230 6.

terials, including easiness of use and applicability in the field, but with higher strength, especially under wet condition, and lower brittleness.

In order to minimize material costs, it is imperative to use the least possible amount of polymer in REC materials to achieve maximum strength and/or stiffness, depending on application. Mechanical properties (both compressive and flexural) of resin concrete as a function of polymer content, show an increase up to a certain limit of polymers content, at which it exhibit maximum strength.

The aim of the present study is to investigate the mechanical properties and microstructure of different mixtures of polymer cement mortar (PCM) and resin mortar (REM) composite materials. Results are compared with plain cementitious mortars (CS) and the effect of UP resin on interfaces of the final composites materials is evaluated.

2 Materials

- 425 Portland cement (Italcementi) conforming to the Italian Standard, with a Blaine fineness of 3910 cm^2/g.
- 'Torre del Lago' (Sisa) normal sand, specific gravity 2.67, size 1.7-0.08 mm.
- calcium carbonate (Craital-N, Umbria Mineraria) specific gravity 2.7, average size 2.5.
- three kinds of commercially available orthophthalate-type unsaturated polyester resins, whose properties are shown in Table 1, together with 50% DMP (dimethyl phthalate) solution of methyl ethyl ketone peroxide (MEKPO) as catalyst.

Table 1. Properties of the polyester resin (UP-St).

RESIN (UP-St)	VISCOSITY 25°C (mPa.s)	STYRENE (%)	TENSIL MODULUS (N/mm^2)	FLEXURAL MODULUS (N/mm^2)
FS4230-X	500	35	3600	3900
FS956-VAM	330	31	3500	3900
FS940-AR	150	38	3200	3800

3 Experimental

3.1 Preparation and curing of specimens.

Prismatic samples (40x40x160 mm) were prepared from mortars of the following composition:

- plain cement mortar as control specimens (CS), with a mix proportion water:cement:sand = 1:2:3, prepared and cured according to the Decreto Ministeriale 03.06.1968 (Italian Standards), i.e. in calcium hydroxide saturated water solution.

- polymer cement mortar (PCM), prepared by mixing cement, water and sand in a Hobarth mixer to obtain a homogeneous mortar, then adding the usaturated polyester resin premixed with the accelerator MEKPO. They were susequently moulded an cured at room temperature in semi-dry condition (40% R.H.).
- unsaturated polyester resin mortar (UP-REM) prepared by mixing firstly sand and calcium carbonate in a Hobarth mixer, and later adding the unsaturated polyester resin premixed with the accelerator MEKPO. They were moulded and cured at room temperature in semi-dry (40% R.H.) atmosphere.

Mix proportions for PCM and REM are reported in Table 2.

Table 2 Mix proportions of PCM and REM

COMPOSITES	MIX	W/C	SAND (%)	$CaCO_3$ (%)	RESIN (%)
Polyester Cement Mortar	--	0.5	64	--	15
Polyester Resin Mortar	A	--	80	14	6
	B	--	79	13	8
	C	--	78	12	10
	D	--	74	11	15

3.2 Experimental Measurements

Measurements of the mechanical strength (both compressive and flexural) were carried out on all samples in an Amsler testing machine (30 tons): at least two measurements for each composition were made. Microstructural observations were performed by a Philips 501 B scanning electron microscope (SEM). Pore size distribution measurements were carried out by a mercury porosimeter Mod. Carlo Erba 200.

4 Test Results and Discussion

Figures 1 and 2 shows the values of the mechanical strength of the specimens: all polyester resin mortars (REM) show a higher compressive and flexural strength, with respect to the normal and polyester mortar at all curing times.

Figure 3 shows the effect of the unsaturated polyester resin content on flexural and compressive strength of REM samples. The mechanical characteristics increase as the resin content increases, reaching a maximum for a resin content of about 8%, and subsequently sligtly decrease.

The FS940-AR resin with 38% styrene shows a higher mechanical resistance, in comparison with the FS956-VAM (31% styrene).

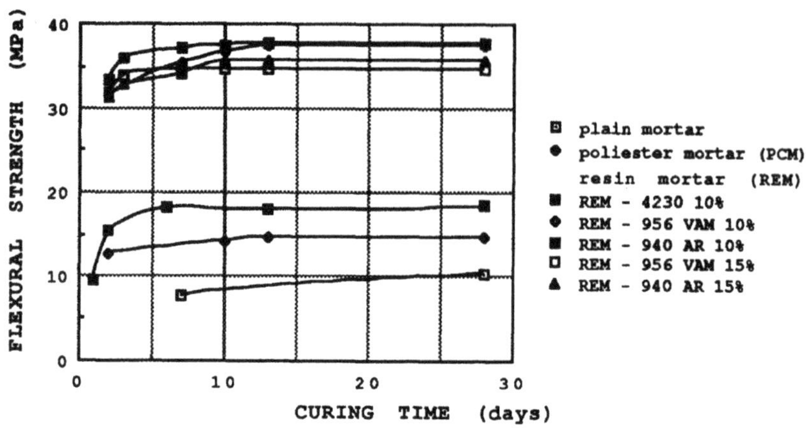

Fig.1. Time of curing vs. flexural strength of plain mortar (CS), polyester cement mortar (PCM) and resin mortar (REM).

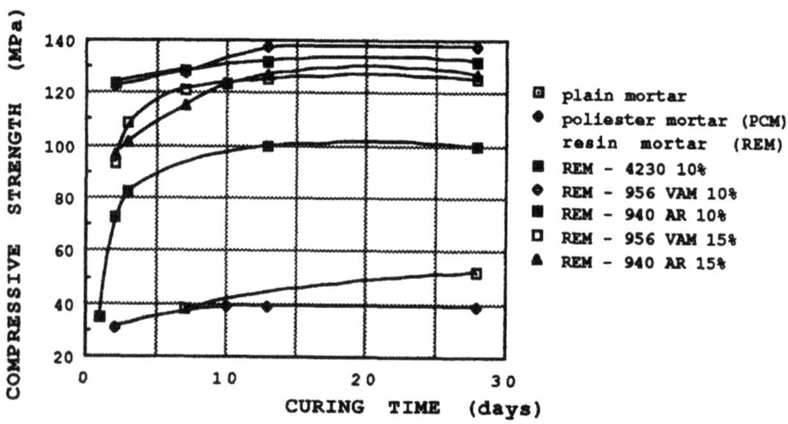

Fig.2. Time of curing vs.compressive strength of plain mortar (CS), polyester cement mortar (PCM) and resin mortar (REM).

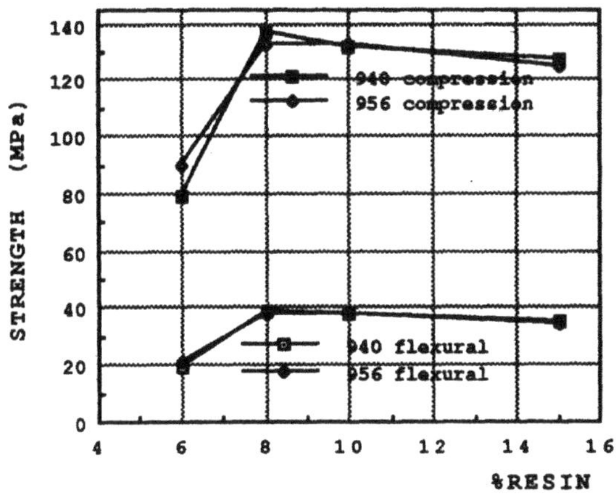

Fig.3. Resin % vs. flexural and compressive strength of composites polyester resin mortar (REM).

Table 4 reports the density and void fraction (ε) parameters for each composite material. For all the investigated materials the mechanical strength decreases, as well known, with the material porosity.

Table 4. Specific gravity and void fractions of composites

COMPOSITES	Apparent Spec.gravity	Average	ε	Average
PLAIN MORTAR	2.24	2.24	10.120	10.120
PCM	1.92	1.92	11.811	11.811
REM 4230 X	2.38	2.38	4.944	4.944
REM 956 /VAM	2.52		2.204	2.705
	2.52	2.52	3.206	
REM 940 /AR	2.77		1.622	2.442
	2.60	2.69	3.262	

5 Microstructural observation

SEM observations were carried out on the undisturbed fractured surfaces of PCM and REM.

General microstructure of PCM samples appears like that reported in Figure 4, where large regions with uniformly distributed microporosity of almost spherical shape can be seen surrounding large aggregates grains. Only a few regions with a more compact microstructure, included in the previously described regions, can be seen (Figure 5). Since the UP resins added to PCM is neither soluble nor disperdible in water, previous investigation showed that water tended to

Fig.4. General view of PCM microstructure (160x).

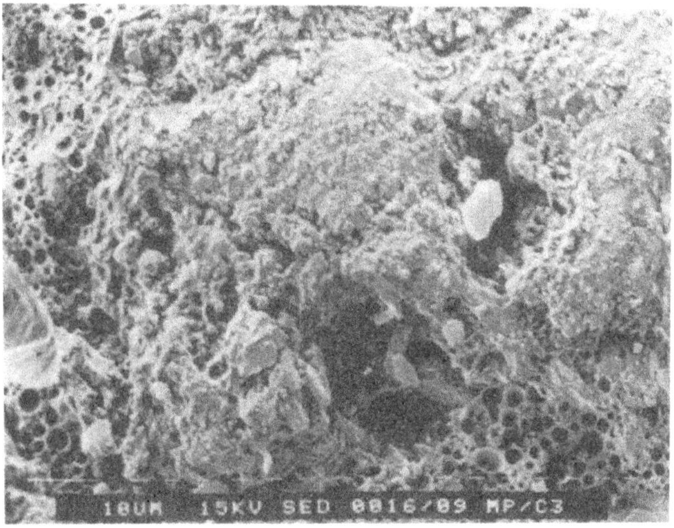

Fig.5. Regions of PCM matrix presenting a more compact microstructure (320x).

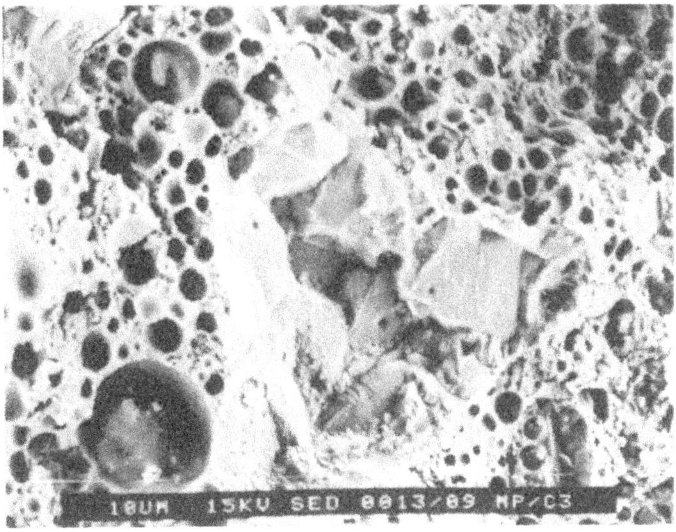

Fig.6 Unhydratated cement grains englobated in the polymeric matrix of PMC composite (640x).

Fig.7 Hydration products in plain mortar (5000x).

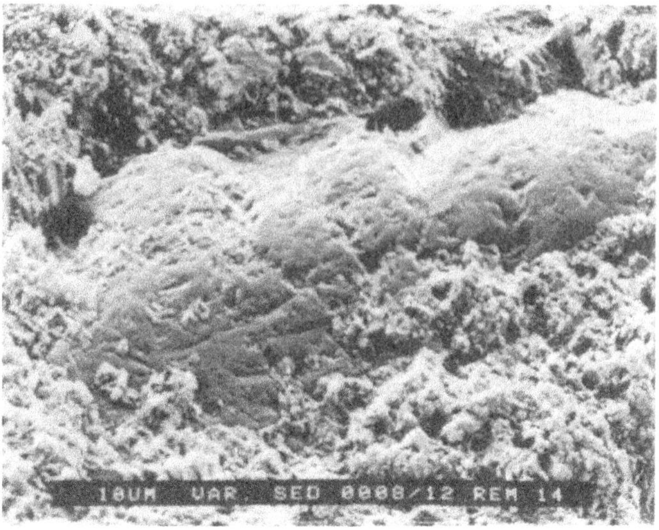

Fig.8. Compact matrix in REM composite (640x).

Fig.9 Fractured aggregates in REM composite (80x).

separate in little spherical droplets as it was mixed with the resin. Thus, the microstructure of the porous phase shown in Figure 4 can be considered as the result of the unhomogeneous mixing of PCM components: the UP resin does not uniformly disperse in the system, and separates in large porous regions; the porosity being probably due to water droplets dispersion after mixing. Indeed, as the mortar is fractured, some little moisture appears on the fractured surface. Cement grains can however be identified within the porosity, as can be seen in Figure 6. This microstructure is completely different from that obtained using water disperdible epoxy resins, as shown elsewhere by Zappia and Sandrolini. The aggregate always presents a somewhat poor adhesion with the polymeric and/or cementitious matrix. It is worth noting that the resin seems to exert a retarding effect on the hydration of the cement, which exhibits, in plain mortars after the same curing time, hydration products, as shown in Figure 7.

On the other hand, REM samples show a compact matrix characterized by low porosity, capable of forming a strong, intimate contact with the aggregate as shown in Figure 8. The presence of fractured aggregates in the matrix, showed in Figure 9, again proves the strong resin-aggregate adhesion. The almost complete absence of pores in this microstructure again confirms that pores present in PCM materials are formed by the englobation of small droplets of water in the resin.

6 Conclusions

a) Polyester resin mortars (REM) showed high mechanical characteristics (flexural and compressive), which are 220-250% higher than that of plain mortars, at shorter curing times.

b) The optimum resin content which provides the best mechanical properties for these polyester resin mortar system (REM) has been evaluated about 8% wt. per cent on the whole composite.

c) PCM mortars show lower mechanical strengths, expecially compressive, compared to plain mortars, on account of their highly porous matrix and of the low level of hydration of the cement particles, at least at the curing times so far investigated.

d) Again, the nature of components and the technological factors greatly affect microstructure and hence properties of PCM composites.

7 Acknowledgment

Scientific support of Research Center of SIR INDUSTRIALE S.p.A.(Macherio, VA) is gratefully acknowledged.

8 References

Decreto Ministeriale 03.06.1968 (Gazzetta Ufficiale No.180, 17.07.1968)

Dharmarajan, N. and Vipulanandan, C. (1991) Fracture toughness of particle filler fiber reinforced polyester composites. Journal of Applied Polymer Science, 42, 601-607.

Italian Standard UNI 6687-73.

Ohama, O., Demura, K. and Komiyama, M. (1979) Properties of polyester resin concrete with various styrene contents. 22nd Japan Congress on Materials Research, Japan, 360-363.

Sandrolini, F. (1990) Concrete polymer materials developments in Italia. Proceedings of the 6th International Congress, Shanghai, China, 36-43.

Zappia, G., Sandrolini, F. and Motori, A. (1990) Premix PCC materials: mechanical properties as a function of technological factors. Materials and Structures, 23, 436-441.

Zhiyong, Y. and Keru, W. (1990) Effect of proportion and curing condition on the mechanical strength of reactive unsatured polyester cement mortar. Proceedings of the 6th International Congress, Shanghai, China, 237-241.

23 ETUDE DES INTERFACES DANS DES COMPOSITES CIMENT-FIBRES DE POLYESTER

(Study of the interfaces of polyester fibre reinforced cement composites)

A. JELIDI, M. CHABANNET, J. AMBROISE and J. PERA
Laboratoire des Matériaux Minéraux, INSA de Lyon, Villeurbanne, France

Résumé *(English abstract, page 308)*
La fibre de polyester est détruite dans un milieu aussi alcalin que la phase interstitielle du ciment portland hydraté. Afin de préserver la durabilité de ce type de fibre, les auteurs ont mis au point une matrice cimentaire comportant du ciment portland artificiel, du métakaolin, de la montmorillonite calcinée et du chlorure de calcium. Ils obtiennent ainsi un composite durable et ductile. Cette ductilité s'explique par la modification des interfaces : le métakaolin réduit la teneur en portlandite, la montmorillonite calcinée piège la potasse et le chlorure de calcium empêche la précipitation de cristaux de gehlénite hydratée à la surface des fibres par formation de chloroaluminates hydratés. La fibre peut alors glisser dans la matrice et l'indice de ténacité du composite est maintenu jusqu'à 360 jours de conservation.
La compréhension des phénomènes a été possible en associant étroitement les analyses physico-chimiques (microscopie électronique à balayage, diffraction des rayons X, analyse élémentaire en dispersion de longueur d'onde et d'énergie, spectrométrie infra-rouge) et l'étude du comportement mécanique.
Mots-Clés: Fibres de polyester, Composite Ciment-Fibres, Ténacité, Durabilité, Dégradation, Pouzzolanes.

1. Introduction

La fibre de polyester possède des caractéristiques intrinsèques intéressantes qui font d'elle une fibre potentiellement valable dans la voie de substitution de l'amiante. Elle présente cependant l'inconvénient majeur de se dégrader en milieu alcalin. Son utilisation comme renfort dans les composites ciment-fibres nécessite d'une part, que les mécanismes de cette dégradation soient bien expliqués et, d'autre part, que la matrice cimentaire soit modifiée pour empêcher cette destruction. Une solution consiste alors à élaborer un liant pouzzolanique contenant des argiles calcinées susceptible de piéger les alcalins libérés lors de

Interfaces in Cementitious Composites. Edited by J. C. Maso. © RILEM.
Published by E & FN Spon, 2 - 6 Boundary Row, London SE1 8HN. ISBN 0 419 18230 6.

l'hydratation du ciment (Ambroise et al., 1985 ; Jelidi et al., 1991).

2. Matériaux de l'étude

2.1. Les fibres de polyester

Il s'agit d'un polyester haute ténacité produit par la Société RHONE POULENC se présentant sous forme d'un fil plat de 50 brins. Il a un titre de 280 dtex, sa résistance en traction est de 875 MPa et son allongement à la rupture vaut 21 %. D'un point de vue chimique, le fil de polyester est un polyéthylène téréphtalate comportant un anneau aromatique et un groupement ester aliphatique.

2.2. Constituants de la matrice cimentaire

Le ciment utilisé est un ciment CPA 55 à fort taux d'alcalins (équivalent Na_2O: 1,1%); son analyse chimique est présentée dans le tableau 1. Les argiles calcinées, dont l'analyse chimique est également donnée dans le tableau 1, sont respectivement:

- un métakaolin, obtenu par calcination en four industriel à 800°C d'un kaolin comportant 90% de kaolinite, de l'anatase et du quartz.
- une montmorillonite calcique calcinée à 950°C.

Tableau 1. Composition chimique du ciment et des argiles

Matière première	Oxydes (% pondéral) SiO_2	Al_2O_3	Fe_2O_3	CaO	MgO	équiv. Na_2O
Ciment	22,2	4,4	3,4	61,0	1,4	1,1
Kaolin	43,4	35,0	2,1	0,3	0,1	0,1
Montmorillonite	54,1	16,8	2,7	2,1	3	1,1

Le métakaolin est une pouzzolane très réactive susceptible de consommer rapidement la portlandite produite lors de l'hydratation du ciment portland artificiel (Ambroise, 1984) et de piéger le potassium dans les C-S-H résultant de la réaction pouzzolanique (Péra, 1989).

La montmorillonite calcique calcinée à 950°C n'a aucune propriété pouzzolanique (Gniewek, 1987) mais sa structure cristalline proche de l'opale (Novak, 1980) lui permet de capter certains ions comme le potassium (Jelidi, 1991).

2.3. Elaboration des composites ciment-fibres de polyester

Les composites sont des plaques minces (longueur: 280 mm, largeur: 70 mm, hauteur: 15 mm) renforcées par des fibres longues placées à 5 mm du fond du moule. Le taux de renfort adopté est de 3 kg/m^3. La matrice est constituée de pâte pure gâchée à consistance normale. Une fois mises en place par vibration, les plaques sont démoulées après 2 jours d'hydratation et placées dans des sacs étanches pour être conservées à 20°C et à 95% d'humidité relative jusqu'aux essais mécaniques et physico-chimiques.

2.4 Essais réalisés

Des essais de flexion trois points ont été réalisés sur une presse ADAMEL-LHOMARGY de capacité 20 kN, asservie en déplacement, aux échéances suivantes : 28, 90, 180 et 360 jours. L'enregistrement des courbes charge-flèche permet de déterminer l'indice de ténacité du composite comme étant le rapport entre l'énergie totale (aire sous la courbe charge-flèche totale) et l'énergie jusqu'à la limite de proportionnalité (Fig. 1).

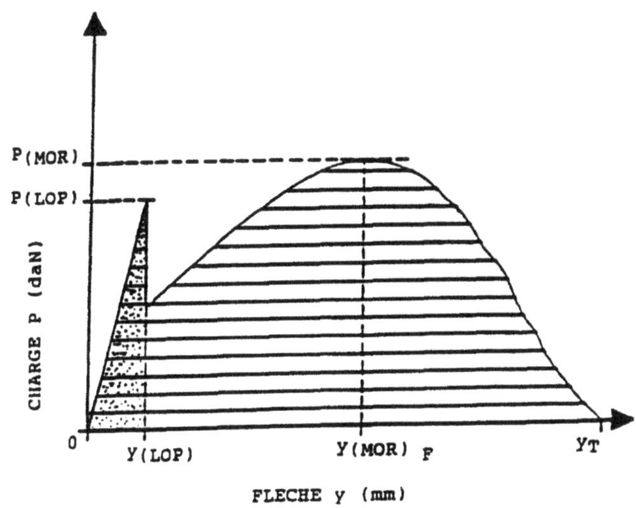

Figure 1. Définition de l'indice de ténacité du composite.

Pour ces différentes échéances, les fibres et leur interface avec la matrice sont examinées en microscopie électronique à balayage (JEOL 35CF) afin d'étudier la variation du diamètre des fibres au cours du temps et de préciser la nature des interfaces par microsonde CAMEX-CABEMAX. Par ailleurs, les fibres extraites de la matrice sont finement coupées et dispersées dans du bromure de potassium à raison de 3 à 6% en masse pour être pastillées et analysées en spectrométrie infra-rouge à transformée de FOURIER (Modèle PERKIN-ELMER 1760X). La nature des hydrates formés dans la matrice est suivie par diffraction des rayons X (Modèle SIEMENS D500, rayonnement CuK_{α}, à monochromateur arrière en graphite) et analyse thermique différentielle (Modèle ADAMEL LHOMARGY, 600 mg de produit, vitesse de montée en température: 10°C/mn) sur des poudres passant au tamis de 100 µm.

3 Comportement des fibres de polyester dans la matrice de ciment portland artificiel

3.1 Comportement mécanique

L'évolution de l'indice de ténacité au cours du temps est donnée sur la figure 2. A 180 jours, le comportement du composite est équivalent à celui de la matrice non renforcée puisque l'indice de ténacité devient égal à 1.

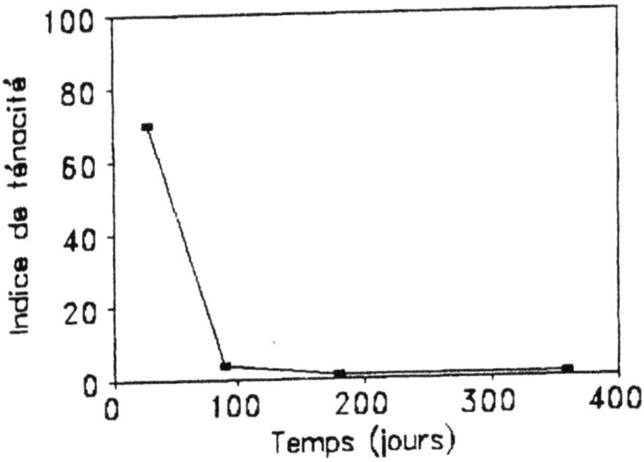

Figure 2. Evolution de l'indice de ténacité du composite CPA-fibres de polyester.

3.2 Examen microscopique

L'examen microscopique permet de constater deux phénomènes:

- la diminution rapide du diamètre de la fibre au cours du temps (Fig. 3),
- la détérioration de la surface de la fibre par des incrustations de cristaux hexagonaux de portlandite (Fig. 4 et 5) et la disparition totale de la fibre après un an d'hydratation (Fig. 6).

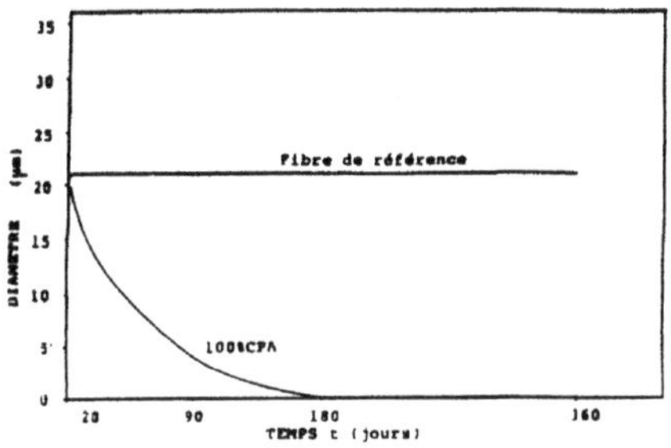

Figure 3. Evolution du diamètre de la fibre dans la matrice de CPA.

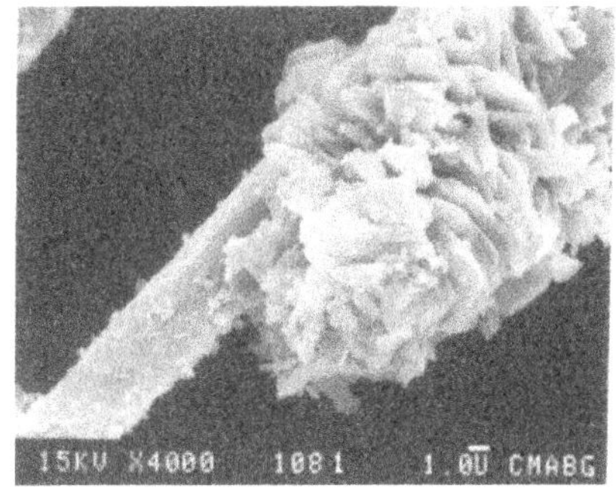

Figure 4. Etat de la fibre après 90 jours d'hydratation dans la matrice de CPA.

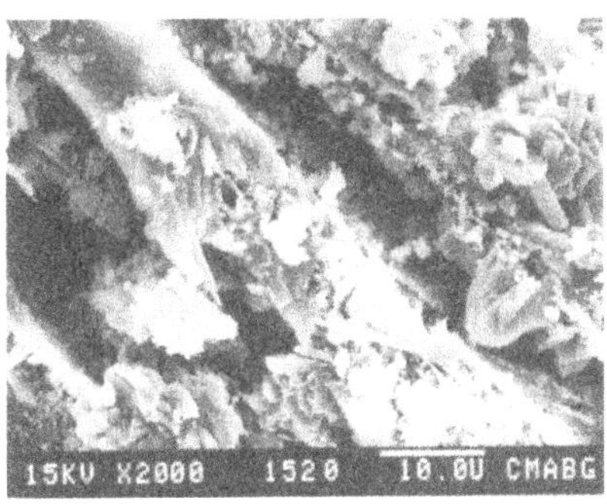

Figure 5. Destruction de la fibre après 180 jours d'hydratation dans la matrice de CPA

Figure 6. Absence de fibre après 360 jours d'hydratation dans la matrice de CPA.

La loi cinétique de variation relative du diamètre peut se mettre sous la forme:

$$\alpha = (d_o - d/d_o = 1 - \exp(kt) \text{ avec } k = 1295 \times 10^{-5} \text{jour}^{-1}$$

3.3 Résultats de l'analyse de spectrométrie infra-rouge

L'analyse de la fibre de polyester extraite des composites et la comparaison de son spectre avec celui de la fibre de polyester de référence (Fig. 7) montre la modification d'un certain nombre de bandes correspondant aux liaisons aliphatiques C-O à 1236 et 1088 cm^{-1}, C-O-C à 500 cm^{-1} et l'anneau aromatique du polyester n'est pas touché. Par ailleurs, de nouvelles bandes apparaissent à 1550, 1400, 850 et 750 cm^{-1}. Elles correspondent à la présence de téréphalate de potassium (Fig. 8): dans la matrice de CPA, la fibre de polyester est attaquée par les ions OH^-, K^+, Na^+ et Ca^{2+}

au niveau des liaisons C-O, C-O-C et O-CH_2-CH_2-CH_2-O (hydrolyse-saponification) entrainant la scission du polyester et la formation de produits alcalino-téréphtaliques et de l'éthylène-glycol. La formation d'éthylène-glycol est difficile à mettre en évidence car ce composé est miscible dans la phase interstitielle du ciment et peut migrer loin de la fibre. Comme le montrent les observations microscopiques, la réaction de dépolymérisation s'opère à la surface de la fibre de manière concentrique et progressive entrainant la diminution du diamètre de la fibre.

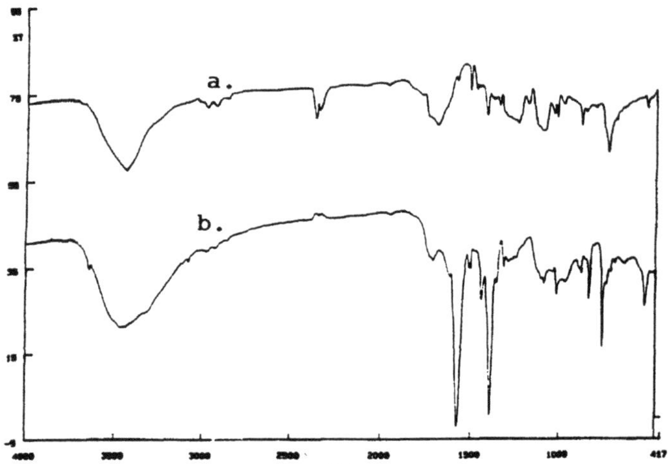

Figure 7. Comparaison entre le spectre de la fibre vierge (a) et le spectre infra-rouge de la fibre de polyester extraite de la matrice CPA à 90 jours (b).

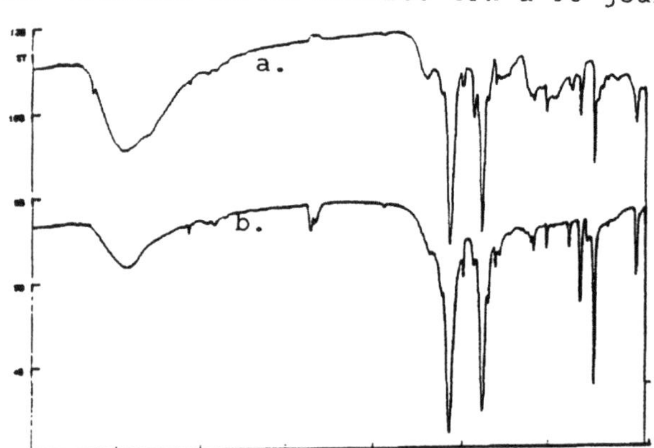

Figure 8. Comparaison entre le spectre I.R. de la fibre de polyester extraite de la matrice de CPA à 90 jours (a) et le spectre du téréphtalate de potassium (b).

L'analyse en microsonde montre une forte concentration de l'élément potassium sur la fibre et à l'interface fibre-matrice qui rend compte de l'extrême mobilité de l'ion potassium dans la matrice cimentaire.

3.4 Conclusion

La dégradation de la fibre de polyester dans une matrice de CPA consiste en une réaction de dépolymérisation du polyester qui se scinde en téréphtalates de potassium (en majorité dans notre cas), de sodium et de calcium, et en éthylène-glycol. Elle s'effectue de manière concentrique sur la surface de la fibre, provoquant ainsi la diminution de son diamètre au cours du temps jusqu'à sa disparition totale. Parallèlement à cette réaction de dépolymérisation, la destruction de la fibre s'effectue également par une incrustation de cristaux hexagonaux de portlandite dans la fibre causant l'apparition d'entailles qui s'approfondissent au cours du temps. Ces réactions physico-chimiques entrainent la chute progressive des performances mécaniques du composite qui se fragilise au cours du temps.

4 Elaboration d'un composite durable

4.1 Composition de la matrice

La matrice mise au point par Jelidi (1991) a pour rôle d'assurer la durabilité de la fibre de polyester en empêchant les réactions chimiques de dépolymérisation et de l'attaque physique des cristaux de portlandite. Elle comprend:

- 50% de ciment portland artificiel,
- 40% de métakaolin nécessaire pour consommer totalement la portlandite libérée lors de l'hydratation du ciment (Ambroise, 1985),
- 10% de montmorillonite calcique calcinée à 950°C, susceptible d'immobiliser les ions alcalins k^+ et Na^+ et d'empêcher leur migration vers la fibre de polyester.

En outre, 2% de chlorure de calcium sont introduits dans l'eau de gâchage afin de modifier la morphologie de certains hydrates: diminution de la quantité de bâtonnets d'ettringite et de cristaux hexagonaux de gehlénite hydratée (C_2ASH_8); ce dernier hydrate étant issu de la réaction pouzzolanique (Ambroise, 1984). Les cristaux de gehlénite hydratée provoquent en effet une attaque physique des fibres de polyester dans une matrice exempte de portlandite (Fig. 9).

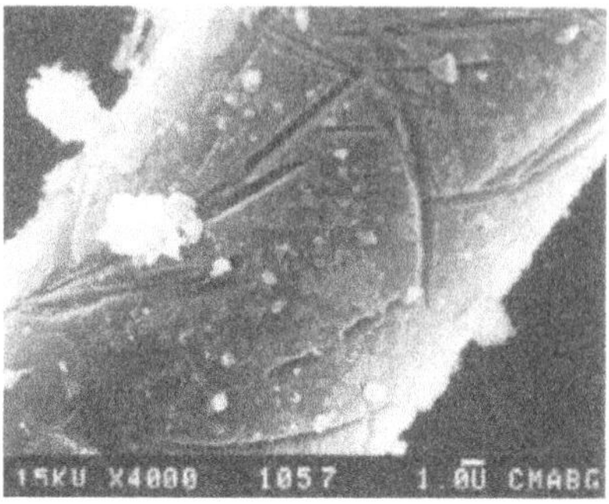

Figure 9. Attaque physique de la fibre de polyester par les cristaux de gehlénite hydratée dans une matrice exempte de chaux à 28 jours (50% CPA + 40% métakaolin + 10% montmorillonite calcinée).

4.2 Comportement mécanique

La ténacité du composite est conservée dans le temps comme le montre la figure 10. Par ailleurs sa valeur initiale est égale à 3 fois la valeur de l'indice de ténacité du composite élaboré avec du ciment pur. Un composite durable et ductile a donc été élaboré.

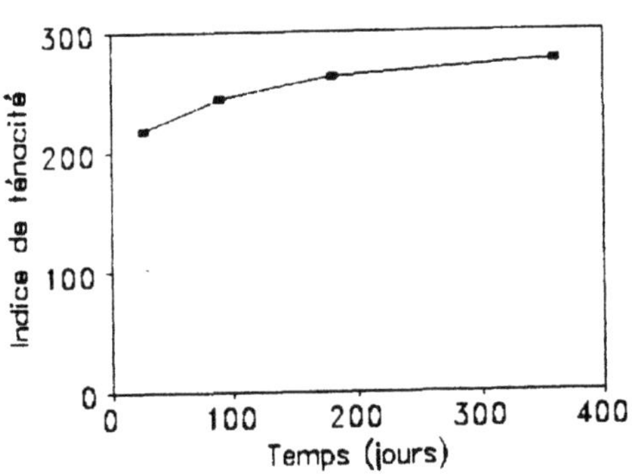

Figure 10. Evolution de l'indice de ténacité du composite ciment-pouzzolanique-fibre de polyester.

4.3 Analyses physico-chimiques

L'examen microscopique de la fibre extraite de la matrice pouzzolanique à différentes échéances montre que la fibre ne subit aucune dégradation au cours du temps; son diamètre demeurant contant (Fig. 11 et 12.)

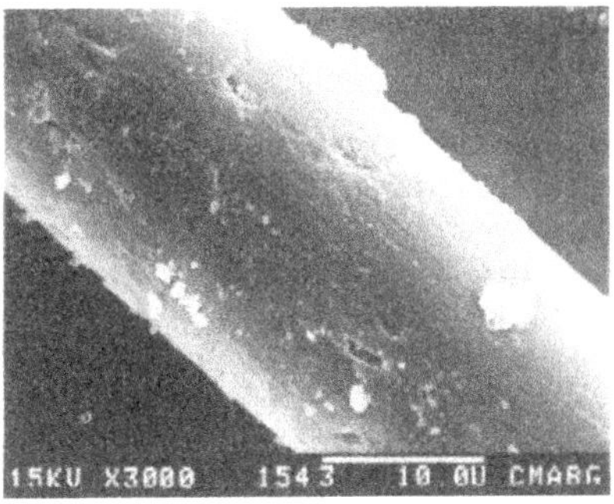

Figure 11. Aspect de la fibre de polyester extraite de la matrice pouzzolanique à 180 jours.

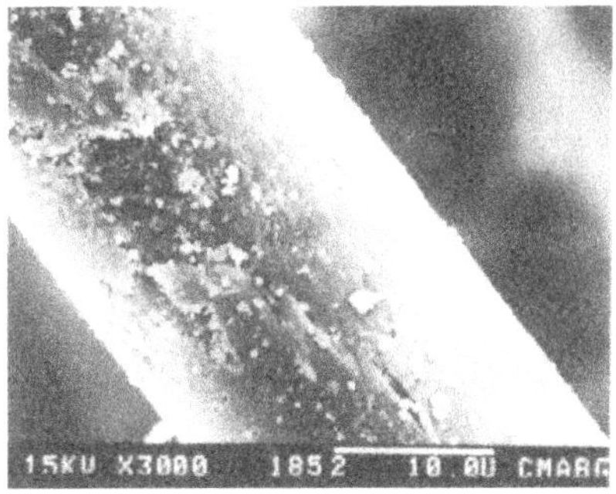

Figure 12. Aspect de la fibre de polyester extraite de la matrice pouzzolanique à 360 jours.

L'analyse en microsonde montre que le potassium est uniformément réparti dans la matrice, confirmant les résultats initialement trouvés par Péra (1989): le potassium est piégé dans les CSH.

L'analyse en spectrométrie infra-rouge des fibres extraites de la matrice pouzzolanique à 90 jours prouve que la fibre est demeurée intacte (Fig. 13): les liaisons et les groupements chimiques, aussi bien de l'anneau aromatique que de l'ester aliphatique, n'ont été touchés ni par les ions hydroxyles ni par les ions alcalins.

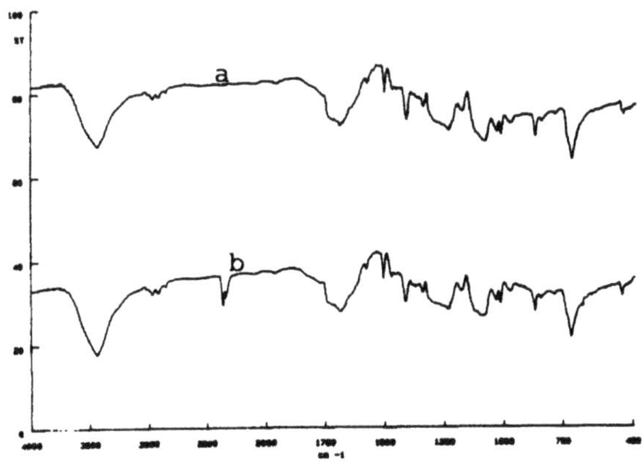

Figure 13. Comparaison entre le spectre I.R. de la fibre de polyester extraite de la matrice pouzzolanique à 90 jours (a) et le spectre de la fibre vierge (b).

5 Conclusion

La saturation en portlandite et la forte concentration en ions potassium et sodium de la phase interstitielle du ciment portland artificiel conduit à une dépolymérisation de la chaîne du polyester qui s'effectue par hydrolyse-saponification. Le diamètre et la masse de la fibre diminuent dans le temps, entraînant la fragilisation du composite.

L'ajout de métakaolin consommateur de portlandite, de montmorillonite calcinée à 950°C fixant les ions alcalins et de chlorure de calcium modifiant la morphologie de certains hydrates (gehlénite hydratée), provoque une modification des interfaces fibre-matrice et permet de garantir la durabilité du composite. Les fibres de polyester peuvent alors concurrencer les fibres de polyamide et de polypropylène pour le renforcement des composites.

6 Références

Ambroise, J. (1984). Elaboration de liants pouzzolaniques à moyenne température et étude de leurs propriétés physico-chimiques. **Thèse de Doctorat ès Sciences**, Institut National des Sciences Appliquées de Lyon, 172 p.

Ambroise, J., Murat, M. et Péra, J. (1985). Durabilité des fibres de verre dans différents milieux cimentaires en cours de consolidation - Etude d'une fibre de verre. **Verreset Réfractaires**, Nov. et Déc. 1985, Vol. 39, N° 6, pp. 905-910.

Gniewek, J. (1987). Réactivité des montmorillonites calcinées et utilisation dans le Génie Civil. **Thèse de Doctorat ès Sciences**, Institut National des Sciences Appliquées de Lyon, 184 p.

Jelidi, A., Chabannet, M., Ambroise, J., and Péra, J. (1991). Development of composites reinforced by polyester fibres. Proceedings **Symposium Materials Research Society** (eds. S. Mindess and J. Skalny), Vol. 211, pp. 275-281.

Jelidi, A. (1991). Conception d'un matériau composite à matrice cimentaire renforcée par des fibres de polyester. **Thèse de Doctorat**, Institut National des Sciences Appliquées de Lyon, 169 p.

Péra, J. et Boudeulle, M. (1989). Connaissance de la microstructure des composites ciment-verre E contenant du métakaolin. **Poster présenté au Colloque IDMAT**, Paris.

24 MODELISATION DU COMPORTEMENT MECANIQUE DE COMPOSITES CIMENT-FIBRES DE VERRE EN LIAISON AVEC LA NATURE DES INTERFACES

(Modelling the behaviour of cement-glass fibre composites and the nature of the interfaces)

J. PERA, J. DEJEAN and J. AMBROISE
Laboratoire des Matériaux Minéraux, INSA de Lyon, Villeurbanne, France

Résumé *(English abstract, page 308)*
L'ajout de métakaolin dans une matrice cimentaire permet de réduire de façon appréciable et, dans certains cas, d'éliminer les cristaux de portlandite qui bloquent les fibres de verre constituant le renfort. Les fibres ont alors la possibilité de glisser dans la matrice ce qui conduit à l'obtention de composites ductiles.
Les méthodes de calcul classiques reposant sur l'hypothèse d'adhérence parfaite fibre-matrice ne s'appliquent plus pour décrire le comportement mécanique de tels composites. Les auteurs proposent une nouvelle modélisation de ce type de comportement.
Analysant un essai de traction directe d'un composite, ils évaluent d'abord l'effort transmis par une fibre à partir d'une description déterministe de sa position par rapport à une fissure. Puis ils calculent l'effort de traction transmis par une unité de largeur de fissure à partir d'une description probabiliste de l'action de K fibres pontant la fissure sur cette largeur. Le nombre de ces fibres est supposé être distribué suivant une loi de Poisson de paramètre K. Ils peuvent ainsi déterminer la valeur limite du frottement interfacial si l'on veut assurer un comportement ductile au composite et ceci, quelle que soit la nature de la matrice cimentaire.
Mots-clés: Composites Ciment-Verre, Fibres, Métakaolin, Interface, Frottement, Modélisation.

1 Introduction

De façon très générale, le comportement mécanique d'un composite dépend des comportements de la matrice, des fibres et de l'interface fibre-matrice. Dans le cas particulier des composites ciment-verre (CCV), des données spécifiques sont à prendre en compte:

- la matrice a un comportement fragile,
- les fibres ont un comportement fragile,
- leurs caractéristiques, et en particulier leur résistance, évoluent dans le temps,

Interfaces in Cementitious Composites. Edited by J. C. Maso. © RILEM.
Published by E & FN Spon, 2 - 6 Boundary Row, London SE1 8HN. ISBN 0 419 18230 6.

- les propriétés de l'interface fibre-matrice évoluent dans le temps (Bentur, 1985 et 1986, Cohen, 1975, Daniel, 1986).

L'étude des propriétés mécaniques de CCV contenant du métakaolin, pouzzolane très réactive susceptible de consommer la portlandite produite lors de l'hydratation du ciment, montre que, malgré la fragilité de la matrice et du renfort, on peut obtenir un matériau globalement tenace grâce à la possibilité de glissement des fibres dans la matrice (Ambroise, 1987, Dejean, 1989). A cause de ce glissement, l'hypothèse d'adhérence parfaite fibre-matrice n'est pas satisfaisante pour décrire un tel comportement. Une nouvelle modélisation s'avère donc nécessaire.

2 Les différentes phases de comportement des CCV contenant du métakaolin

La courbe charge-flèche d'un CCV contenant du métakaolin peut être schématisée en trois phases, comme le montre la figure 1.

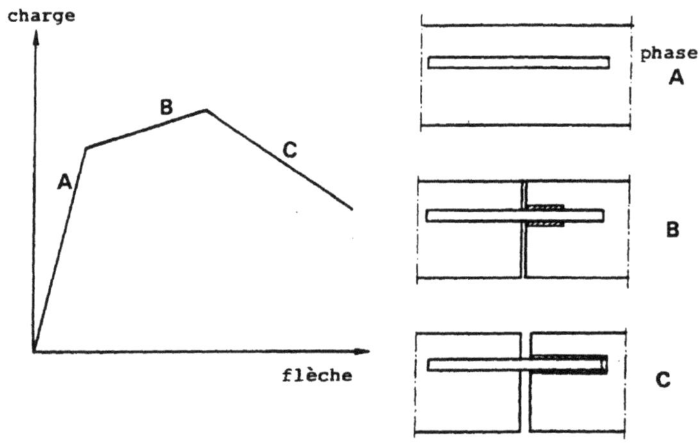

Figure 1. Différentes phases du comportement d'un CCV contenant du métakaolin.

La première phase, notée A sur la figure 1, correspond quasiment au comportement de la matrice non renforcée. De nombreux auteurs ont étudié cette phase en utilisant la loi des mélanges (Hannant, 1978, Swamy, 1974) et ont obtenu de bons résultats. Les phases de grandes déformations, notées B et C, sont essentiellement gouvernées par le comportement de l'interface fibre-matrice. La loi frottement-glissement qui caractérise mécaniquement cette interface a fait l'objet de nombreuses études, le plus souvent fondées sur des essais de type "pull-out" (Shah, 1985). Des essais de pelage ont aussi été réalisés (Cheick Larbi, 1985). Tous ces essais présentent l'inconvénient de ne pas recréer les conditions réelles de déformation dans la matrice, en particulier au niveau de l'interface. On ne peut donc pas assurer que les efforts mesurés dans ces essais correspondent à ceux transmis à l'interface dans les composites.

Ces raisons nous ont incité à développer une autre voie pour établir une valeur moyenne des frottements interfaciaux. Notons que la recherche d'une résistance maximale du composite passe par l'obtention de frottements maximaux. Mais, comme le font remarquer Bartos (1986) et Mindess (1987), pour les CCV, cette résistance s'obtient au détriment de la ductilité car des ruptures de fibres se produisent au cours de la phase de ruptures locales d'adhérence que nous avons notée B, pouvant entrainer une rupture totale du composite. Bentur (1985) a également montré que, pour obtenir une énergie de rupture maximale, la valeur du frottement devait être limitée. On conçoit aussi qu'elle ne doive pas chuter brutalement pour de grands glissements afin de développer la phase d'arrachement des fibres notée C sur la figure 1.

3 Description analytique du transfert des efforts dans une section fissurée

Nous supposons que la section est entièrement fissurée; seules les fibres assurent la transmission des efforts au niveau de la fissure. Sous sollicitation de traction pure, l'effort transmis est normal à la section fissurée. Nous supposerons, par la suite, que cet effort est porté par l'axe OY, le plan de fissure est donc XOZ. Le frottement est supposé constant sur la longueur de la fibre; celle-ci ne peut transmettre, de par sa géométrie, que des efforts normaux. L'effort transmis dans la section apparait donc comme la somme des projections sur OY des efforts normaux des fibres pontant la fissure (Fig. 2).

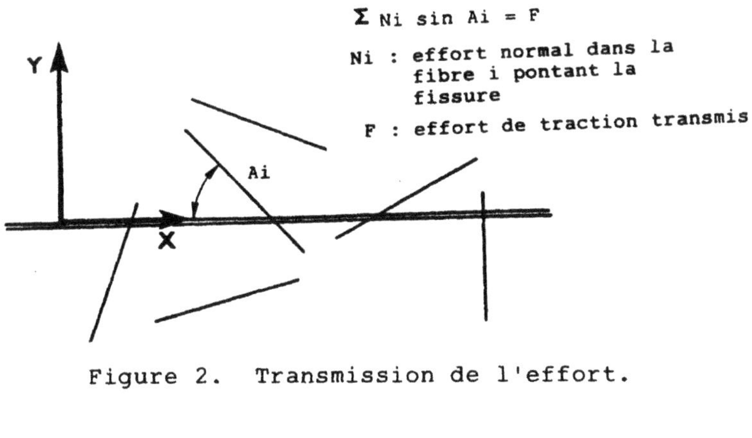

Figure 2. Transmission de l'effort.

Comme le propose Fukuda (1982), nous allons évaluer l'effort transmis par une fibre, à partir d'une description déterministe de sa position par rapport à la fissure. Puis, nous déduirons l'effort de traction transmis par unité de largeur de fissure à partir d'une description probabiliste de l'action de K fibres pontant la fissure sur cette largeur. Nous estimerons pour cela que le nombre de fibres pontant la fissure est un nombre aléatoire distribué selon une loi de Poisson de paramètre K. Naaman (1973) a présenté une justification de cette hypothèse.

3.1 Effort normal moyen transmis par une fibre

Les fibres sont supposées avoir toutes la même longueur:L. Chaque fibre est caractérisée par deux paramètres indépendants: la position d'un de ses points (extrémité supérieure) et son orientation (angle formé par rapport à la fissure: A). Les fibres qui pontent la fissure appartiennent à l'une des trois populations suivantes :

a) les fibres dont l'extrémité est située à une distance comprise entre L et L/2 de la fissure et dont l'angle est compris entre $-\pi/2$ et -Arc sin (y/l) pour les orientations comprises entre $-\pi/2$ et $+\pi/2$. La rupture de l'adhérence concerne, pour ces fibres, une longueur L + y/sin A (Fig. 3).

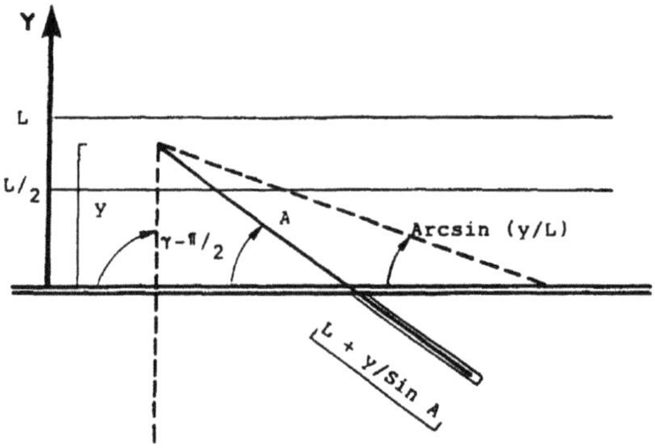

Figure 3. Fibres de type I.

b) les fibres dont l'extrémité est située à une distance inférieure à L/2 de la fissure et dont l'orientation est comprise entre -Arc sin (2y/L) et -Arc sin (y/L) (Fig. 4). La longueur déchaussée s'exprime encore ici par L + y/sinA.

c) les fibres dont l'extrémité se situe dans la même zone mais qui sont déchaussées "de l'autre côté" de la fissure. Leur orientation peut varier (Fig. 5) de $-\pi/2$ à Arcsin (2y/L) pour une longueur déchaussée y/sin A.

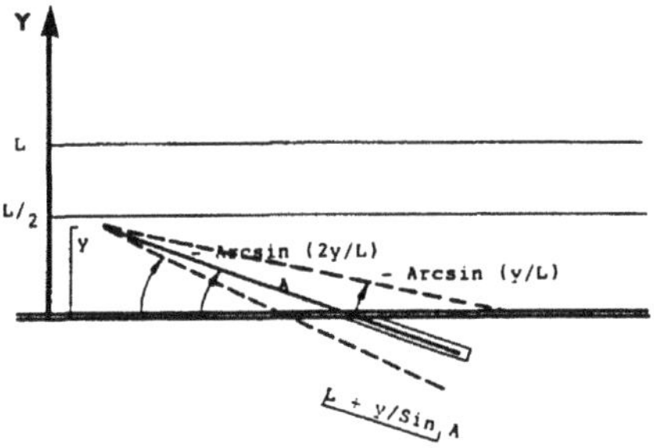

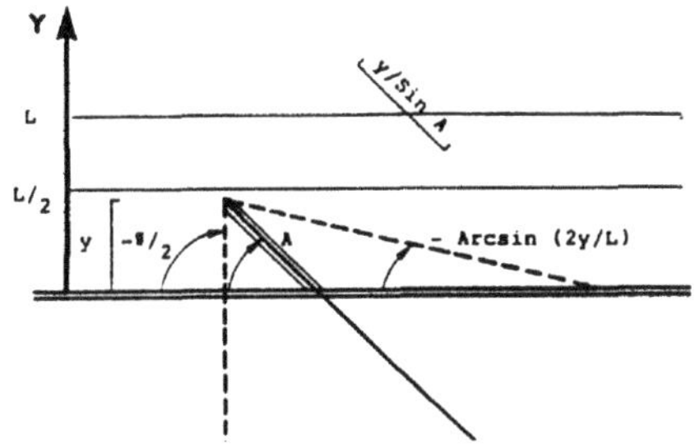

Figure 4. Fibres de type II. Figure 5. Fibres de type III.

En supposant que le frottement est constant sur toute la longueur déchaussée, on obtient l'effort normal transmis par une fibre "moyenne" (diamètre $\varnothing$) de chaque population:

$$F_1 = 0{,}034\ L\varnothing\tau \qquad (1)$$
$$F_2 = 0{,}153\ L\varnothing\tau \qquad (2)$$
$$F_3 = 0{,}049\ L\varnothing\tau \qquad (3)$$

Pour obtenir l'effort transmis, on doit aussi tenir compte des fibres d'orientation comprise entre $\pi/2$ et $3\pi/2$, c'est-à-dire doubler la somme $F_1 + F_2 + F_3$; l'effort total transmis par la fibre "moyenne" vaut donc:

$$F = 0{,}196\ L\varnothing\tau \qquad (4)$$

3.2 **Effort transmis sur une large donnée**

Supposons qu'il y ait une moyenne de K fibres sur la surface définie par la largeur b de plaque fissurée et la longueur L de part et d'autre de la fissure. Considérons le nombre N de fibres ayant une de leurs extrémités dans cette surface. En faisant abstraction des effets de bord et de largeur réelle d'éprouvette, le milieu est homogène. La bibliographie relative aux problèmes de comptage en milieu fini (Duncunha, 1972, Johnson, 1962), pour des variables discrètes, montre l'adaptation de la loi de Poisson à cette étude. Le nombre de fibres N apparait alors comme une variable aléatoire distribuée selon une loi de Poisson de paramètre K:

$$E\ (N = n) = e^{-k} K^n / n!$$

Pour des fibres cylindriques de diamètre $\varnothing$, en appelant n% le pourcentage volumique de fibres, e l'épaisseur de la plaque, le paramètre K s'exprime par:

$$K = n\%\ 8\ eb/\pi\varnothing^2 \qquad (5)$$

et l'effort transmis sur une largeur b:

$$T = 0{,}5\ n\%\ eb\,\tau\, L/\varnothing \qquad (6)$$

De même, pour des fibres en ruban de largeur W et d'épaisseur t, K s'écrit :

$$K = n\% \ 2be/Wt,$$

d'où l'effort transmis:

$$T = 0{,}26 \ n\% be\tau L/t \qquad (7)$$

3.3 Application aux essais de traction sur C.C.V.

Dans les C.C.V., l'élément de base du renfort est le fil constitué par l'assemblage d'environ 200 fibriles. Sa section réelle est de l'ordre de $2{,}8 \times 10^{-8} m^2$. L'observation microscopique des fibres émergeant d'une fracture montre que le fil reste le plus souvent sous forme de ruban, de largeur voisine de 1 mm et d'épaisseur faible devant la largeur ($2{,}8 \times 10^{-5}$ m).

Pour des composites, notés MK40:
Matrice: 60% CPA-40% Métakaolin-Sable/Liant = 0,2,
Fibre E: 5% en poids (4% en volume), longueur 30 mm,
Largeur: 5 cm,
Epaisseur: 1,2 cm,

vieillis 120 jours en atmosphère saturée, nous avons obtenu un comportement ductile au cours d'un essai de traction pure, de déplacement contrôlé (Fig. 6).

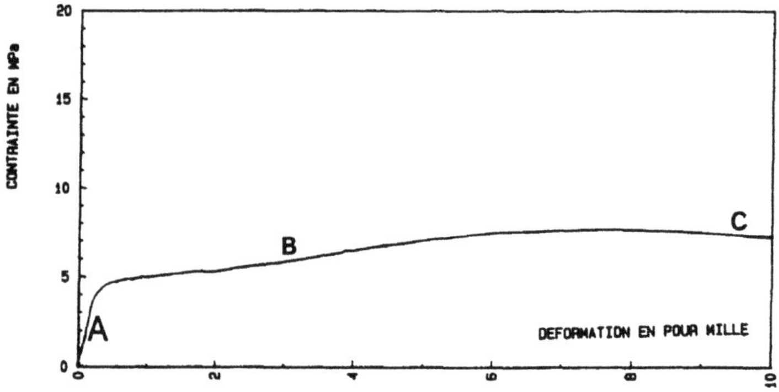

Figure 6. Loi de comportement du composite MK40 après 120 jours d'hydratation.

Au-delà de la rupture de la matrice, qui intervient pour une déformation de l'ordre de 200 à 300 10^{-6} m/m correspondant à une contrainte de l'ordre de 4 MPa, nous notons que la phase B de mobilisation progressive des frottements interfaciaux est très étendue. Elle s'achève pour une ouverture de fissure de l'ordre de 0.3 à 0.4 mm. La transition entre les phases B et C est très progressive. A ce niveau, on peut estimer que le frottement interfacial est entièrement mobilisé. L'application de l'expression (7) pour l'effort maximal de 462 daN conduit à un frottement interfacial global de 0,69 MPa. Au cours de la phase C, cet effort décroît progressivement sous l'effet de deux fac-

teurs conjugués:
- la dégradation de l'interface sous l'effet des glissements,
- la diminution de la longueur encastrée des fibres.

Ce composite est le seul sur lequel nous avons pu obtenir, après vieillissement, des comportements ductiles en traction. Il parait donc difficile de déduire à partir des essais quelle pourrait être la valeur limite, à long terme, de ce frottement interfacial conduisant à un comportement ductile des composites étudiés. On ne peut pas séparer l'effet de la baisse de résistance des fibres (due à leur dégradation) et l'effet de l'augmentation de ce frottement. Nous pouvons seulement calculer les valeurs de ces contraintes, pour différentes matrices aux jeunes âges (7 jours) afin de s'affranchir de l'effet de dégradation des fibres.

Matrice	CPA	CPA + 15 % polymère	0.75 CPA + 0.25 MK	0.6 CPA + 0.4 MK
(MPa)	0.9	1.0	0.8	0.6

Les lois de comportement correspondantes sont représentées sur la figure 7.

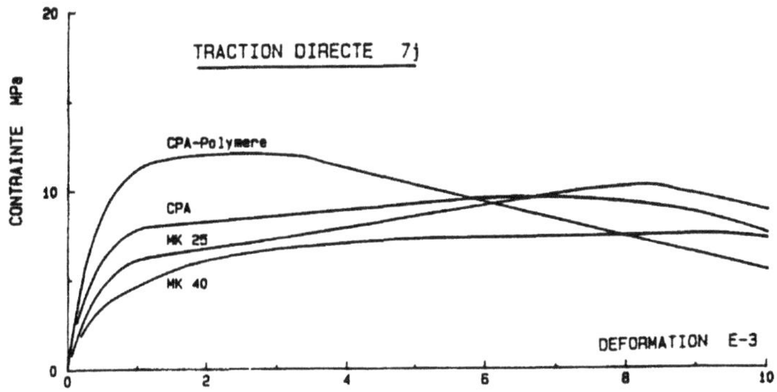

Figure 7. Lois de comportement en traction. Comparaison.

La rapide décroissance de l'effort transmis dans le composite à matrice CPA + polymère suggère que de nombreuses fibres sont rompues alors qu'à cet âge, aucune dégradation physico-chimique des fibres n'a été notée.

Nous pouvons donc conclure qu'en dehors de toute dégradation des fibres, la valeur limite du frottement interfacial, tel que nous l'avons défini, se situe autour de 1 MPa, si l'on veut assurer un comportement ductile.

3.4 Limites de cette modélisation

La modélisation proposée repose sur l'hypothèse que le composite périt par rupture d'adhérence des fibres et non par rupture des fibres. C'est donc la comparaison entre la résistance propre des fibres et celle de l'interface qui va conditionner la limite de la modélisation, c'est-à-dire, la limite du comportement ductile.

Les fibres les plus sollicitées, c'est-à-dire celles qui se rompront les premières, sont celles qui sont encastrées de la longueur maximale de part et d'autre de la fissure, soit L/2. Sous l'effet du vieillissement, leur résistance $\sigma(j)$ va diminuer alors que la liaison interfaciale, paramétrée par $\tau(j)$ va se renforcer. Mais on n'aura pas de rupture tant que:

$$\tau(j) \;.\; 2\,W \;.\; L/2 < \sigma(j) \;.\; X.\; t$$

soit:

$$\tau(j) < \sigma(j)\; t/L. \qquad (8)$$

La résistance propre des fibres peut être évaluée par un essai S.I.C. (traction simple d'un fil) soit en viellissement naturel, soit en vieillissement accéléré. Nous avons procédé à de tels essais, pour des fibres de verre E ou AR et pour des matrices CPA ou CPA-MK. Les valeurs de résistances obtenues jusqu'à 180 jours correspondent à des vieillissements en atmosphère saturée, celles à très long terme à un vieillissement sous eau à 50° pendant 6 semaines. Les résistances obtenues sont représentées sur la figure 8.

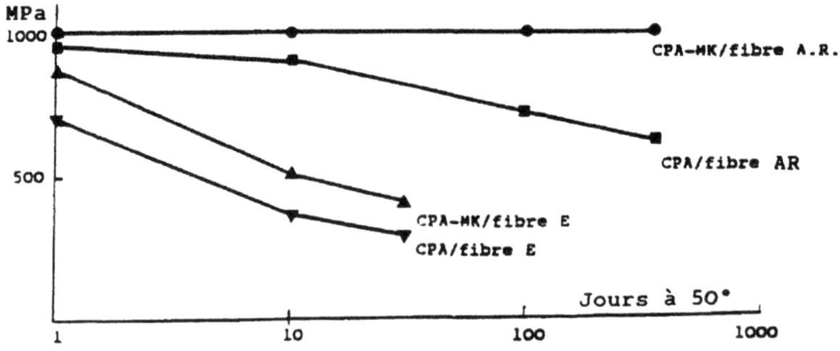

Figure 8. Résultats des essais S.I.C.

Pour des composites qui ne subissent pas de dégradation des fibres, tels les composites 40 MK/fibres AR, nous obtenons une valeur limite de $\sigma(j)$ de 1000 MPa. Pour des fibres de géométrie courante ($t = 2{,}8\ 10^{-5}$ m, L = 30 mm) la valeur maximale du frottement serait donc de l'ordre de 0,9 MPa, à comparer à la valeur 0,69 MPa, obtenue expérimentalement. La ductilité est donc assurée.

Par contre, pour des composites dont les fibres subissent des pertes de résistance ou (et) dont les efforts interfaciaux croissent, la seule solution pratique envisageable pour préserver la ductilité consiste à utiliser des fibres plus courtes :

$$L < \sigma(j)/\tau(j)\ .t. \qquad (9)$$

Bien évidemment cette conservation de la ductilité s'effectue au détriment de la résistance du composite.

En reprenant le cas de la matrice dont le liant comporte 40% de métakaolin mais renforcée de fibres E, dont la résistance à long terme chuterait autour de 500 MPa, pour la même valeur du frottement interfacial (0,69 MPa) nous obtenons une longueur maximale de fibres de 2 cm. Cette rupture de certaines fibres explique la perte de déformabilité que nous avons constatée lors de l'étude du comportement de CCV CPA-MK renforcés par du verre E (Dejean, 1989).

La ductilité ne sera préservée que si l'on utilise un liant conduisant à des frottements interfaciaux limités à:

$$\tau < \sigma(j)\ t/L \qquad (10)$$

soit pour les géométries de fibres considérées (L = 3 cm, $t = 2{,}8\ 10^{-5}$ m)

$$\tau < 9{,}3.10^{-4}\sigma j$$

Les deux valeurs de $\sigma(j)$ évoquées - 1000 ou 500 MPa - correspondent à des valeurs limites de τ de 0,93 et 0,47 MPa. On voit donc que dès les jeunes âges, les valeurs de τ obtenues pour des matrices CPA sont critiques. La nécessité d'utiliser une matrice mixte apparait clairement puisque même en l'absence de dégradation, l'accroissement des efforts interfaciaux avec le temps mènera à des ruptures de fibres. Nous retrouvons là, sous l'aspect mécanique, l'intérêt de fibres A.R., intérêt que nous avions évoqué précédemment (Ambroise, 1987).

4 Conclusion

L'approche analytique du comportement des CCV montre que l'association verre-matrice cimentaire est critique sur le plan mécanique. L'utilisation de matrice mixte est donc en ce sens pleinement justifiée. Cette méthode pourra permettre d'évaluer et d'optimiser d'autres composites, de matrice ou fibres de nature différente, et ainsi de garantir la sécurité des éléments à travers la résistance et la ductilité.

5 Références

Ambroise, J., Dejean. J. et Péra, J. (1987). Study of fifer matric interfaces in metakaolin OPC blended cement GRC composites. Proceedings of The **Material Research**

Society, Boston. (Edit. Mindess). Boston, V.114, p.175-180.

Bartos, P. (1986). Bond in Cement reinforced with bundles of fibers. Proceedings **Congrès de la Chimie des Ciments**, Rio, Edit. ICCC, Rio, ICCC 1986. V.3, p. 485-490.

Bentur, L. (1986). Ageing process of glass fiber reinforced cements with different cementitious matrix. Proceedings of The **Rilem Symposium on Development in FRC**, Sheffield, July, 1986, V.2 § 7.3.

Bentur, L. (1985). Mechanism of potential embrittlement and strength loss of GFRC. Proceedings **Durability of GFRC Symposium**, Chicago, 1985. Edited by S. Diamond 1985, p. 109-123.

Cheick Larbi, A. (1985). Etude de l'interface fibre-matrice dans les composites à matrice cimentaire. **Thèse Doct**: U.T. Compiègne, 120 p.

Cohen, E. (1975). Validity of flexural strength reduction as an indication of alkali attack on GRC. Proceedings of The **Developments of FRC, Rilem Symposium**, Paris, Edit. Neville.

Daniel, J.I. and Schultz, D.M. (1985). Durability of glass fiber reinforced concrete systems. Proceedings of **The Symposium on Durability of GFRC**, Chicago, Nov. Edited by S. Diamond, 1985. p. 174-197.

Dejean, J. (1989). Amélioration de la durabilité des composites ciment-verre par ajout de métakaolin. **Thèse Doct.**, INSA Lyon, 163 p.

Fukuda, F. (1982). A probabilistic theory of the strength of short fibers composites. **Journal of Material and Science**, N° 17. p. 1003-1011.

Hannant, D.J. (1978). **Fibre Cements and Fibre Composites**, John Wiley and Sons, 219 p.

Mindess, S. (1987). Bonding : How important is it ? Proceedings of The **Material Research Society**, Boston, 1987. Edit. S. Mindess. Boston: MRS, V. 114, p. 3-10.

Naaman, A.E. (1973). A fracture model for fiber reinforced cementitious materials. **Cement and Concrete Research**, Vol. 3., p. 397-41.

Shah, S.P. (1985). Reinforcement mechanisms in GFRC composites. Proceedings **Durability of GRC Symposium**, Chicago, Edit. S. Diamond, 1985. p. 90-107.

Swamy, R. (1974). A theory for the flexural strength of fiber reinforced concrete. **Cement and Concrete Research** V 4. p. 313-325.

25 NUMERICAL STUDY OF THE INFLUENCE OF INTERFACIAL PROPERTIES ON THE MECHANICAL BEHAVIOUR OF CEMENT-BASED COMPOSITES

E. SCHLANGEN and J. G. M. van MIER
Stevin Laboratory, Delft University of Technology,
Delft, The Netherlands

Abstract
Numerical simulations with a simple lattice model in which the properties of the individual components in the concrete can be varried are shown. Fracture in laboratory scale specimens is simulated. It is shown that the fracture mechanism is influenced by the strength of aggregates, matrix and interfacial zone. Varying the bond strength leads to a change in the amount of microcracking outside the main crack, but has only a slight influence on the overall specimen strength.
Keywords: Lattice Model, Interfacial Properties, Concrete Fracture.

1 Introduction

Fracturing in concrete structures is often simulated with Finite Element Methods (FEM) based on continuum mechanics. Following these procedures satisfying predictions are only found if homogeneous materials are concerned or if the disorder in the material is weak.

Cracks in a heterogeneous material like concrete propagate in a direction which is perpendicular to the maximum tensile stress, but they also follow the weakest link in the material. The disorder in concrete is a consequence of the grain structure in the material. In a normal weigth concrete the interfacial zone between the aggregates and the matrix is the weakest link and will therefore crack first. This was first recognised by Hsu et al. (1963).

In the present investigation a two dimensional lattice type model is used for simulating the fracture process in concrete, Schlangen & Van Mier (1991a+b). The model is based on a linear elastic finite element code. Crack growth is obtained by removing elements from the lattice that exceed their tensile strength.

With this model fracture in laboratory scale specimens subjected to different loading situations is simulated. The results of the simulations are compared with experiments. By varying the strength of the respective elements in matrix, aggregates and interfacial zone a qualitative study of the fracture mechanism in a single-edge-notched uniaxial tensile specimen is carried out.

Interfaces in Cementitious Composites. Edited by J. C. Maso. © RILEM.
Published by E & FN Spon, 2 - 6 Boundary Row, London SE1 8HN. ISBN 0 419 18230 6.

Quantitative results of the influence of the bond strength on the overall specimen strength are obtained from simulations of four point shear experiments on double-edge-notched beams.

2 Lattice Model

In lattice model the medium is discretized such that all sites are equivalent, i.e. the grid is not refined in regions of higher stress as is often done in FEM. Each site has the same number of neighbours so that the lattice is regular. In the present study a triangular lattice of beam elements is chosen as shown in Figure 1. Different procedures for crack growth through the medium can be used, see for example Herrmann & Roux (1990). The current model is based on a linear elastic finite element code (DIANA). Fracturing of the material takes place by removing in each load step the beam element with the highest stress (relative to its strength f_t). The stress is derived from the fracture law:

$$\sigma_t=\beta*(F/A+\alpha*\max(|M_i|,|M_j|)/W)$$

where F is the normal force in the beam element, M_i and M_j are the bending moments in nodes i and j of the beam element respectively, A=b*h is the cross-sectional area of the beam, and $W=b*h^2/6$. The coefficient α is introduced in order to select a failure mode where bending plays either a dominant or a restricted role. The coefficient β is a scaling factor for the global maximum stress.

The normal force F and the bending moments M_i and M_j are obtained from a linear elastic analysis of a lattice with a prescribed external load P_{pre} or prescribed boundary displacement u_{pre}.

The external load that can be carried by the lattice is in each step simply determined by multiplying P_{pre} by the factor (f_t/σ_t) of the beam element that will be removed in that step. The same procedure has to be followed to derive the boundary displacement in the case of a prescribed u_{pre}.

After removing a beam element a linear analysis on the new mesh is carried out followed by a next removal. This procedure is repeated until the lattice is completely fractured. Following this procedure an exact solution is obtained.

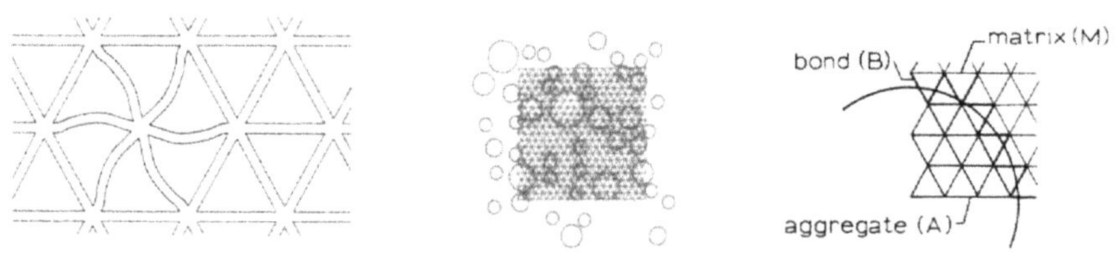

Fig.1. Triangular lattice and projection on generated grain structure.

The heterogeneity of the material is implemented by generating the grain structure of concrete. In order to generate the grain structure, an assumption has to be made for the distribution of the diameters of the intersection circles (aggregate particles are assumed to be perfect spheres) in a cross section of a concrete body. For the distribution of the aggragate particles in the concrete mix a Fuller curve has been chosen. Using a cumulative distribution function derived by Walraven (1980), see also Schlangen & Van Mier (1991b), for a certain concrete mix the distribution of circle diameters in a certain cross section can be generated. The circles are randomly positioned in an area starting with the largest circles. In the simulations shown in this paper the maximum aggregate size in the concrete is 8 mm.

The triangular lattice is projected on top of the generated grain structure, and different strengths are assigned to the respective beam elements, see Figure 1. When a beam element is situated inside an aggregate particle, the failure strength of the aggregate will be assigned to this element; a beam element located on the boundary between aggregate and matrix will get a low bond strength when fracturing of normal concrete is to be simulated. Matrix strength will be assigned to those beams projected on the cement matrix in the generated two-phase material.

In the simulations of laboratory scale specimens only the part of the mesh where cracks are expected to grow is modelled with the lattice. For the remaining part of the specimen isoparametric eigth-noded plane stress elements available in DIANA are used. To get the same linear elastic behaviour for the lattice part as for the continuum part the following procedure is adopted to derive the input paramaters for the beams. A lattice with beam elements and a mesh with continuum elements having the same dimensions are loaded in tension. The Young's modulus E and the thickness b of the beams is taken equal to the Young's modulus and thickness of the continuum elements. The length l and the height h of the beams are chosen such that the overal Young's modulus of the lattice is equal to the Young's modulus of the mesh. This results in $h=0.68*l$ for the beam elements. The length of the beams strongly depends on the smallest aggregate size in the grain structure.

3 Influence of beam strength on fracture mechanism

Experiments of single-edge-notched plates loaded in uniaxial tension between non-rotating end platens are simulated. In the experiments a vacuum impregnation technique was adopted for determining the extent of internal cracking. The specimen dimensions are shown in Figure 2a. Some main results of the experiments will be presented here, for a complete overview see Van Mier (1991). After a specimen was loaded to a prescribed axial deformation, it was unloaded until P=0. At this moment the specimen was taken out of the machine and vacuum impregnated with a low viscosity fluorescenting epoxy. After hardening of the epoxy the specimen was cut into slices. The internal surfaces were photographed under Ultra Violet light.

In Figure 3a the load-displacement diagrams for normal weight concrete and mortar loaded up to different crack openings are shown. From the photos of the internal surfaces, see Van Mier (1991), it can be concluded that up till a crack opening of about 50 μm the crack is propagating through the specimen cross section. At larger opening,

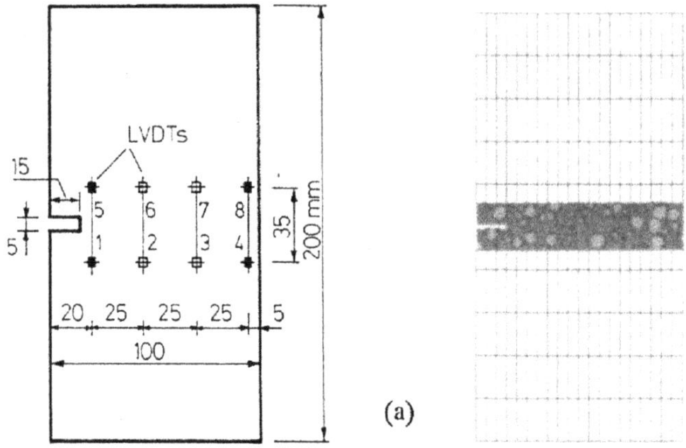

(a) (b)

Fig. 2. Single-edge-notched specimen used for impregnation experiments (a); Element mesh with generated grain skeleton (b).

the crack has traversed the specimen completely but is not continuous: overlaps with intact ligaments in between called crack face bridges can be observed (Fig. 3b). This implies that still load can be transferred between the crack faces. The long tail in the softening diagram can probably be explained by a large deformation between the crack faces through bending of the bridges.

The experiments described above have been simulated using the lattice model. As shown in Figure 2b only the part where cracks are expected to grow is modelled with beam elements. The bottom edge of the specimen is supported in the vertical direction; the specimen is loaded by giving the upper edge a vertical displacement. Aggregates smaller than 2 mm were excluded from the grain structure. The length of the beams was taken 1.25 mm in this simulation. For α an arbitrary value of α=1/3 was chosen. Four

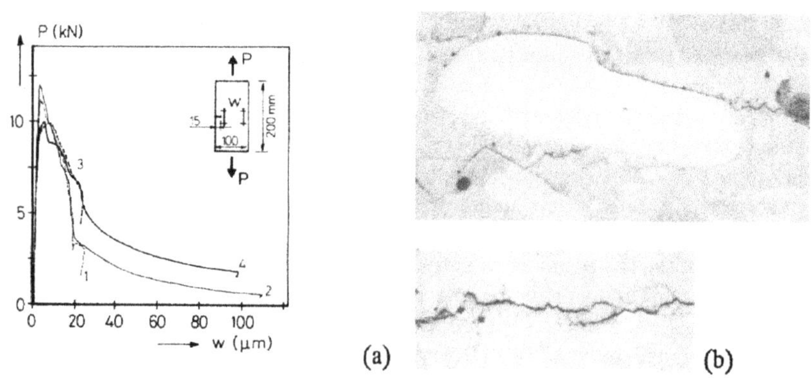

(a) (b)

Fig. 3. Load-displacement curves for normal weight concrete (curves 3 and 4) and mortar (curves 1 and 2) specimens loaded to a prescribed deformation of 25μm and 100μm (a); Crack face bridges in concrete and mortar (b).

simulations are carried out with different properties for the elements in aggregate, matrix and bond zone as shown in Table 1. For the properties of the individual components in the grain structure realistic values are taken according to experiments by Rehm et al. (1977).

Table 1. Input parameters for simulations of uniaxial tensile experiment.

Analysis	I	II	III	IV		I	II	III	IV
E_A [GPa]	70	25	70	70	σ_A [MPa]	10	1	10	10
E_M [GPa]	25	25	25	25	σ_M [MPa]	5	5	5	5
E_B [GPa]	25	25	25	25	σ_B [MPa]	0.5	2	0.5	1

E is the Young's modulus; σ is the failure strength of the lattice beam elements.
The subscripts A,M and B stand for aggregate, matrix and bond respectively.

The crack history for simulation I is shown in Figure 4. The σ-w curve is plotted in Figure 5. It can be concluded that microcracking in the entire cross section starts already before the maximum load is reached (Fig. 4a). After peak load the crack starts to localize in one main crack (Fig. 4b). At the beginning of the long tail in the descending branch the cross section is completely cracked (Fig. 4c). Small intact material pieces bridging the crack faces are the reason for the long tail (Fig. 4d).

In simulation II and III higher strengths for the elements in the bond zone are taken. From Figure 6 where the final crack patterns for the four simulations are plotted it is clearly visible that, with increasing bond strength less microcracking occurs outside the main crack path, and the number of bridges decreases.

In simulation IV the strength of the beams inside the aggregates is taken smaller in order to simulate fracture in lightweight concrete. In this case, the aggregates crack first, see Figure 6d. In this analysis also crack face bridges inside the aggregates are observed, which is in agreement with experimental observations, see Van Mier (1991).

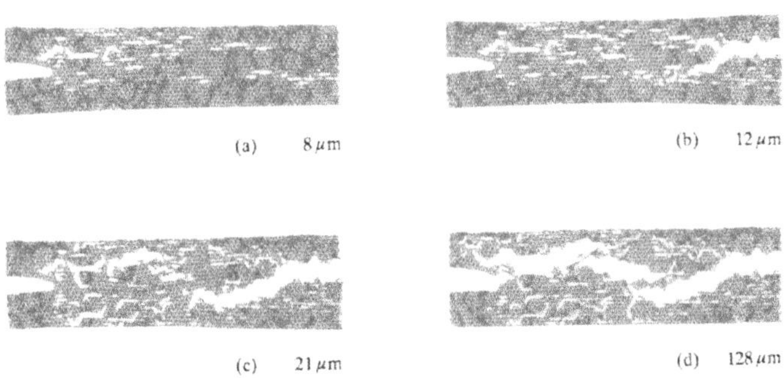

Fig. 4. Crack history for simulation I.

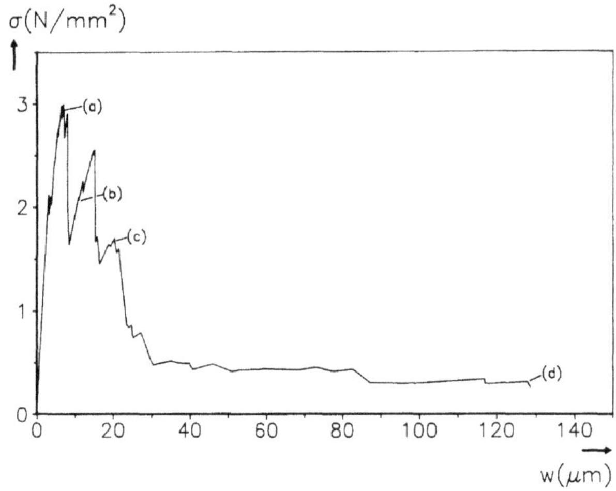

Fig. 5. σ-w response for simulation I.

4 Influence of bond strength on overall specimen response

A quantitative investigation on the influence of the strength of the elements in the interfacial zone on the overall specimen strength is carried out by simulating four point shear experiments on double-edge-notched beams. The specimen geometry is according to the tests of Bažant & Pfeiffer (1986). They claimed that a 'true shear crack' (defined as a regular array of inclined microcracks) developed in these beams.

The tests of Baž ant & Pfeiffer have recently been duplicated, using the same geometry but a different loading device with well known boundary conditions, see Van Mier et al.

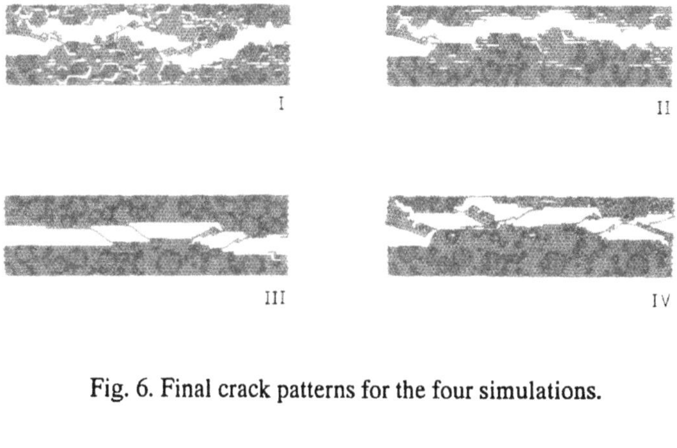

Fig. 6. Final crack patterns for the four simulations.

(1992). From these experiments it is concluded that not a shear crack but a curved mode I crack will develop in the four point shear specimen (Fig. 7.).

In Schlangen & Van Mier (1992) the experiments are compared with simulations with the lattice model for different beam sizes. In this paper only a single beam size (d=75 mm, thickness=37.5 mm) is simulated taking nine different strength values for the elements in the interfacial zone in order to study the influence of the bond strength on the specimen strength.

The input parameters (Table 2) that are used in the simulations have been determined from a parameter study. For this purpose simulations were carried out of a uniaxial tensile experiment on double edge notched prisms, see Schlangen & Van Mier (1992). The parameters that gave the best fit with the experiments are listed in Table 2.

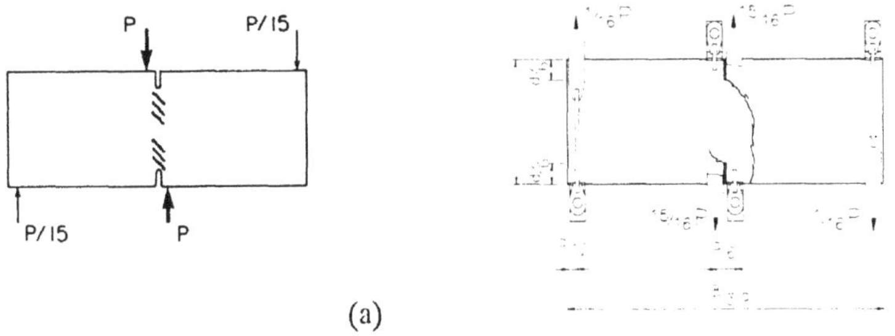

(a) (b)

Fig. 7. (a) Shear crack according to Bažant & Pfeiffer (1986); (b) Curved mode I crack according to Van Mier et al. (1992)

Table 2. Input parameters for simulations of four point shear beams.

default parameters for the beam elements:
l=5/3 mm, b=37.5 mm, h=1.133 mm; fracture law: α=0.005, β=2.0
E_A=70 GPa, E_M=25 GPa, E_B=25 GPa, σ_A=10 MPa, σ_M=5 MPa

Analysis No.	1	2	3	4	5	6	7	8	9
σ_B [MPa]	1.25	0.5	0.75	1.0	1.5	1.75	2.0	3.0	5.0
P_{max} [kN]	14.3	14.6	14.7	12.6	14.3	15.6	16.4	17.8	19.4

E is the Young's modulus; σ is the failure strength of the lattice beam elements.
The subscripts A,M and B stand for aggregate, matrix and bond respectively.

Simulation 1 has the default input parameters with σ_B=1.25 MPa; for the other eight simulations all the input parameters including the generated grain structure are kept the same excepting the strength of the beams in the interfacial zone. In Figure 8 the crack patterns are shown. In Figure 9 the load-displacement curve for simulation 1 is plotted. The displacement in Figure 9 is the average of the crack mouth opening at the top and the bottom of the beam. The shape of the load-displacement curves of the other simulations are identical. The maximum loads of all simulations are listed in Table 2.

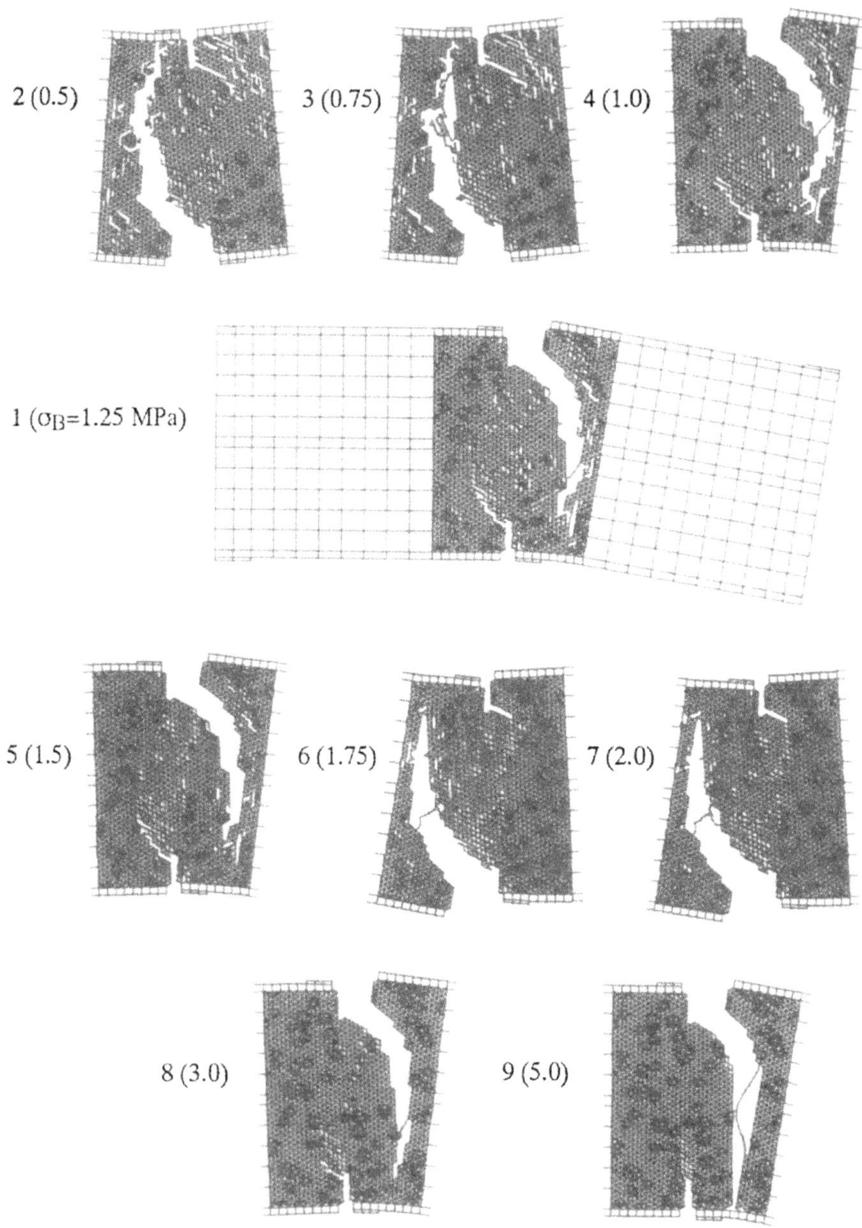

Fig. 8. Simulated crack patterns in double-edge-notched four point shear beams.

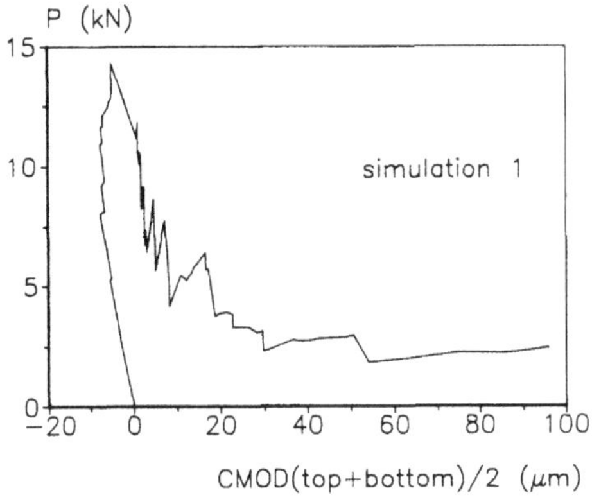

Fig. 9. Load-displacement curve for simulation 1 of double-edge-notched four point shear beam.

Crack propagation is similar in all the simulations: before the peak load two curved cracks initiate from the two notches, beyond peak one curved mode I crack propagates leading to failure. Either the crack at the top notch or the crack at the bottom notch opens after peak load depending on the bond strength. Also in this specimen an increase in microcracking with decreasing bond strength can be observed. A comparison of the maximum loads shows that with changing bond strength, a relatively small difference in specimen strength is obtained. From this it can be concluded that the maximum load is mainly defined by the strength of the matrix.

5 Conclusions

In the paper the effect of the interfacial strenght between aggregates and cementmatrix on the overall mechanical behaviour of concrete has been studied. A recently developed simple lattice model was used. The following conclusions can be drawn:

(1) With increasing bond strength the amount of microcracking outside the main crack will decrease.

(2) The number of bridges connecting the two crack faces will decrease if the strength of the interfacial zone between matrix and aggregates increases. This leads also to a more brittle specimen behaviour, see Schlangen & Van Mier (1991a).

(3) An increase of strength of the interfacial zone with a factor 10 leads to an increase of specimen strength of 30-35%. Therefore it may be concluded that the specimen strength is mainly governed by the matrix strength. This conclusion seems in agreement with experimental data, see for instance Mindess (1989).

(4) The random character of the model is demonstrated in the paper. A small change in bond strength leads to random crack shapes, with the final failure crack nucleating either from the top or the bottom notch.

6 Acknowledgement

The work was sponsored through a grant from the Dutch Technology Foundation (STW). The authors are indebted to Mr. A.S. Elgersma for his expert help in the experiments.

7 References

Bažant, Z.P. and Pfeiffer, P.A. (1986) Shear fracture test of concrete. *Materials and Structures*, 19, 111-121.

Herrmann, H.J. and Roux, S. (1990) *Statistical Models for the Fracture of Disordered Media*. Elsevier Applied Science Publishers B.V. (North-Holland).

Hsu, T.T.C. Slate, F.O. Sturman, G.M. and Winter, G. (1963), Microcracking of plain concrete and the shape of the stress-strain curve. *Journal of the American Concrete Institute*, 60, 2, 209-224.

Mindess, S. (1989) Interfaces in concrete, in *Materials Science of Concrete I* (ed. J.P. Skalny), The American Ceramic Society, Inc., Westervill (OH), pp. 163-180.

Rehm, G. Diem, P. and Zimbelmann, R. (1977) Technische Möglichkeiten zur Erhöhung der Zugfestigkeit von Beton. *Deutscher Ausschuss für Stahlbeton*, 283, 1-76.

Schlangen, E. and Van Mier, J.G.M. (1991a) *Experimental and Numerical Analysis of Micromechanisms of Fracture of Cementbased Composites*. Report No. 25.5-91-1/VFC, Delft University of Technology, to appear in *J. Cem. & Conc. Comp.*, special issue on micromechanics of fracture of cementbased composites (ed. V.C. Li).

Schlangen, E. and Van Mier, J.G.M. (1991b) Lattice model for numerical simulation of concrete fracture, in *Proc. Int'l. EPRI Conference on Dam Fracture* (eds V.E. Saouma, R. Dungar and D. Morris), Boulder, Colorado, Sept. 11-13, 1991, Electric Power Research Institute, Palo Alto, Ca 94304, USA, pp. 511- 527.

Schlangen, E. and Van Mier, J.G.M. (1992) Shear fracture in cementitious composites, part II: numerical simulations, in *Proc. FRAMCOS I* (ed. Z.P. Bazant), Breckenridge, Colorado, June 1-5, 1992, Elsevier Publishers, London.

Van Mier, J.G.M. (1991) Mode I fracture of concrete: discontinuous crack growth and crack interface grain bridging. *Cement & Concrete Research*, 21, 1-15.

Van Mier, J.G.M. Schlangen, E. and Nooru-Mohamed, M.B. (1992) Shear fracture in cementitious composites, part I: experimental observations, in *Proc. FRAMCOS I* (ed. Z.P. Bazant), Breckenridge, Colorado, June 1-5, 1992, Elsevier Publishers, London.

Walraven, J.C. (1980) *Aggregate Interlock: a Theoretical and Experimental Analysis*. PhD-thesis, Delft University of Technology.

26 UNE METHODE INTEGRALE DE FRONTIERE TENANT COMPTE DU COMPORTEMENT MECANIQUE DE LA LIAISON HETEROGENEITE-MATRICE

(A boundary integral method taking into account the mechanical behaviour of the matrix-inhomogeneity interface)

L. ULMET and S. CAPERAA
Laboratoire Génie Civil, University Limoges, Egletons, France

Résumé *(English abstract, page 308)*
Ce travail montre l'apport de la méthode des Eléments de Frontière pour la détermination du champ de contraintes avoisinant des hétérogénéités, en prenant en compte diverses conditions de liaison interfaciales. La méthode a été également utilisée pour des problèmes de propagation de fissures dans un milieu composite.
Mots-clés: Eléments de Frontière, Modélisation, Composites, Hétérogénéités, Fissuration.

1 Introduction

Les progrès en matière de mécanique des milieux composites tels que les bétons sont largement conditionnés par une meilleure connaissance des lois de comportement des zones interfaciales entre la matrice et les granulats qu'elle enrobe. Toutefois, quel que soit le degré de sophistication des modèles de comportement caractérisant la liaison, une étape difficile est à franchir pour injecter le comportement micro-mécanique local dans un modèle plus large, incluant les champs de déformations et de contraintes propres à la matrice et aux granulats. Ces derniers présentant des formes et des dispositions très variées, les solutions analytiques sont inexistantes, et il est alors généralement fait appel à des méthodes d'éléments finis. Dans ce cas, le caractère fastidieux de la saisie de la géométrie des maillages est le plus souvent rédhibitoire.

La méthode des équations intégrales de frontière constitue une alternative intéressante, dans la mesure où la discrétisation est confinée aux frontières des domaines étudiés. Ses principaux points forts résident dans une plus grande souplesse en phase de saisie des géométries, mais aussi et surtout dans une possibilité de génération automatique de maillages, pour la mise en oeuvre d'algorithmes itératifs ou incrémentaux. Nous montrons ici quelques applications du code de calcul "MUZO", développé au Laboratoire de Génie Civil de l'Université de Limoges.

Interfaces in Cementitious Composites. Edited by J. C. Maso.
Published by E & FN Spon, 2 - 6 Boundary Row, London SE1 8HN. ISBN 0 419 18230 6.

2 Méthode des éléments de frontière

Cette méthode est désormais largement utilisée dans le domaine de la mécanique {BREBBIA}. Nous nous contenterons ici d'en rappeler les grandes lignes, ainsi que les caractéristiques du code de calcul "MUZO" que nous avons développé {CAPERAA, ULMET}.

Considérons un domaine élastique linéaire isotrope Ω, borné par une frontière Γ. L'identité de SOMIGLIANA exprime le déplacement $u_j(P)$ en tout point P du domaine en fonction uniquement des déplacements $u_i(Q)$ et $t_i(Q)$ sur la frontière. Dans le cas particulier de l'absence de forces de volume, cette identité s'écrit :

$$u_j(P) = \int_\Gamma U_{ij}(P,Q).t_i(Q).d\Gamma(Q) - \int_\Gamma T_{ij}(P,Q).u_i(Q).d\Gamma(Q)$$

Les tenseurs $U_{ij}(P,Q)$ et $T_{ij}(P,Q)$ représentent la solution fondamentale de l'élasticité, permettant de connaître respectivement les déplacements et les contraintes en tout point P d'un milieu infini, sous l'effet d'une charge ponctuelle appliquée en Q. Si le point P tend vers la frontière Γ, le passage à la limite conduit à l'équation intégrale de frontière, qui fait uniquement intervenir des déplacements et tensions sur Γ. La résolution d'un problème d'élasticité consiste, dans un premier temps, à résoudre l'équation intégrale de frontière afin de déterminer les déplacements et tensions inconnus sur la frontière, en fonction des conditions aux limites imposées. Le calcul peut ensuite s'étendre à tout point intérieur de Ω, par application de l'identité de SOMIGLIANA.

Le code de calcul "MUZO" utilise une discrétisation de la frontière en éléments isoparamétriques à trois noeuds de contrôle. L'interpolation des coordonnées géométriques, déplacements et tensions s'effectue au moyen de fonctions de forme quadratiques. Une méthode de collocations est utilisée pour la détermination des inconnues sur la frontière : le point P pour lequel est écrite l'équation intégrale de frontière occupe successivement tous les noeuds de contrôle du contour, ce qui conduit à une égalité de la forme :

$$[A].\{u\} = [B].\{t\},$$

liant déplacements et tensions sur la frontière. Les conditions aux limites imposées sont alors prises en compte, afin de constituer le second membre d'un système linéaire. La résolution, par une méthode de GAUSS classique, conduit à la détermination des déplacements et tensions inconnus sur la frontière.

3 Prise en compte des lois interfaciales pour des matériaux composites

3.1 Traitement numérique

Le code permet de modéliser des problèmes multi-zones, et notamment le cas d'hétérogénéités inclues dans une matrice, chacune des zones étant continue, homogène, isotrope et élastique linéaire. D'un point de vue général, la résolution consiste à appliquer la méthode des collocations pour chacune des zones, et à tenir compte du couplage entre zones par des relations traduisant la loi de comportement interfaciale.

Nous nous intéressons ici en particulier au cas d'une hétérogénéité qui, avant chargement de la structure, "remplit" complètement une cavité de la matrice. L'interface autorise toute forme de discontinuité de déplacement et ne supporte aucune contrainte de traction. En cas de compression de l'interface, le contact obéit à la loi de frottement de COULOMB.

Une solution analytique est établie par HUSSAIN et PU dans le cas d'une hétérogénéité circulaire, de mêmes propriétés élastiques que la matrice, cette dernière étant infinie et soumise à une traction uniforme à l'infini. Au cours d'une phase monotone de chargement, l'interface présente des zones de décollement, de contact glissant, et de contact adhérent dont l'étendue est constante. La difficulté majeure réside dans la détermination de la répartition de ces zones, qui ne peut être connue *a priori*. Hussain et Pu obtiennent une solution approchée grâce à une formulation variationnelle.

La résolution du problème par éléments de frontière fait appel à un processus itératif : partant d'une hypothèse "raisonnable" concernant la répartition des différents états d'interface, nous calculons les inconnues sur la frontière. Les conditions aux limites sont les suivantes, et sont résumées sur la figure 1 :

- éléments en contact adhérent : continuité des déplacements et des tensions, les inconnues étant communes aux deux zones,

- éléments en contact glissant : continuité de tension, continuité de déplacement normal, relation caractéristique du frottement : $\sigma_t = \pm \mathrm{tg}\, \varphi . \sigma_n$, σ_t désignant la tension tangentielle au contour, et σ_n la tension normale. Le signe est fonction du sens de glissement relatif supposé.

- éléments en décollement : discontinuité des déplacements, tensions nulles sur les deux zones.

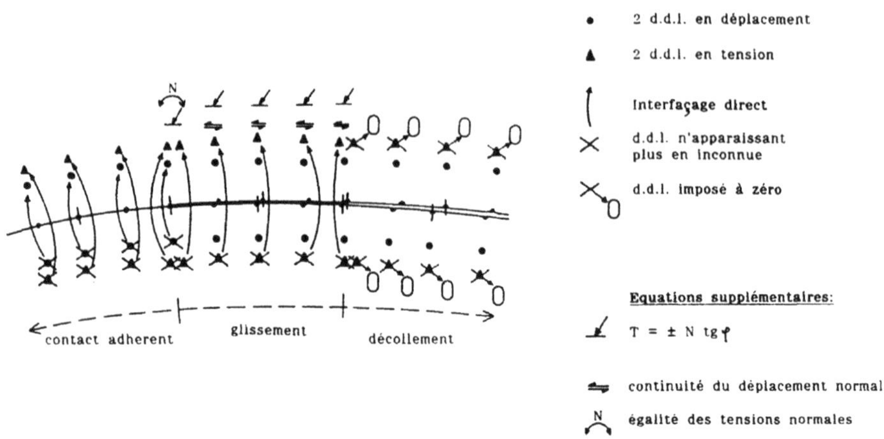

Fig.1. Traitement des conditions aux limites sur l'interface

Après résolution, l'examen des tensions d'interface permet de décider de l'évolution à donner au maillage pour l'itération suivante. Il convient alors de dilater, contracter, ou faire "ramper" certains éléments le long du contour d'interface. Les critères d'obtention de limites convenables pour chaque zone d'état d'interface sont les suivants :

- limite contact adhérent / glissement : la tension tangentielle σt doit être continue entre les deux éléments adjacents,
- limite glissement / décollement : la tension normale σn doit être nulle au dernier noeud de l'élément glissant.

Une solution suffisamment précise s'obtient en général en moins de dix itérations. Des utilitaires de visualisation graphique des tensions et de génération automatique des maillages de frontière facilitent les interventions de l'opérateur. Dans l'état actuel des connaissances, un tel processus serait aisément automatisable.

3.2 Exemples

Contrairement aux hétérogénéités parfaitement adhérentes à la matrice, celles décrites plus haut présentent un comportement fortement dissymétrique entre traction et compression, les zones d'état d'interface n'étant pas les mêmes (voir Fig.2.). Nous nous intéressons à une hétérogénéité circulaire incluse dans une matrice infinie soumise à un chargement lointain uniforme.

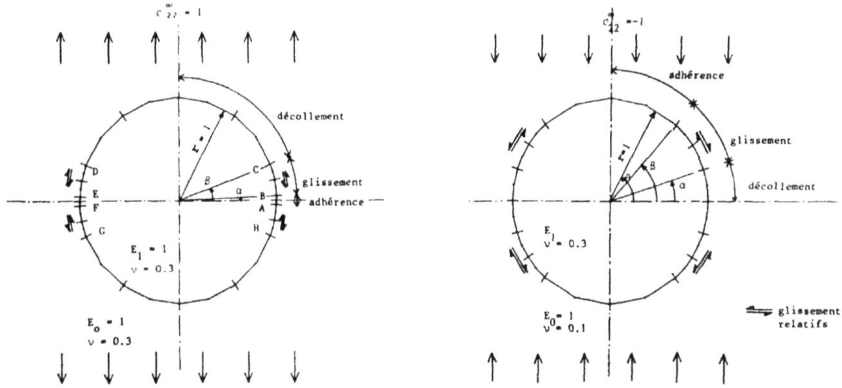

Fig.2.a. Traction Fig.2.b. Compression

Fig.2. Maillages d'éléments de frontière et conditions de contact sur l'interface

3.2.1 Traction

Il apparaît que l'hétérogénéité n'est en contact avec la matrice que sur deux faibles portions de son contour : l'angle ß caractérisant la limite des surfaces en contact (voir Fig.2.a.) est de l'ordre de 20°, et ne dépend pratiquement pas du coefficient de frottement tg φ . Les deux zones d'adhérence sont très peu étendues pour des valeurs courantes de tg φ . Bien que le maillage soit relativement grossier (20 éléments), ces résultats, ainsi que la répartition des tensions le long des surfaces de contact sont en bonne corrélation avec les travaux de HUSSAIN et PU. L'examen des perturbations du champ de contraintes dans la matrice montre une grande similitude entre ce type d'hétérogénéité et une cavité.

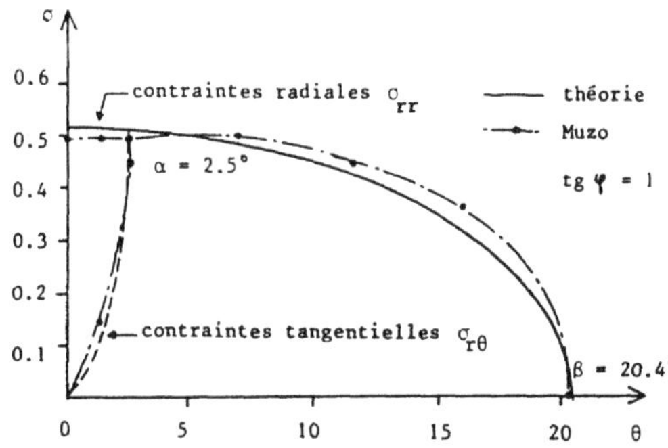

Fig.3. Hétérogénéité dans un champ de traction : tensions le long de la zone de contact

3.2.2 Compression

L'examen des perturbations introduites par l'hétérogénéité sur le champ de contraintes (Fig. 4) met en évidence les points suivants :

- en A : amplification des contraintes de compression σ_{22} normales à l'interface, apparition de contraintes de traction σ_{11} sur des facettes perpendiculaires à l'interface
- en B : amplification des contraintes de compression σ_{22} dans la matrice, près de la zone de décollement,
- en C : compression bi-axiale de l'hétérogénéité près de la zone de contact.

Le tableau (Fig.5) résume les influences du coefficient de frottement et du rapport des modules entre hétérogénéité et matrice, d'une part sur les angles limites α et β des différentes zones (voir Fig.2.b), et d'autre part sur les amplifications de contraintes dans la matrice.

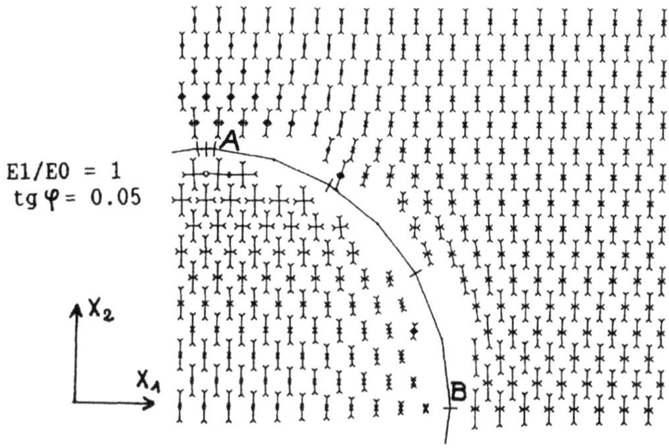

Fig.4. Hétérogénéité comprimée : champ de contraintes

	tg φ = 0.05		tg φ = 0.3		tg φ = 1		adhér.
	E1/E0=1	E1/E0=∞	E1/E0=1	E1/E0=∞	E1/E0=1	E1/E0=∞	E1/E0=∞
α β	31.5° 88°	35° 88°	28° 84°	30.5° 82.5°	17° 50°	18° 48°	
$\sigma_{11}(A)$ $\sigma_{22}(A)$ $\sigma_{22}(B)$	+0.90 -1.25 -2.00	+0.73 -1.90 -1.52	+0.31 -1.11 -1.89	+0.30 -1.58 -1.38	0.00 -0.99 -1.59	+0.12 -1.39 -0.92	-0.38 -1.53 -0.02

Fig.5. Angles et contraintes maximales

A l'examen de ces résultats, il apparait clairement que le comportement de l'interface peut induire des phénomènes assez inattendus : ainsi, un glissement interfacial est susceptible de provoquer des contraintes de traction dans la matrice, bien que la sollicitation externe soit une compression pure. Ceci constitue probablement un élément d'explication des mécanismes de fissuration et d'endommagement de matériaux hétérogènes tels que le béton.

4 Traitement des problèmes de fissuration

4.1 Fissures interfaciales

Le cas de la fissure "interfaciale" située sur une partie de la frontière hétérogénéité-matrice apparait comme un cas particulier du problème précédent, pour lequel ne subsistent que deux types de conditions de contact, non évolutifs : adhérence parfaite ou décollement au niveau de la fissure.

L'exemple considéré sur la figure 6 concerne une fissure en arc de demi-ouverture θ. La répartition des contraintes normale et tangentielle le long de l'interface, établie par PERLMAN et SIH , est retrouvée avec une bonne corrélation par notre modèle (moins de 5% d'écart).Dans le cas du milieu homogène, MUSKHELISHVILI propose l'expression analytique des facteurs d'intensité de contrainte K1 et K2 caractérisant cette sollicitation en mode mixte; la figure 7 montre la comparaison avec cette solution, de l'évolution de ces facteurs en fonction de l'angle θ. Il convient de noter que la méthode "rudimentaire" utilisée pour la détermination de K (méthode cinématique basée sur l'exploitation des déplacements relatifs des lèvres de la fissure), peut être améliorée par diverses techniques spécifiques aux méthodes d'éléments de frontière: élément spécial, fonctions de poids...

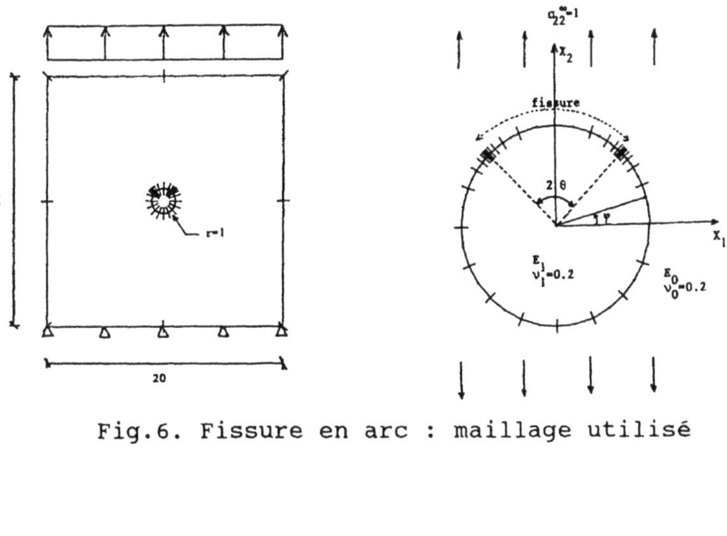

Fig.6. Fissure en arc : maillage utilisé

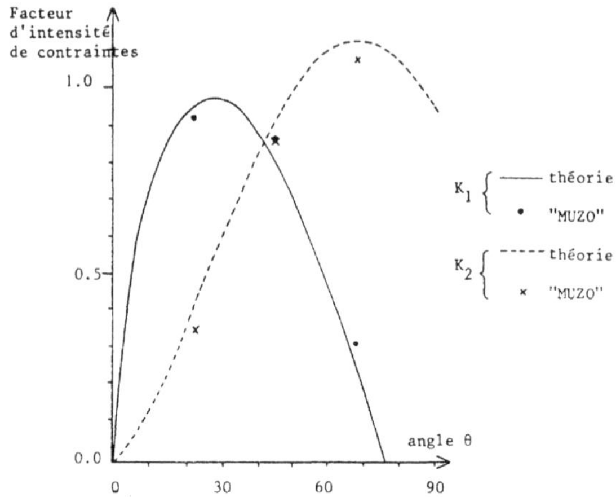

Fig.7. Fissure en arc : facteurs d'intensité de contrainte en milieu homogène

4.2 Interaction entre hétérogénéités et fissures

Un autre problème important pour l'étude de l'endommagement d'une zone fissurée est l'influence d'une hétérogénéité sur le champ de contrainte au-devant du front d'une fissure existante. La figure 8 montre un solide sollicité en traction, en déformations planes, comportant une hétérogénéité rigide (rapport des modules hétérogénéité-matrice égal à 10) située à une distance d d'une fissure non débouchante, assimilée à une ellipsoïde.

Les résultats, présentés sous forme de l'évolution du critère de RANKINE (figure 9), confirmés par les études théoriques de KUNIN et GOMMERSTADT , montrent que la présence de l'hétérogénéité confine l'endommagement, tendant à stopper la propagation et provoquant une bifurcation.

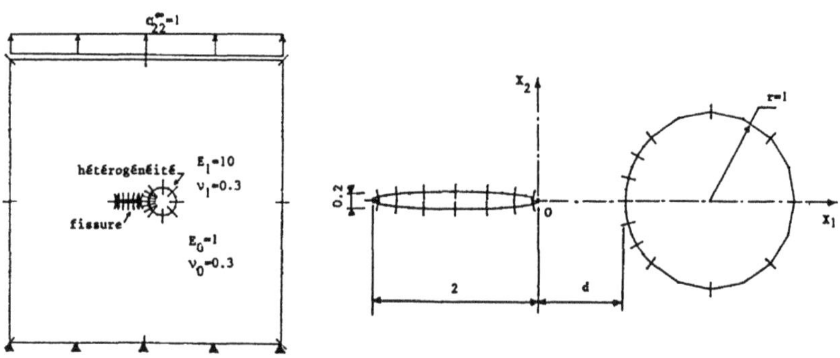

Fig.8. Interaction fissure-hétérogénéité : maillage.

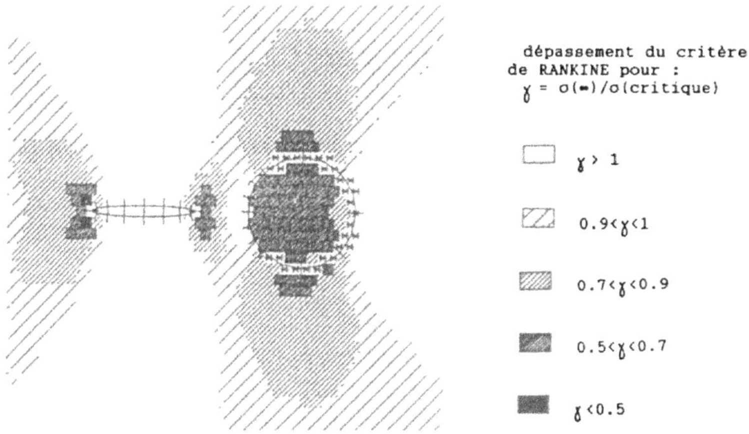

Fig.9. Critère de Rankine.

Le programme MUZO permet également de traiter des problèmes plus complexes, tel que celui qui est présenté sur la figure 10, montrant la cartographie du facteur de concentration de contraintes autour d'une "fissure" et de 4 hétérogénéités de forme elliptique, en déformations planes. Mentionnons que la saisie complète du fichier de données n'a nécessité que 16 minutes de travail; la modification de la forme et de la position d'une hétérogénéité se fait en moins d'une minute de travail. Il parait difficile d'envisager de telles performances par Eléments Finis...

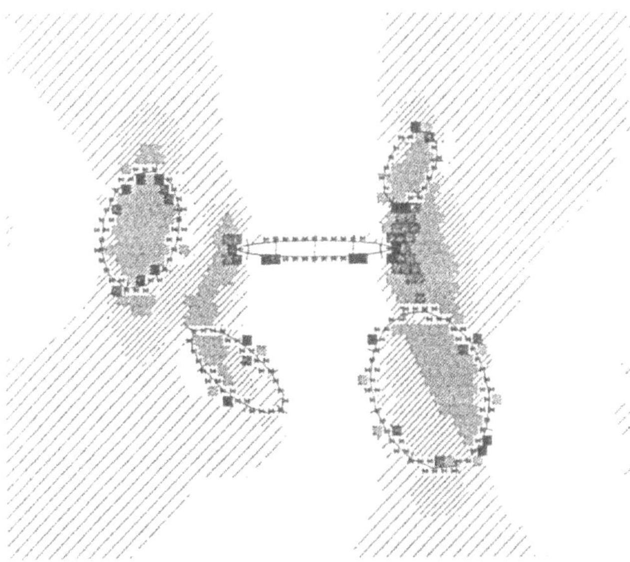

Fig.10. Exemple à plusieurs hétérogénéités.

5 Conclusion

Cette étude avait pour objectif de présenter l'apport de la technique des Eléments de Frontière, associée à un algorithme de traitement non-linéaire, pour l'étude du champ mécanique local au sein d'un matériau composite. En particulier, les avantages en matière d'évolution de maillage peuvent être mis à profit pour la modélisation de la propagation de fissures.

De nouvelles conditions de liaison, fournies par les spécialistes, peuvent être intégrées; si de plus, les critères les plus significatifs sont introduits, le modèle, ainsi adapté à un matériau particulier, peut constituer un outil en vue d'affiner la connaissance des lois d'endommagement de ce matériau.

6 Références bibliographiques

Brebbia,C.A. et Dominguez,J. (1989).Boundary Elements An Introductory Course.Computational Mechanics Publications, Mc Graw-Hill Book Company.

Caperaa,S. (1985).Modélisation des interfaces et singularités.Thèse d'Etat.Université de Limoges.

Hussain,M.A. et PU,S.L. (1971).Slip phenomenon for a circular inclusion.J.Applied Mech.,p.627-633.

Kunin,I. et GOMMERSTADT (1985).On elastic crack-inclusion interaction.Int. Journ. Solids and Struct.Vol.21, n°7.

Muskhelishvili,N.I. (1963).Some basic problems of the Math. Theory of Elasticity.Noordhof.

Perlman,A.B. et Sih,G.C. (1967). Elastostatic problems of curvilinear cracks.Int.J.Eng.Science, vol. 5.

Ulmet,L. (1989).Perturbation du champ mécanique d'un matériau par la présence d'hété.Thèse Doct. Limoges.

PART 2.2

RELATIONSHIPS BETWEEN MICROSTRUCTURE OF THE INTERFACIAL ZONE AND PROPERTIES OF THE COMPOSITE: TRANSPORT PROPERTIES

(Relations entre la microstructure de l'auréole et les propriétés du composite: les propriétés de transfert)

27 INTERFACIAL ZONE PERCOLATION IN CEMENT-AGGREGATE COMPOSITES

K. A. SNYDER
National Institute of Standards and Technology,
Gaithersburg, Maryland, USA
D. N. WINSLOW
Purdue University, West Lafayette, Indiana, USA
D. P. BENTZ and E. J. GARBOCZI
National Institute of Standards and Technology,
Gaithersburg, Maryland, USA

Abstract
Interfacial zones exist between the aggregates and paste in portland cement mortars and concretes. The microstructure of these interfacial zones consists of both larger pore diameters and higher porosity than the microstructure of the bulk paste. If a sufficient number of the aggregate particles are present, the interfacial zones may overlap and link together, percolating the system. Mercury intrusion experiments performed on mortars indicate that the intrusion characteristics change dramatically at a critical sand content. A computer simulation model has been developed that can predict the volume of sand required for the percolation of the interfacial zone phase given an interfacial zone thickness and an aggregate size distribution. The model was used to simulate $1cm^3$ of mortar, consisting of approximately 10000 aggregate particles. A comparison of the results of the computer simulation model with the results of the mercury intrusion experiment yields an estimate of the interfacial zone thickness which is in agreement with published values. The model is also applied to $27cm^3$ systems of concrete in order to predict the critical aggregate content required for percolation.
Keywords: Percolation, Composite, Hard Core-Soft Shell, Microstructure, Computer Model, Interfacial Zone, Mercury Intrusion.

1 Introduction

The role aggregates play in the transport properties of the paste phase of concrete and mortar has received little attention. Although the transport properties of concrete strongly depend upon the properties of the paste, the effect of aggregates upon the properties of the paste must not be overlooked. Zimbelmann (1985) has demonstrated that the paste immediately surrounding an aggregate has a different microstructure than the paste away from the aggregate. This region of transition is referred to herein as the "interfacial zone". Computer simulation models by Garboczi and Bentz (1991a) and Bentz and Garboczi (1991) suggest that the interfacial zone is due to both inefficient packing of the cement particles near the aggregate and a "one-sided growth" effect of the hydration products.

Interfaces in Cementitious Composites. Edited by J. C. Maso. © RILEM.
Published by E & FN Spon, 2 - 6 Boundary Row, London SE1 8HN. ISBN 0 419 18230 6.

Within the interfacial zone, the porosity is highest near the aggregate and decreases with increasing distance from the aggregate, finally decreasing down to the bulk porosity value. Scrivener and Gartner (1988) have shown that the porosity at the surface of the aggregate can be up to three times the value of the bulk porosity. Since diffusivity depends upon porosity, one expects that the diffusivity of the interfacial zones will be greater than the bulk diffusivity. The interfacial zones may act like "short circuits" for the diffusion of ions. Therefore, the presence of aggregates could increase the overall diffusivity of the composite. One expects similar behavior for permeability, which depends upon both porosity and pore size.

Individual, isolated interfacial zones should affect the bulk transport properties of the composite very little. At low aggregate densities, transport should be dominated by the properties of the bulk microstructure. However, if a sufficient density of aggregates are present to allow the interfacial zones to percolate, see Stauffer (1985), the transport properties could become affected by the continuous path of high porosity that exists through the system. Expected service life is also affected; Clifton and Knab (1989) have shown that service life prediction is based extensively upon knowledge of the transport properties.

The potential for percolation of the interfacial zone phase is related to the *hard core-soft shell* problem. Hard cores are impenetrable spheres, representing aggregate particles, which can not overlap one another, and are randomly parked (a particle intersecting another is removed and a new random location is chosen) into the system until a predefined volume fraction is achieved. Soft shells, representing interfacial zones, are placed concentrically about each of the hard cores and may overlap one another.

The critical volume fraction of hard cores is that volume fraction required for interfacial zones to percolate the paste. These quantity depends upon the size distribution of the hard cores and the thickness of the soft shells. Rikvold and Stell (1985a) (1985b) have derived approximate analytical solutions to the soft shell phase fraction for monosized hard core spheres. However, the volume fraction of the hard cores or the soft shell phase at percolation is not known analytically for an arbitrary hard core size distribution.

A continuum computer model program was written to determine the critical aggregate fraction numerically. Results from the computer model were compared with the results of a mercury intrusion experiment performed on mortar specimens containing varying sand fractions. The mercury intrusion results indicate the existence of a critical sand volume fraction. The computer model was applied to different interfacial zone thickness and sand contents, and the results were compared with the mercury intrusion data to estimate an interfacial zone thickness. The computer model was then applied to concretes to compare critical sand fractions for two different sand gradations.

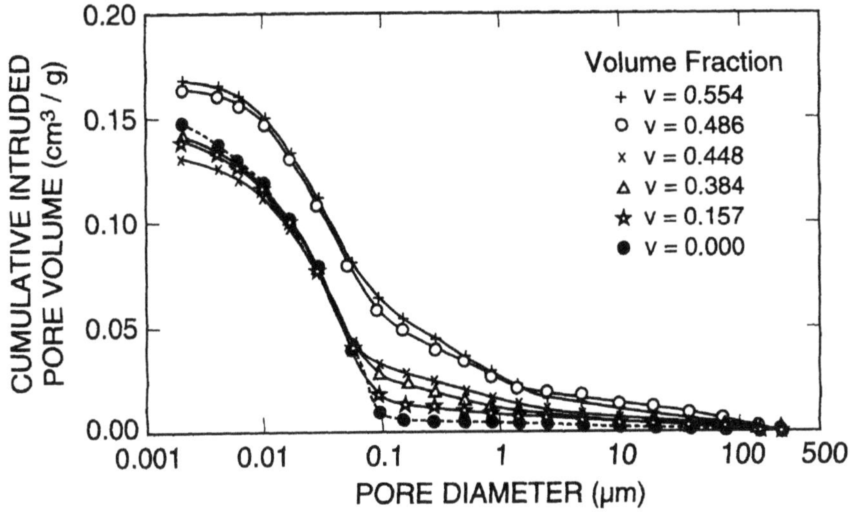

Figure 1: Mercury intrusion porosimetry for mortars with varying sand contents.

2 Mercury Intrusion Experiment

A mercury intrusion porosimetry (MIP) experiment was performed on mortars in which the volume fraction of sand was varied while keeping the water:cement ratio (w/c) constant at 0.40. Materials and methods of preparation were the same as those previously described by Winslow and Liu (1990). The non-reactive sand had zero porosity, insuring that all pores intruded by mercury were cement paste pores. Fig. 1 shows the cumulative volume of mercury intruded, normalized per gram of cement, versus the effective pore diameter, determined from the intrusion pressure via the Washburn equation as described by van Brakel (1981).

Figure 1 contains two groups of curves exhibiting differing behavior. For sand contents below 45% by volume, the intruded volume increased steadily with increasing sand content for pore sizes larger than 0.1μm. If the sand particles did not disturb the microstructure, all of the curves should superpose upon one another since the results are normalized per gram of cement paste. Therefore, the monotonic increase in volume intruded with increasing sand content suggests that the sand particles are altering the cement paste microstructure.

The intrusion curves for sand contents greater than 45% appear to be nearly identical. Both curves show a disproportionate increase in the volume intruded for pore sizes between 0.1 and 10 μm. The dramatic difference between the results for an increase in sand volume fraction from 45% and 49%, relative to the other sand volume changes, indicates a critical phenomenon may be occurring in regard to the connectivity of the larger pores in the interfacial zones.

The results for aggregate volumes less than 45% can be explained, at least in

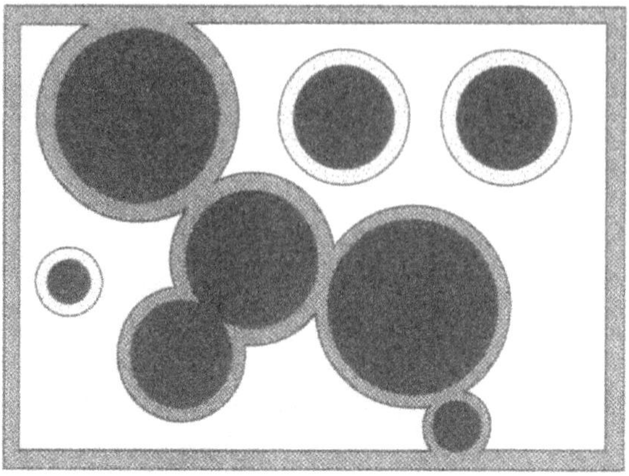

Figure 2: Schematic of mercury intrusion at percolation of the soft shell phase.

part, by the fact that the presence of the sand alters the development of the cement paste microstructure immediately surrounding the sand particles, producing the interfacial zones. As more aggregates are placed in the system, the interfacial zone volume fraction increases, allowing more mercury to intrude pores larger than 0.1μm. However, this does not explain the jump in intruded volume for sand contents between 45% and 49%.

If the volume of aggregate in mortar or concrete were sufficiently great, one would expect that the aggregate particles would be close enough to one another that the interfacial zones could percolate the system, forming a continuous path through the system. Interfacial zones ranging in thickness from 25 to 50 μm have been observed by Scrivener et al. (1988) and by Scrivener, Bentur, and Pratt (1988) using scanning electron microscopy (SEM) imaging techniques. Work performed by Diamond, Mindess and Lovell (1982) found the distance between aggregate surfaces in a typical concrete mix, estimated from two-dimensional images, to be on the order of 100 μm. Since the three-dimensional distance would be closer than this, percolation of interfacial zones less than 50 μm thick is quite plausible.

3 *Hard Core-Soft Shell* Model of Mortar

A macrostructure model of the interfacial zones was used because the quantity of information required to incorporate the microstructure would make computational requirement prohibitive. Instead, a macro structure model could convey large amounts of information from a small amout of structural information. For example, consider Fig. 2. The artificial system shown in Fig. 2 represents a 2D slice through a very small eight sand particle mortar, with each sand particle (black) surrounded by an interfacial zone; the white region represents bulk paste. Note the overlapping that occurs between the interfacial zones. The figure depicts mercury intruding first through the larger pores of the interfacial zones accessible from the outside surface. The dark gray areas represent the presence of mercury. The lighter gray interfacial zones are not yet accessible to the intruding mercury. The mercury has penetrated deeper into the sample because some of the interfacial zones are linked (percolated) into a continuous pathway. If the interfacial zones were isolated then intrusion could only occur at the sample surface.

4 Continuum Computer Simulation Model

A computer program was written to study the percolation of overlapping interfacial zones in mortars containing varying sand volume fractions. The aggregate particles are represented by spherical hard cores and the interfacial zones are represented by concentric, spherical soft shells. The soft shells have a constant thickness, independent of aggregate size. Garboczi and Bentz (1991a) and Bentz and Garboczi (1991) have shown that this is physically valid as long as the aggregate is much larger than the cement particles. The hard core centers are placed randomly in a cubic cell, employing periodic boundary conditions; that is, the portion of a sphere, hard or soft, extending beyond an exterior cube surface is moved onto the opposite surface, protruding into the cube. Maintaining periodic boundary conditions helps to alleviate any artificial effects arising from the finite size of the cubic cell. Hard cores are placed into the system until a predefined hard core volume is achieved, representing the aggregate volume fraction in a mortar. After the addition of the soft shells, a burning algorithm, as described by Stauffer (1985), is used to determine whether percolation has occurred. The volume fraction of the soft shell phase is determined via a systematic point-count method, see Cochran (1963).

This model is a continuum model, rather than a digital image model. This means that each hard core can be specified by four numbers, (x, y, z, r), where (x, y, z) are the coordinates of the hard core center and (r) is the hard core radius. A continuum model allows many more particles to be used than would be possible in a digital image model because only four numbers are needed to represent each particle, instead of hundreds or thousands of pixels required in a digital model. Also, shells that are thin compared to the hard core size are difficult to represent accurately with pixels, and the digital "roughness" would probably affect the results since the small volume of interfacial zone overlap would be inaccurately represented by pixels.

The model system was representative of the mortar used in the MIP experiment. The system size was $1 cm^3$, about the same size as a MIP specimen, and contained on the order of 10000 particles, depending upon the volume fraction of sand desired. The program used approximately the same sand size distribution that was used in the MIP experiment and shown in Table 4. In the intervals between sieve sizes, the particle volume distribution was assumed to be uniform. Therefore, the number of particles chosen from a single interval is the product of the total sand volume desired and the ratio of the volume fraction of the interval to the average particle volume of the interval. The volume fraction of the interval is calculated from the mass differences between intervals in Table 4. The average particle volume of the interval is simply the midpoint of the limits of particle volume for the interval.

Table 1: Sand size distribution used in computer model.

Sieve No.	Sieve Size μm	% Passing (mass)	Sieve No.	Sieve Size μm	% Passing (mass)
#4	4750	100.0	#30	600	41.4
#8	2360	94.4	#50	300	19.0
#16	1180	64.6	#100	150	0.0

Hard cores, representing sand particles, were randomly parked in the system. The locations of the hard core centers were determined from a uniform random number generator and the largest particles were placed first. Equal thickness soft shells, or interfacial zones, were then placed around each sand particle, concentric to the hard core. Once all of the hard cores and soft shells were placed, the program would begin at the top surface of the cube and search for a continuous path of soft shells to the bottom surface using a burning algorithm. Any path that led to a side surface of the cube was continued on the opposite side; this path was assured due to the periodic boundary conditions.

Upon completion of the test for interfacial zone percolation, the program reported the volume fraction of soft shells, determined from a $100 \times 100 \times 100$ point count grid. As a check of the program, a test case was run where the hard core radius was zero, so the soft shells became monosized, perfectly penetrable spheres. The critical volume fraction of soft shells, or equivalently, interfacial zones, at percolation, 0.29, agreed with the published results of Balberg (1987) and of Balberg and Binenbaum (1987).

5 Model Results

Results of the continuum model analysis for 1 cm^3 mortar samples appear in Fig. 3. The plot displays the fraction of the total soft shell volume that is part of the percolation backbone, as a function of sand volume fraction, for a specified interfacial zone thickness. If all of the soft shells are connected together then the fraction connected is equal to one. Prior to percolation, the fraction connected is defined to be zero. At percolation, not all of the interfacial zones are interconnected, only those forming the percolation backbone. After percolation of the soft shell phase, there is a gradual increase of the fraction connected with increasing sand content until all of the interfacial zones become members of the backbone at high sand contents. The different curves in Fig. 3 represent the fraction connected for a single interfacial zone thicknesses. Each point on the graph represents the results of a single system.

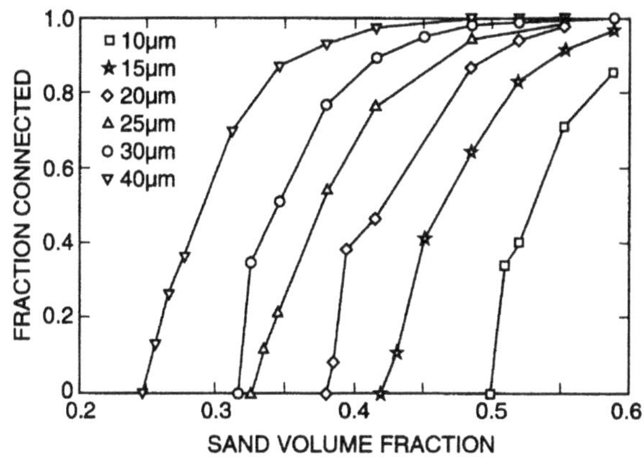

Figure 3: Results of continuum model used to determine interfacial zone percolation in mortars with varying sand contents and interfacial zone thicknesses (μm), using the sand gradations in Table 1.

The number of sand particles required for interfacial zone percolation ranged from 8000 to 15000, with smaller numbers required for the thicker interfacial zones. The number of hard cores at percolation of the interfacial zone phase was determined to the nearest 100 particles. This was sufficient to estimate the volume fraction of hard cores at percolation to the nearest percent volume.

An estimation of the interfacial zone thickness for the mercury intrusion data can be found by comparing the computer model results with the MIP results. The MIP results indicate that the intrusion volume increased very little from a sand content of 49% to a sand content of 55%. This implies that for 49% sand volume and greater, the interfacial zones are totally percolated, with a fraction connected in Fig. 3 nearly equal to one. This eliminates the 10μm curve. Also, the 45% sand content must either be unpercolated, or much less percolated, in terms of the value of fraction connected, than the 49% sand content data point. This eliminates the 40,30, and 25μm curves as well. Therefore, the apparent interfacial zone thickness that best explains the MIP data is 15–20μm.

These continuum model results may also be extrapolated to concrete. Consider a mortar that is percolated. If a portion of the mortar is replaced with a large aggregate, which is also surrounded by an interfacial zone, the mortar must remain percolated. Averaged over many systems, the addition of large aggregates will always reduce the sand fraction of the mortar volume required for percolation because the addition of large aggregates helps to percolate a mortar's interfacial zone phase that is not already percolated. Since the mortar fraction of concrete typically has a sand content of about 50%, one expects the interfacial zone phase in concrete to be percolated for interfacial zones as small as 10μm.

The computer model was applied to two concretes with different aggregate

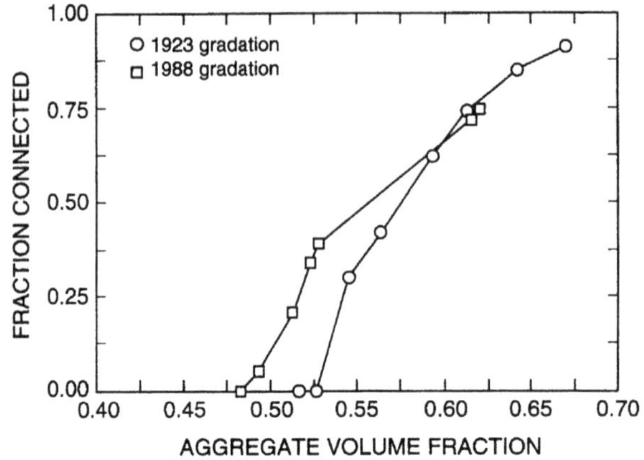

Figure 4: Percolation of two concrete mixtures containing different aggregate gradations.

gradations taken from Shilstone (1990), and are dated 1988 and 1923. Qualitatively, the 1923 aggregate gradation is a uniform blend of aggregate sizes while the 1988 gradation appears bi-modal.

The results from 27cm^3 model samples containing on the order of 500000 particles are shown in Fig. 4. Percolation occurs at aggregate fractions of 54% and 49% respectively, for the older and newer gradations, for an interfacial zone thickness of 20μm. The percolation data indicate that gradation can affect the aggregate fraction required for percolation.

The mortar phases of the concretes were analyzed using the concrete aggregate gradation for particles smaller than a #8 sieve. For the percolated concrete specimens, the sand fraction of the mortar phase at percolation was 38% and 30% for the old and new gradations, respectively. Using the same gradations, 1cm^3 mortar systems were created as described in the previous section. The mortars percolated at sand fractions of 42% and 37% for the old and new gradations, respectively. The small decrease in critical sand volume for the old gradation suggests that the addition of the coarser aggregate to the mortar of the 1923 gradation should have less affect upon transport properties than the newer gradation.

6 Discussion and Summary

The computer simulation model suggests that the mercury intrusion data are consistent with an interfacial zone thickness of 15–20 μm, which is smaller than the 25–50μm typically measured by SEM. The apparent discrepancy can be explained by examining the two different procedures. SEM results define the interfacial zone thickness by measuring capillary porosity as a function of distance from an aggregate surface, and then defining the edge of the zone as the point where the porosity has decreased from its high value at the aggregate surface down to its value in the bulk paste. It is almost certain that the biggest interfacial zone pores are contained in the highest porosity regions, closer to the aggregate. The interfacial zone pores that contribute the most to the threshold diameter seen in the mercury intrusion experiments would be closer to the aggregate, suggesting a thinner interfacial zone thickness than that measured by the SEM technique.

Percolated interfacial zones may have a subtle effect on the diffusivity and fluid permeability of concrete. For a homogenous matrix, assuming no interfacial zones, the addition of impenetrable inclusions decreases the diffusivity and permeability of the composite because the path for transport has become more restrictive. The presence of interfacial zones will increase the diffusivity and permeability by an amount that depends upon the relative diffusivities of the bulk and the interfacial zones. Recent work by Ping, Beaudoin and Brousseau (1991) on model cement-aggregate composites shows that the electrical conductivity is enhanced by the presence of the interfacial zone porosity in ordinary portland cement systems. Although the diffusivity of the composite may not increase with the addition of aggregates, the diffusivity is greater than a homogeneous matrix (no interfacial zones) containing solid, impermeable inclusions at the same volume fraction as the aggregates.

In the past, the requirements for suitable aggregate gradations were based upon considerations such as economy, workability, and finishability. This study indicates that composite transport properties are an additional consideration.

7 Acknowledgements

The authors would like to thank the National Science Foundation's Science and Technology Center for Advanced Cement-Based Materials (ACBM) for support of this work.

8 References

Balberg, I. (1987) Recent developments in continuum percolation, **Philo. Mag. 56** (6), 991–1003.

Balberg, I. and Binenbaum, N. (1987) Invariant properties of the percolation thresholds in the soft-core–hard-core transition, **Phys. Rev. A 35**, 5174–5177.

Bentz, D.P. and Garboczi, E.J. (1991) Simulation studies of the effects of mineral admixtures on the cement paste-aggregate interfacial zone, **Amer. Conc. Inst. Mater. J. 88**, 518–529.

Clifton, J.R. and Knab, L.I. (1989) Service Life of Concrete, NIST Internal Report, **NISTIR 89–4086**.

Cochran, W.G. (1963) **Sampling Techniques**. Wiley, New York.

Detwiler, R.J. and Mehta, P.K. (1989) Chemical and physical effects of silica fume on the mechanical and behavior of concrete, **Amer. Conc. Inst. Mater. J. 86**, 609.

Diamond, S., Mindess, S. and Lovell, J. (1982) On the spacing between aggregate grains in concrete and the dimension of the aureole de transition, **RILEM, Liaisons Pates De Ciment Materiaux Associes**, C42–C46.

DeSimone, T., Stratt, R.M., and Demoulini, S. (1986) Continuum percolation in an interacting system: Exact solution of the Percus-Yevick equation for connectivity in liquids, **Phys. Rev. Letts. 56** (11), 1140–1043.

Garboczi, E.J. and Bentz, D.P. (1991a) Digital simulation of the aggregate–cement paste interfacial zone in concrete, **J. Mater. Res. 6**, 196–201.

Garboczi, E.J. and Bentz, D.P. (1991b) Computer simulation of the diffusivity of cement-based materials, **J. Mater. Sci.** (in press).

Lee, S.B. and Torquato, S. (1988) Porosity for the penetrable–concentric–shell model of two–phase disorded media: Computer simulation results, **J. Chem. Phys. 89** (5), 3258–3263.

Ping, X., Beaudoin, J.J., and Brousseau, R. (1991) Flat aggregate–portland cement paste interfaces, I. Electrical conductivity models, **Cem. and Conc. Res. 21**, 515–522.

Rikvold, P.A. and Stell, G. (1985a) Porosity and specific surface for interpenetrable-sphere models of two-phase random media, **J. Chem. Phys. 82** (2), 1014–1020.

Rikvold, P.A. and Stell, G. (1985b) D–dimension interpenetrable-sphere models of random two–phase media: Microstructure and an application to chromatography, **J. of Coll. and Inter. Sci. 108** (1), 158–173.

Scrivener, K.L. and Gartner, E.M. (1988) Microstructural gradients in cement paste around aggregate particles, in **Bonding in Cementitious Composites** (eds S. Mindess and S.P. Shah), Materials Research Society, Pittsburgh, PA, 77–85.

Scrivener, K.L., Bentur, A. and Pratt, P.L. (1988) Quantitative characterization of the transition zone in high strength concretes, **Adv. in Cem. Res. 1** (4), 230–237.

Shilstone, J.M. Sr. (1990) Concrete mixture optimization, **Concrete International**, June 1990, 33–39.

Stauffer, D. (1985) **Introduction to Percolation Theory**. Taylor and Francis, London.

Torquato, S. (1986) Two–point distribution function for a dispersion of impenetrable spheres in a matrix, **J. Chem. Phys. 85** (10), 6248–6249.

van Brakel, J. (1981) A special issue devoted to mercury porosimetry, **Powder Technology 29**(1).

Winslow, D.N. and Liu, D. (1990) The pore structure of paste and concrete, **Cem. and Conc. Res. 20**, 227–235.

Zimbelmann, R. (1985) A contribution to the problem of cement-aggregate bond, **Cem. and Conc. Res. 15**, 801–808.

28 DIFFUSIVITE DES IONS CHLORE DANS LA ZONE DE TRANSITION ENTRE PATE DE CIMENT ET ROCHE GRANITIQUE

(Diffusivity of chloride ions in the transition zone between cement paste and granite)

D. BRETTON
CRIB, Sherbrooke University, Canada
J.-P OLLIVIER
LMDC, INSA-UPS, Toulouse, France
G. BALLIVY
CRIB, Sherbrooke University, Canada

Résumé *(English abstract, page 308)*
La perméabilité ou la diffusivité de la zone de transition roche-pâte de ciment sont importantes pour l'efficacité et la durabilité des injections de fissures. Pour déterminer si les propriétés de transfert de la zone de transition sont différentes de celles du reste de la pâte, nous avons utilisé deux types d'essais basés sur le flux de migration des ions chlore: par diffusion et par influence d'un champ électrique. Les résultats indiquent que la zone de transition possède une plus forte diffusivité et perméabilité par rapport à la pâte de ciment. De plus, il semble que ces techniques permettent d'observer les effets du retrait endogène de la pâte dans la fissure injectée.
Mots clés : Zone de transition, perméabilité, ions chlore, diffusion, champ électrique, pâte de ciment, roche.

1 Introduction

La mise en évidence d'une plus forte perméabilité ou d'une plus forte diffusivité de la zone de transition par rapport au reste de la pâte de ciment n'a pas encore été démontrée clairement. L'existence d'une telle zone peut être nuisible et même néfaste pour la durabilité des injections de fissures par des coulis à base de ciment soit dans des massifs rocheux ou de vieilles structures de béton.

À l'aide de la technique de l'imagerie par électrons rétrodiffusés (BEI) sur des échantillons polis et recouverts d'un film de carbone, Scrivener et coll. (1987, 1988) concluent que la zone de transition est plus poreuse que le reste de la pâte. De même, Tognon et Cangiano (1980), à partir d'études faites au porosimètre à mercure, arrivent aux mêmes conclusions. Ces travaux de recherche proposent que, si les pores de cette zone sont interconnectés, la zone de transition aurait une plus forte perméabilité par rapport au reste de la pâte de ciment.

Cependant, d'autres chercheurs mentionnent que cette zone n'est pas plus poreuse et ni plus perméable. En effet, à partir d'études faites au porosimètre à mercure, Kayyali (1987) conclut que la porosité de la zone de transition est plus faible. Finalement, Wakeley et Roy (1982) concluent, à partir d'essais de perméabilité à l'eau, que la zone de transition ne semble pas être plus perméable pour fournir un chemin préférentiel à l'écoulement de l'eau.

Ces divergences d'opinion peuvent provenir des méthodes d'analyses utilisées. Lorsque l'on fait des observations au microscope électronique à balayage (MEB), lors de la préparation des échantillons, le polissage peut engendrer l'arrachement des grains de ciment anhydres et fausser les analyses. De plus, les observations sont souvent effectuées sur quelques sections qui, en fait, ne représentent pas réellement l'ensemble de l'échantillon. Pour les analyses réalisées au porosimètre à mercure, les échantillons, préalablement séchés, développent souvent des microfissures qui n'appartiennent pas

Interfaces in Cementitious Composites. Edited by J. C. Maso. © RILEM.
Published by E & FN Spon, 2 - 6 Boundary Row, London SE1 8HN. ISBN 0 419 18230 6.

réellement à la pâte de ciment durcie, ce qui peut alors biaiser les résultats. Dans le cas des observations au perméamètre à l'eau, il est possible que l'apport du débit engendré par l'épaisseur de la zone de transition soit négligeable par rapport au débit total pour l'ensemble de l'échantillon.

2 MÉTHODES DE MESURE

Pour déterminer si les propriétés de transfert de la zone de transition sont différentes de celles du reste de la pâte de ciment, nous utilisons deux types d'essais basés sur le flux de migration des ions chlore traversant des échantillons de roche injectés dont, pour une même section, la quantité de pâte de ciment reste toujours la même; seuls le nombre et l'étendue des contacts roche-coulis changent (Fig.1). Si la zone de transition a une diffusivité plus importante, il en résultera une plus forte migration des ions chlore pour les échantillons ayant le plus grand nombre de contacts roche-pâte de ciment. De plus, afin de connaître l'apport exact du flux traversant la section de roche par rapport à celui de la pâte, des échantillons constitués essentiellement de pâte ou de roche ont été analysés (Fig.2).

Les deux types d'essais utilisés pour nos analyses sont la migration des ions chlore par diffusion et la migration des ions chlore par champ électrique.

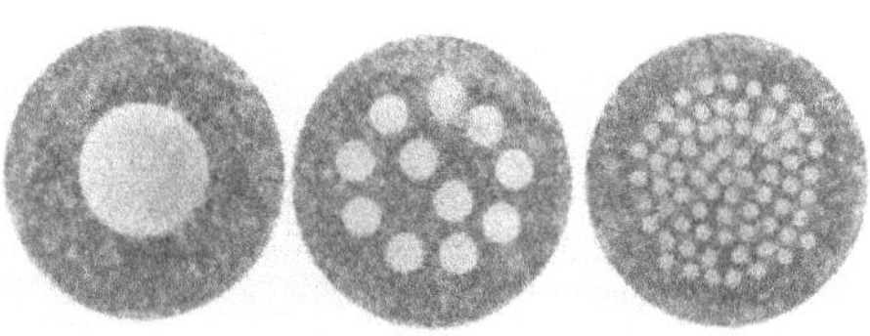

Fig. 1. Échantillons de roche injectée

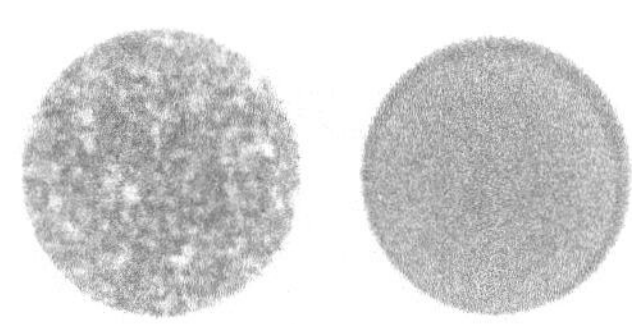

Fig. 2. Échantillons de roche et de pâte de ciment

2.1 Migration des ions chlore par diffusion

La migration des ions chlore par diffusion est un phénomène relié à un gradient de concentration. Ce phénomène est exprimé par la première loi de Fick :

$$F_x = -D\frac{dC}{dx} \quad (1)$$

où Fx : flux des ions Cl^- dans la direction x, exprimé en mole $cm^{-2}\ s^{-1}$ de Cl^-

$\frac{dC}{dx}$: gradient de concentration exprimé en mole cm^{-4}

D : coefficient de diffusion apparent exprimé en $cm^2\ s^{-1}$

Pour connaître le coefficient de diffusion apparent, nous utilisons des cellules à double réservoirs (de volumes V_1 et V_2), contenant des solutions de concentrations C_1 et

C_2 en ions Cl^-, séparés au centre par un échantillon de section (A) et de longueur (L) connues (Fig.3). Selon Page et coll. (1981), le flux F_x pénétrant dans le réservoir V_2, est donné par:

$$F_x = \frac{V_2}{A} \frac{dC_2}{dt} \qquad (2)$$

en incluant l'équation (1) dans (2), on obtient, en régime établi, avec la condition aux limites $C_1 >>> C_2$:

$$\Delta C_2 \cong DAC_1 \frac{\Delta t}{V_2 L} \qquad (3)$$

où ΔC_2 est l'accroissement de concentration en ion Cl^- dans le compartiment aval pendant l'intervalle de temps Δt.

Le coefficient de diffusion apparent (D) est la caractéristique des propriétés de transfert que nous utiliserons comme base de comparaison entre les échantillons.

2.2 Migration des ions chlore sous l'effet d'un champ électrique

Le deuxième type d'essais utilisé est basé sur la migration des ions Cl^- par l'action d'un champ électrique constant. La relation mathématique entre le flux de migration et le champ électrique est donnée par l'équation de Nernst-Planck (Bockris, Reddy; 1970) :

$$F_x = D\, C \frac{zFX}{RT} - D \frac{dC}{dx} \qquad (4)$$

où F_x : flux des ions Cl^- dans la direction x, exprimé en mole $cm^{-2}\, s^{-1}$ de Cl^-

$\frac{dC}{dx}$: gradient de concentration exprimé en mole cm^{-4}

D : coefficient de diffusion apparent exprimé en $cm^2\, s^{-1}$

C : concentration des ions Cl^-, au point x, exprimée en mole cm^{-3}

zF : charge portée par une mole d'ions exprimée en Coulomb $mole^{-1}$

X : champ électrique exprimé en volt cm^{-1}

R : constante des gaz parfaits exprimé en Joule K^{-1} $mole^{-1}$

T : température exprimée en degré K

Des cellules à deux réservoirs contenant des électrodes d'acier inoxydable ont été fabriquées (Fig.4)

La particularité de ce type d'essais est que, sous l'influence d'un champ électrique (X) constant, un courant électrique (i) est produit par la migration des ions à l'intérieur de l'échantillon. Puisque le flux de migration est en partie fonction de la diffusivité du milieu, nous caractérisons les échantillons à partir de ce courant électrique. De plus, afin de quantifier la contribution de la zone de contact par rapport au reste de la pâte, comme paramètre de comparaison entre les échantillons, nous utilisons la densité de courant (I) :

$$I = \frac{i}{S} \qquad (5)$$

I : densité de courant exprimée en A cm^{-2}

i : courant électrique exprimé en A

S : surface d'écoulement exprimée en cm^2.

C'est-à-dire que, plus la densité de courant est élevée, plus le milieu est perméable.

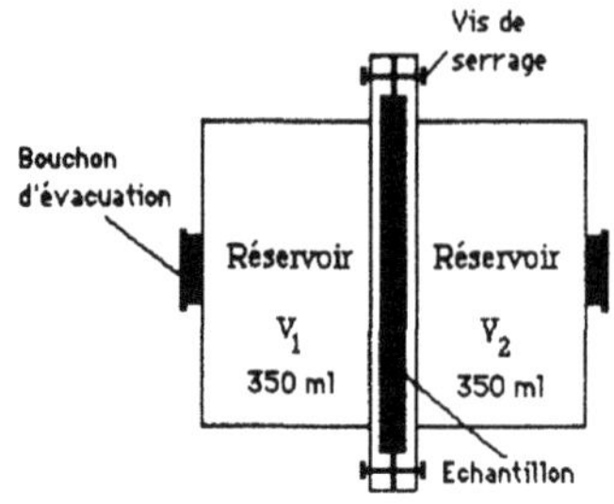

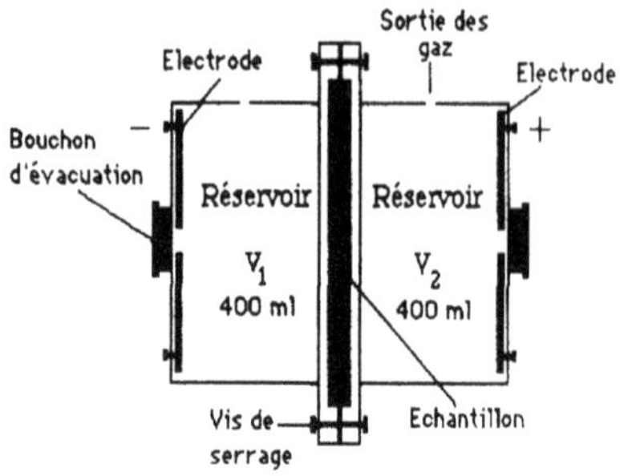

Fig. 3. Cellule de diffusion

Fig. 4. Cellule de champ électrique

3 Essais de migration des ions Cl^- par diffusion

3.1 Préparation des échantillons

Huit échantillons ont servi à cet essai :

- un disque de roche granitique;
- un disque de pâte de ciment;
- deux disques de roche préalablement percés d'un trou de 50 mm de diamètre puis injectés de pâte de ciment (T_1);
- deux disques de roche préalablement percés de 11 trous de 15 mm de diamètre puis injectés de pâte de ciment (T_{11});
- deux disques de roche préalablement percés de 69 trous de 6 mm de diamètre puis injectés de pâte de ciment (T_{69}).

Tous les échantillons étaient sous forme de disque circulaire de 110 mm de diamètre et d'environ 3,50 mm d'épaisseur. Dans les cellules de diffusion, la section (A) qui permet un écoulement des ions est de 90 mm. Le tableau 1 donne les caractéristiques physiques de chacun de ces échantillons.

TABLEAU 1 - CARACTÉRISTIQUES DES ÉCHANTILLONS POUR L'ESSAI DE DIFFUSION

Type d'échantillon	Nombre de trou	Diamètre des trous (mm)	Épaisseur (L) des échantillons (mm)	Surface de la roche disponible à l'écoulement (cm^2)	Surface de la pâte disponible à l'écoulement (cm^2)
roche (R)	--	--	3,5	63,61	--
pâte (P)	--	--	3,2	--	63,61
T1A	1	50	3,6	43,98	19,63
T1B	1	50	3,9	43,98	19,63
T11A	11	15	3,5	44,17	19,44
T11B	11	15	3,65	44,17	19,44
T69A	69	6	3.95	44,10	19,51
T69B	69	6	3,65	44,10	19,51

Le coulis de ciment utilisé est à base de ciment Portland de type 10 ordinaire (ASTM) ayant un rapport E/C de 0,50. L'analyse chimique du ciment est donnée dans le tableau 2.

TABLEAU 2 - COMPOSITION CHIMIQUE DU CIMENT PORTLAND DE TYPE 10 (%)

SiO_2	Al_2O_3	Fe_2O_3	CaO	MgO	SO_3	Na_2O	K_2O	Perte au feu	Insolubles	Chaux libre	Finesse Blaine
20,8	4,1	3,1	63,4	2,5	2,9	0,24	0,80	1,6	0,7	0,5	338,6

L'injection des échantillons s'est faite à l'aide d'un réservoir, fermé hermétiquement, préalablement rempli de coulis, dans lequel on retrouve une entrée (partie supérieure) permettant d'appliquer une pression d'air et une sortie (partie inférieure) permettant au coulis de s'écouler librement vers les moules d'injection (Fig.5). Les échantillons sont disposés à l'intérieur des moules de sorte que les trous dans la roche soient à la verticale pour favoriser un remplissage parfait de ceux-ci. La pénétration du coulis à l'intérieur des moules se fait par la base pour minimiser l'entraînement de bulles d'air et favoriser leur évacuation s'il y a lieu.

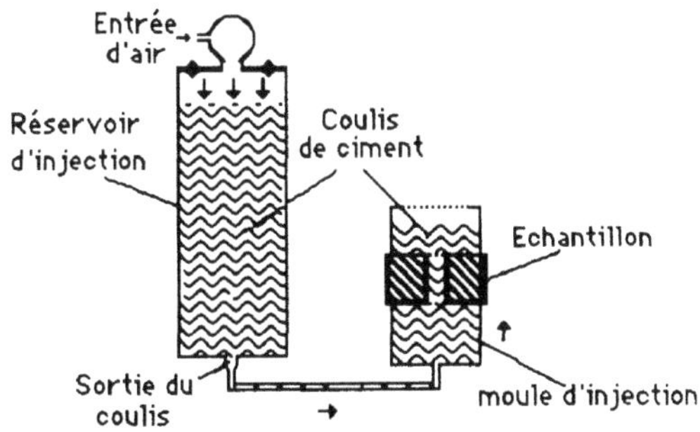

Fig. 5. Schéma d'injection

Après injection des échantillons, ceux-ci sont mûris 7 jours à 100% h.r. puis placés dans l'eau saturée en chaux pendant 3 jours. Par la suite, ils sont désaérés sous l'eau pendant 16 heures avant leur mise en place dans les cellules de diffusion. Au début de l'essai, les deux compartiment sont remplis avec une solution aqueuse de NaOH (1 g/L) et de KOH (20 g/L), solution à pH 13 utilisée pour éviter un lessivage trop prononcé des minéraux constituant la pâte de ciment. De plus, dans le compartiment amont (V_1) on introduit des chlorures (sous forme de NaCl) à une concentration de 20g de Cl^- par litre.

Durant l'essai, d'une durée d'environ 3 mois, le compartiment aval est vidangé à plusieurs reprises pour analyses, puis rechargé avec une nouvelle solution de même composition initiale de KOH et de NaOH. De plus, jamais la solution amont n'est changée lors des essais.

Toutes les mesures des concentrations en ions chlore des solutions sont faites par potentiométrie.

3.2 Analyses des résultats

Selon les résultats obtenus (Fig.6 à 10), on constate que la quantité d'ions Cl^- ayant migré, pendant environ 3 mois, à travers l'échantillon de roche est presque nulle comparativement à celles des autres échantillons.

TABLEAU 3 - COEFFICIENT DE DIFFUSION APPARENT ($10^{-9}cm^2/s$)

(R)	3,99
(P)	59,8
T1A	78,6
T1B	73,6
T11A	79,1
T11B	87,0
T69A	96,6
T69B	117,0

Le coefficient D étant calculé à partir des droites, les résultats présentés dans le tableau 3 montrent que le coefficient de diffusion apparent de la pâte de ciment est plus élevé que celui de la roche et qu'il augmente dans les autres échantillons avec l'étendue de la zone de transition.

Afin de quantifier la contribution de la zone de transition par rapport à la pâte de ciment, nous avons supposé que l'épaisseur de cette zone ayant une plus forte porosité est de 100 µm. Connaissant la surface que représente cette zone (Ai) et la surface totale qu'occupe la pâte de ciment (At), nous avons tracé un graphique représentant le coefficient de diffusion (D) en fonction du rapport (Ai/At), exprimé en % (Fig.11).

Selon ce graphique, on constate qu'il y a une bonne linéarité entre les points, sauf pour l'échantillon ayant un trou de 50 mm. À partir de l'équation obtenue de cette droite :

$$D = 6{,}81 \frac{Ai}{At} + 63$$

Pour une valeur de Ai/At égale à 100%, ce qui correspondrait à un échantillon dont la microstructure serait celle de la zone de transition , on aurait un coefficient de diffusion apparent de 744 cm^2/s, ce qui correspondrait à une valeur environ de 12 fois supérieure à celle de la pâte de ciment.

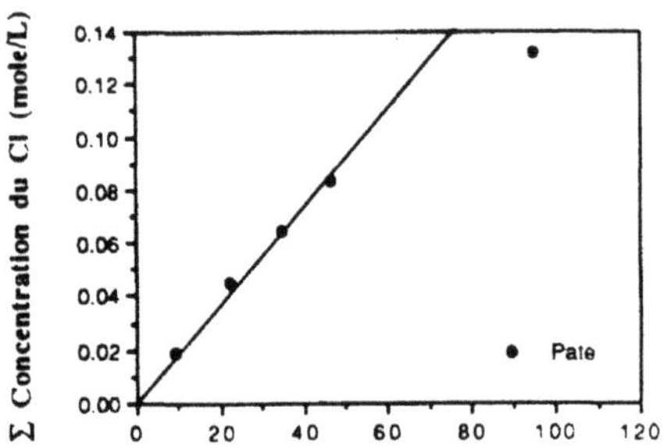

Fig. 6.Migration du Cl^- (pâte) en fonction du temps

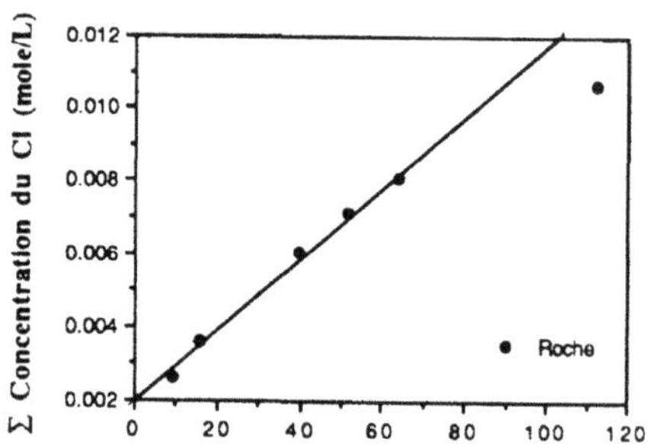

Fig. 7. Migration du Cl^- (roche) en fonction du temps

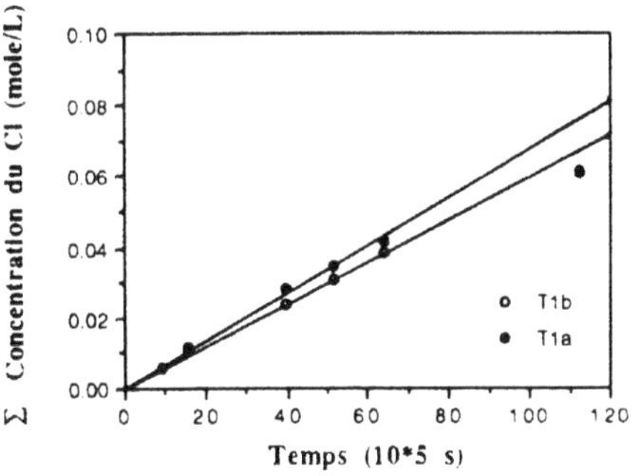

Fig. 8.Migration du Cl⁻ (T1) en fonction du temps

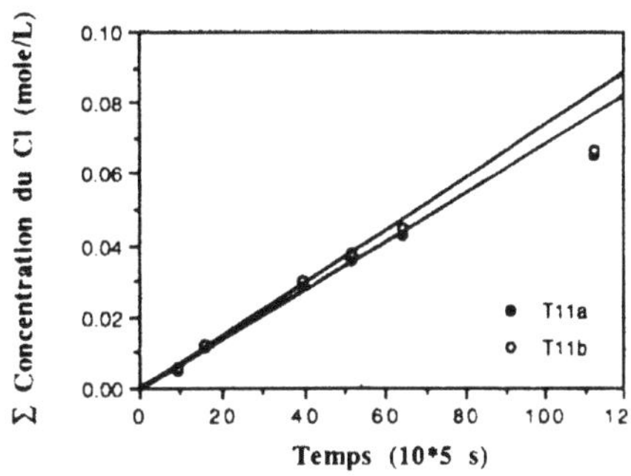

Fig. 9. Migration du Cl⁻ (T11) en fonction du temps

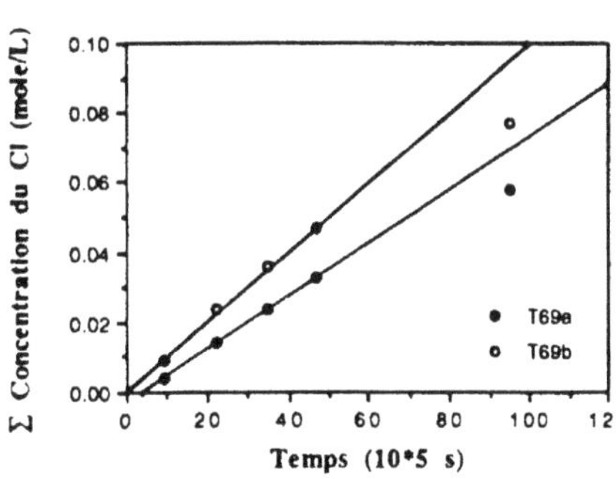

Fig. 10. Migration du Cl⁻ (T69) en fonction du temps

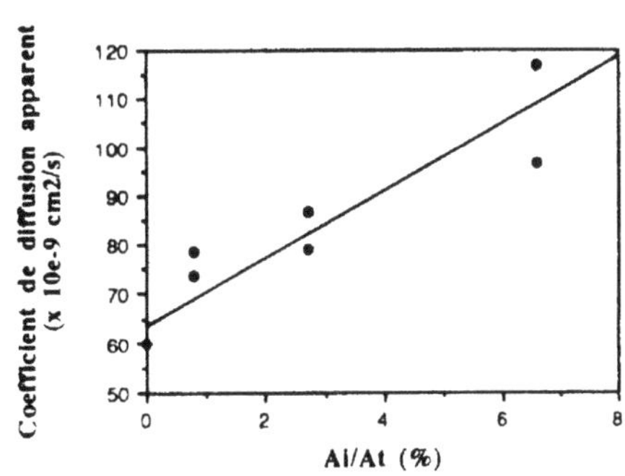

Fig. 11.Coefficient de diffusion apparent en fonction de (Ai/At)

4 Migration des ions Cl⁻ sous l'influence d'un champ électrique

4.1 Préparation des échantillons

Cinq échantillons ont été préparés pour cet essai :

- un disque de roche granitique;
- un disque de pâte de ciment;
- un disque de roche préalablement percé d'un trou de 50,8 mm (2 pouces) de diamètre puis injecté de pâte de ciment (T_1);
- un disque de roche préalablement percé de 16 trous de 12,7 mm (1/2 pouce) de diamètre puis injectés de pâte de ciment (T_{16});
- un disque de roche préalablement percé de 64 trous de 6,35 mm (1/4 pouce) de diamètre puis injectés de pâte de ciment (T_{64}).

Tous les échantillons sont des disques circulaires de 95 mm de diamètre et de 10 mm d'épaisseur. Dans la cellule, la section (A) qui permet un écoulement des ions est de 85 mm de diamètre. Le tableau 4 donne un résumé des caractéristiques des échantillons.

TABLEAU 4 - CARACTÉRISTIQUES DES ÉCHANTILLONS POUR L'ESSAI SOUS CHAMP ÉLECTRIQUE

Type d'échantillon	Nombre de trou	Diamètre des trous (mm)	Épaisseur (L) des échantillons (mm)	Surface de la roche disponible à l'écoulement (cm^2)	Surface de la pâte disponible à l'écoulement (cm^2)
roche (R)	--	--	10	56,74	--
pâte (P)	--	--	10	--	56,74
T1	1	50,8	10	36,47	20,27
T16	16	12,7	10	36,47	20,27
T64	64	6,35	10	36,47	20,27

Les caractéristiques pour le type de coulis de ciment et pour le type de roche utilisés pour cet essai sont les mêmes que celles employées pour l'essai par diffusion, de même que les procédures d'injection et de mûrissement. Au début de l'essai, le compartiment amont V_1 (Fig.4), situé du côté de la cathode, est rempli avec une solution chargée en chlorure (sous forme de NaCl), à une concentration de 18,20 g/L de Cl^-. Le deuxième compartiment V_2, situé du côté de l'anode, est rempli avec une solution aqueuse de NaOH (30 mL/L). Tout au long de l'essai, la différence de potentiel appliqué aux bornes des électrodes est de 5 volts, ce qui, selon la configuration du montage, représente un champ électrique de 0,237 volt/cm. Le choix d'un faible champ électrique a pour but de diminuer les possibilités d'augmentation de température qui fausserait les résultats. De plus, afin de maintenir le plus constant possible le gradient de concentration dC/dx à l'intérieur des cellules, on vidange les deux réservoirs durant l'essai.

4.2 Analyse des résultats

Selon les résultats obtenus (Fig.12 à 14), on constate que le courant électrique circulant à l'intérieur des échantillons contenant de la pâte de ciment diminue avec le temps, contrairement à l'échantillon de roche. Nous pensons que ce phénomène peut provenir d'une désaturation ionique des eaux interstitielles contenues dans les pores de la pâte de ciment. Dans le cas de la roche, on observe plutôt l'effet inverse, c'est-à-dire une saturation ionique des pores, provoquée par la migration des ions à l'intérieur de l'échantillon. La présence de pics sur les courbes correspond aux périodes de vidange des échantillons. À mesure que l'essai avance, il semble que ceux-ci soient moins apparents.

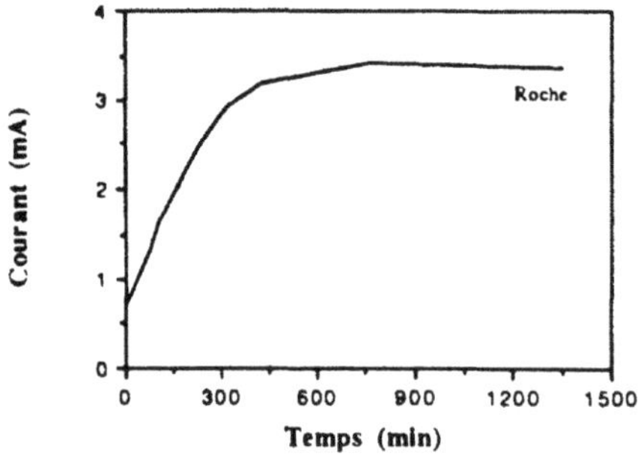

Fig. 12.Courant électrique en fonction du temps (Roche)

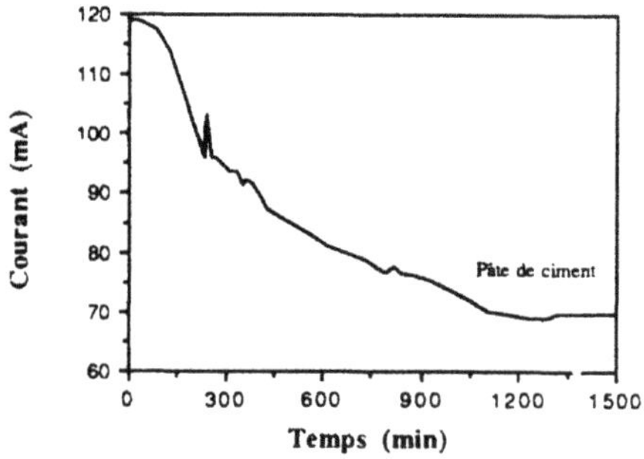

Fig. 13.Courant électrique en fonction du temps (Pâte)

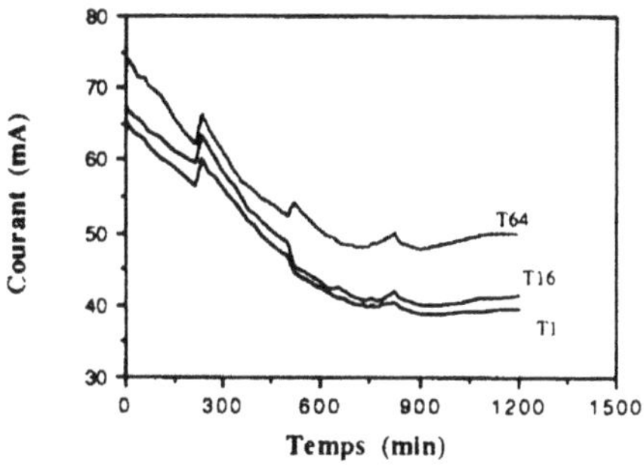

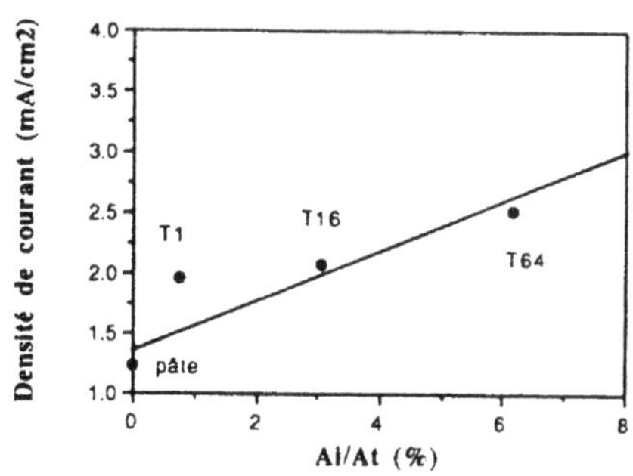

Fig. 14.Courant électrique en fonction du temps (Roche injectée)

Fig. 15.Densité de courant en fontion de (Ai/At)

Pour ce qui est du courant, on constate que l'échantillon de roche laisse passer peu de courant alors que, pour les échantillons injectés de coulis de ciment, on constate que, plus le contacts roche-coulis est étendu, plus le courant électrique est élevé.

Afin de quantifier l'importance de la zone de transition par rapport au reste de la pâte, nous avons tracé un graphique de la densité de courant (I) en fonction du rapport (Ai/At), exprimé en %. Étant donné le faible courant circulant à l'intérieur de la roche, nous négligeons sa valeur dans les calculs. Selon la figure 15, on constate qu'il y a une bonne linéarité entre les points, sauf pour l'échantillon ayant un trou de 50 mm. À partir de l'équation obtenue de cette droite :

$$I = 0{,}187 \frac{Ai}{At} + 1{,}35$$

en posant Ai/At égal à 100%, on obtient une densité de courant de 20,05 A/cm^2, ce qui correspond à une valeur environ de 15 fois supérieure à celle de la pâte de ciment pour la zone de transition.

5 Discussion des résultats

L'utilisation de la migration des ions chlore, soit par diffusion ou par champ électrique, présente des résultats intéressants. Les deux techniques d'analyses utilisées nous montrent sensiblement les mêmes résultats. On constate que le coefficient de diffusion apparent (D) et la densité de courant (I) sont plus élevés pour la pâte de ciment que pour la roche et qu'ils augmentent dans les échantillons composites roche-pâte de ciment avec l'étendue de la zone de transition.

Après avoir estimé l'épaisseur de la zone de transition ayant une plus forte porosité à 100 µm, lorsque l'on trace D ou I en fonction du rapport aire de l'interface (Ai) sur aire totale (At), exprimé en %, on obtient une bonne linéarité entre les points. Seuls les points représentant les échantillons ayant un trou d'environ 50 mm sont nettement supérieurs aux droites tracées sur les graphiques. Ce comportement peut provenir du retrait endogène de la pâte de ciment dont les effets seraient plus prononcés pour la dimension de ce trou par rapport aux autres. Il est possible qu'il y ait eu de la décohésion entre la pâte de ciment et la roche, sur une très faible épaisseur, ce qui favorise une migration plus facile des ions le long de ce contact.

Étant donné que, pour les autres échantillons ayant des trous de dimensions inférieures, on retrouve une bonne linéarité, on peut prétendre que seule la zone de transition provoque une hausse de diffusivité. À partir des équations de ces droites, en posant que Ai égale At, les résultats indiquent que la zone de transition aurait une diffusivité environ de 12 à 15 fois supérieure à celle de la pâte. Cependant, dans le cas où l'épaisseur de la zone de transition serait de 10 µm, plutôt que de 100 µm, on aurait une diffusivité environ de 120 à 150 fois supérieure à celle de la pâte, ce qui ne serait pas négligeable.

Dans un milieu poreux à température constante, le coefficient de diffusion apparent du Cl^- est fonction de l'intéraction de ces ions avec le milieu et de la nature des pores de ce milieu. Même s'il n'existe pas de relation bien établie entre la perméabilité et la diffusivité, il n'en demeure pas moins que pour un même matériau, plus la diffusivité est élevée plus la perméabilité est élevée, étant donné que seule la nature des pores en est responsable. Cela indique que la zone de transition serait un chemin préférentiel à l'écoulement et qu'elle pourait devenir un lieu favorable à la détérioration de la fissure injectée.

6 Conclusion

Les techniques d'analyse utilisant la migration des ions chlore par diffusion ou par champ électrique, sur des échantillons de roche injectée dont, pour une même section, la quantité de pâte reste toujours la même et où seuls le nombre et l'étendue des contacts roche-coulis changent, permettent de caractériser la diffusivité de la zone de transition par rapport au reste de la pâte. De plus, il semble que ces techniques permettent d'observer les effets du retrait endogène de la pâte de ciment à l'intérieur d'une fissure injectée.

Les résultats obtenus indiquent, pour un coulis de ciment de type 10 (E/C = 0,50) injecté dans une roche granitique, que la zone de transition, par rapport au reste de la pâte de ciment, possède une diffusivité plus grande. Dans le cas où celle-ci aurait une épaisseur de 100 µm, la diffusivité de cette zone serait environ de 12 à 15 fois supérieure à celle de la pâte de ciment.

De plus, étant donné que les résultats sont obtenus à partir de la même pâte de ciment, la zone de transition possède également une plus forte perméabilité que le reste de la pâte de ciment.

7 Références

Bockris, J. O'M, I., Reddy, A.K.N. (1970) Modern electrochemistry, **Plenum Press**, New York, 1, 622 p.

Kayyali, O.A. (1987) Porosity of concrete in relation to the nature of the paste-aggregate interface, **Matériaux et Constructions**, RILEM, 20, p. 19-26.

Page, C.L. et coll. (1981) Diffusion of chloride ions in hardened cement paste, **Cement and Concrete Research**, 11, p. 395-406.

Scrivener, K.L. et coll. (1987) Microstructural gradients in cement paste around aggregate particles. **Symp. on bounding in cementitious composites**, MRS, Boston, p. 77-85.

Scrivener, K.L. et coll. (octobre 1988) Quantitative characterization of the transition zone in high strength concretes. **Advances in Cement Research**, 1, n° 4, p. 230-237

Tognon, G.P., Cangiano, S. (1980) Interface phenomena and durability of concrete. **7th Int. Congr. on the Chemistry of Cement**, Paris, 3, p. VII-133-VII-137.

Wakeley, L.D., Roy, D.M. (1982) A méthod for testing the permeability between grout and rock, **Cement and Concrete Research**, 12, p. 533-534.

29 INFLUENCE OF AGGREGATE CONCENTRATION ON THE DIFFUSION OF CO_2 AND O_2

Y. F. HOUST
Federal Institute of Technology (EPFL), Lausanne, Switzerland
H. SADOUKI and F. H. WITTMANN
Federal Institute of Technology (ETHZ), Zurich, Switzerland

Abstract
Gas diffusion process through cementitious materials must be influenced, first of all, by the aggregate concentration, and, secondly, by the interfacial zones between aggregates and bulk cement paste. Indeed, an increase in the aggregate concentration reduces, on the one hand, the porous volume through which gases diffuse, and, on the other hand, augments the volume of the interfacial zones which is more porous than the bulk.

The microstructure of carbonated mortars with increasing sand concentrations has been studied by mercury intrusion porosimetry and water adsorption. The diffusion coefficients of CO_2 and O_2 through the same mortars have been measured at three different relative humidities. In addition, water vapor permeability has been measured through the same mortars at four humidity gradients.

The diffusion coefficients of CO_2 and O_2 are little influenced by water content in the range 55%-86% R.H.. It is also shown that at low sand concentration the diffusion coefficients decrease with the sand content. Above 50% sand by volume, a steep increase is observed partly due to a high volume of interfacial zones which become interconnected.
Keywords: Interfacial zones, CO_2 diffusion, O_2 diffusion, Carbonation, Microstructure, Mortar.

1 Introduction

It is well known that the porosity is of the greatest importance for the durability of cementitious materials. Aggressive substances are transported into the material in the liquid or in the gaseous state through the complex porous system. The carbonation is related to the diffusion of atmospheric carbon dioxide through cementitious materials. Its unfavorable consequences on the durability of reinforced concrete have been already largely discussed. According to Tuuti (1982), the service life of a structure, from the point of view of reinforcement corrosion, can be subdivided into an initiation stage and a propagation stage. The carbonation of the concrete cover takes place during the initiation stage and the steel corrosion during the propagation stage. The corrosion rate, which is practically zero during the initiation stage, is generally limited by the rate of diffusion of oxygen, specially in water saturated concrete. From the surface of a carbonating concrete, one can observe, after a certain time, a fully carbonated concrete part, a reaction front, where carbonation is tacking place, and the bulk of non-carbonated concrete. The carbonation proceeds by the displacement of the reaction front. In this simple model, carbon dioxide diffuses through carbonated material. During the propagation stage, oxygen diffuses also through carbonated material. That is why diffusion coefficients of carbon dioxide and oxygen have to be determined through carbonated material. Water, which is produced by the carbonation reaction, diffuses from the reaction front to the surface, where the moisture is lower. The rates of carbon dioxide

Interfaces in Cementitious Composites. Edited by J. C. Maso. © RILEM.
Published by E & FN Spon, 2 - 6 Boundary Row, London SE1 8HN. ISBN 0 419 18230 6.

and oxygen diffusion are dependent on numerous factors like porosity, pore-size distribution, water, cement and aggregate content, curing, time, as well as on the environment: temperature, CO_2 and O_2 concentrations, relative humidity, etc.

The bulk of the experimental data available cannot be understood and interpreted fully because so far there is no comprehensive general theoretical concept of the mechanisms involved in carbonation and corrosion of reinforced concrete. Houst, Roelfstra and Wittmann (1983) have developed a numerical model to describe the diffusion of carbon dioxide and oxygen through the concrete structure and the drying process. All parameters of the model can be linked with technological data. However, some of these parameters were unknown. This was true in the case of the influence of moisture content on gas diffusion through porous materials. In a first step, Houst (1991) has determined the influence of the microstructure and moisture content on the diffusion coefficients of CO_2 and O_2 through hydrated and carbonated cement paste. For reasons of time to get an equilibrium with moisture and also of time of measurement of gas diffusion, it was not possible to dermine these parameters on representative samples of concrete. That is why we have determined the influence of fine aggregates on the microstructure and on diffusion coefficients. A composite material model is under development to describe the results.

2 Gas diffusion, microstructure and modelling

2.1 Gas diffusion through porous system

The one-dimensional diffusion of a gas can be described phenomenologically by Fick's first law:

$$J = -D\frac{dc}{dx} \tag{1}$$

where J is the flux of the gas, dc/dx the gradient of concentration and D the diffusion coefficient.

Equation (1) defines the diffusion coefficient of two gases diffusing into one another without hindrance. If the diffusion is hindered by a porous medium, the gases must diffuse through the pores of the material. The distance to cross the material is longer than that in unhindered diffusion. The tortuosity is defined as the distance to cross the material divided by its thickness. Thus only an effective diffusion coefficient can be determined for a specimen having the representative thickness :

$$J = -D_{eff}\frac{\Delta c}{d} \tag{2}$$

where d stands for the thickness of the representative porous material. The representative thickness is defined as the minimum thickness for which D_{eff} becomes independent of d.

In reality, diffusion processes in capillaries are complex and it is possible to distinguish at least three different transport mechanisms. The Knudsen diffusion begins to take place as soon as the pores dimensions get smaller than the mean free path of the gas molecules. In large pores, where the mean free path of the gas molecules is smaller than the pore diameter, normal gas diffusion predominates. In addition to the two mentioned mechanisms, surface diffusion also can occur at the same time. These mechanisms have been described in more detail by Houst (1991) who gives also references.

2.2 The cement-paste/aggregate interface

At least for transport mechanisms, concrete cannot be considered to be a simple composite material consisting of aggregates and hydrated cement paste (hcp). Farran (1956) has demonstrated in some cases the presence of an interfacial zone up to 40 μm wide between aggregates and cement paste. The existence of this interfacial zone is partly due to the bleeding and depends on the reactivity of the aggregate. It was currently assumed that interfacial zones represent only a small volume in concrete until a study by Diamond et al (1982), who observed that the mean spacing between aggregates in concrete is only 75 to 100 μm. With an interfacial zone of about 40 μm, the most of the paste is included in interfacial zones. Therefore, only a small volume of the bulk hcp can be present in concrete. The concentrations of compounds such as calcium hydroxide and ettringite are larger near the aggregate. The existence of microcracks in the interfacial zone due to drying shrinkage is also observed.

The image analysis of backscattered electron measurements has allowed Scrivener and Pratt (1987), Scrivener and Gartner (1988), and Scrivener et al (1988) to determine the variation of porosity with distance from a piece of aggregate. They observed a dramatic encrease of porosity as the interface is being approached. The porosity, which is very high (30%) at 2 μm from the aggregate, drops to a relatively constant level of about 10% after 30 μm. The amount of anhydrous cement grains is very low at the interface and increases almost linearly up to a distance of about 30 μm from the interface. According to Scrivener et al (1988) the particle size distribution of the cement determines the width of the interfacial zone.

The high porosity of the interfacial zone must have a great influence on transport mechanisms such as gas diffusion. An increase of the aggregate content tends to bring interfacial volumes closer together and they become interconnected. This creates continuous channels of high porosity.

It is well known that carbonation reduces the porosity, the specific surface area of hcp and concrete, and, therefore, also gas diffusivity. Houst (1991) has found that the lower the porosity of non-carbonated hcp, the higher the decrease of porosity after carbonation. For example, non-carbonated hcps have porosities of 24.1%, 32.3% and 48%; after carbonation the porosities drop to 13.0%, 23.5% and 42.3%. In the same paper, it is shown that diffusion coefficients of O_2 and CO_2 through carbonated hcp are much less influenced by relative humidity or water content than expected and found for non-carbonated cementitious materials. It seemed us very important to study the influence of the carbonatation on materials with interfacial zones, like mortars, with the aim of determining microstructural changes due to carbonation on carbonated mortars and gas diffusion coefficients.

3 Experimental

3.1 Materials

For the determination of diffusion coefficient in our experimental set-up, we needed samples thin enough to be completely carbonated and equilibrated at different relative humidities in a relatively short time. For this purpose, we prepared Portland cement mortars with a constant water-cement ratio (W/C = 0.6) and three cement-sand ratios (C/S = 0.25, 0.43, 0.67). The cement used was a rapid hardening Swiss Portland cement and the sand was composed fully of quartz grains. The particle-size distributions of the cement and the sand, determined by means of a laser granulometer (Malvern Mastersizer 1002), is given in figure 1. This fine sand was chosen in order to have a large volume of interfacial zones and at the same time to avoid having to cut a large number of aggregates and so modify considerably the properties of interfaces.

After 28 days of curing in a tight cylindrical mold of 160 mm diameter and 320 mm high, discs of mortar of 3 to 5 mm thick were sawn. Then, the specimens have been equilibrated at 76% R.H.. A part of them have been artificially carbonated at 76 % R.H. in an atmosphere of 80% to 90% CO_2. After a complete carbonation checked by the phenolphtalein test, samples were equilibrated at different relative humidities before CO_2 and O_2 diffusion and H_2O permeability measurements. For mercury intrusion porosimetry and water adsorption, we used pieces of the same discs as those used for diffusion measurements. For the determination of total porosity by water absorption, we used discs of about 30 mm thick.

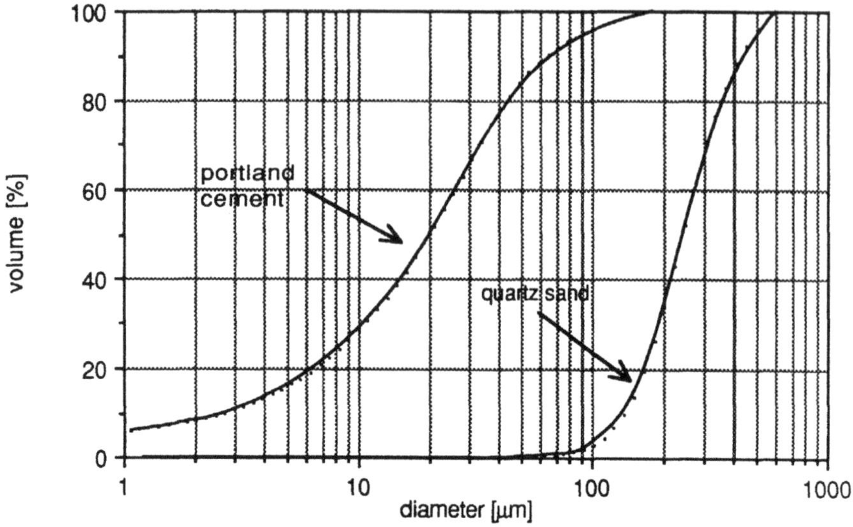

Fig. 1. Particle size distribution of quartz sand and portland cement

3.2 Mercury intrusion porosimetry (MIP)
A standard porosimeter (Micromeritics Autopore 9200) which allows the application of pressures up to 415 MPa has been used. This corresponds to a minimum radius of 1.7 nm. After having introduced a number of simplifying assumptions, the radius r of pores, which can be penetrated by mercury at pressure P, can be related by equation (3):

$$r = -\frac{2\,\sigma \cos\theta}{P} \qquad (3)$$

where σ stands for the surface tension of mercury and θ for the mortar-mercury contact angle. A value of 135° for θ was used in this study. The pore-size measurements have been carried out on small pieces of mortar (3 to 5 mm) pre-dried at 70°C.

3.3 Water adsorption isotherm
The samples (pieces of 3 to 5 mm), dried at 105°C, were first placed in desiccators where different relative humidities are obtained by saturated salt solutions. After the equilibrium was reached, the water uptake was measured by weighing. The water content as a function of the relative humidity allowed us to draw the adsorption isotherms. All adsorption experiments were carried out at 18°C on non-carbonated and carbonated mortar. The results of carbonated samples are not available at the time of writing this paper.

3.4 Water permeability
The wet cup method has been used to determine the water vapor permeability at 18 °C. Four humidity gradients (97%-86% R.H., 86%-76% R.H., 76%-55% R.H. and 55%-33% R.H.) have been obtained by saturated salt solution and have served for the measurements on two discs of each mortar: non-carbonated and carbonated.

3.5 Diffusion of O_2 and CO_2
We have developed equipment which allows us the simultaneous determination of diffusion coefficients of O_2 and CO_2 as a function of the relative humidity, by a non-

steady state method. The heart of the apparatus is a measuring cell which is divided in two rooms by the porous disc of the mortar. At the begining of the experience, the pre-conditioned gas mixture (2% CO_2, 20% O_2 and 78% N_2) enters one of the room whilst the other one contains pure nitogen. Then, the gas mixture diffuses through the mortar into the room filled with N_2 where O_2 and CO_2 are monitored by gas analyzers during one or one and a half day. The diffusion coefficients are deduced from gas concentration-time curves for O_2 concentrations of 2% to 12% and for CO_2 concentrations of 0.2% to 1%. Before and during the measurements, all gases are moistured with water vapor. All measurements were carried out between 20 °C and 25 °C on fully carbonated mortar discs.

The method of determination of the diffusion coefficients is described in more detail in a publication by Houst and Wittmann (1986).

4. Results and discussion

4.1 Microstructure

The pore-size distribution of the non-carbonated and carbonated mortars are shown in figure 2. Each curve of this figure is the average of 3 to 4 runs. The porosity of pores with radii < 300 µm, measured by MIP, and the total porosity are given in table 1. The total porosity of non-carbonated mortars varies between 28.6% and 23%, whilst porosity of pores of radii < 300 µm is quite identical. The total porosity is hardly lowered by the carbonation for mortars of C/S = 0.25 and 0.43. But, that of the mortar

Table 1. Main microstructural properties of the non-carbonated (nc) and carbonated (c) mortars

C/S	Total porosity[1] [% vol./vol.]		Porosity[2] [% vol./vol.]		BET sp. surf. area[3] [m²/g]		Sp. surf. micro- and meso-pores[4] [m²/g]	
	nc	c	nc	c	nc	c	nc	c
0.25	25.4	24.6	15.5	13.4	26.6	14.7	21.2	11.0
0.43	23.6	23.1	14.6	14.9	42.0	19.7	35.2	14.8
0.67	28.6	18.4	15.3	14.9	53.6	23.4	48.5	20.3

[1] Measured by water absorption under vacuum

[2] Measured by mercury intrusion porosimetry

[3] Calculated from water adsorption isotherm data beetwen 9% and 44 % R.H.

[4] Calculated from water adsorption isotherm data beetwen 11.2% and 96.4 % R.H.

of C/S = 0.67 is largely reduced. The pore-size distribution is modified: the volume of pores with radii > 0.03 µm is increased for the two mortars with higher cement contents and is decreased for pores with lower radius. The same modification is observed for the mortar of lower cement content, but the limit is about 1 µm. The pore-size distribution of macro- and micro-pores, computed from adsorption isotherms according to Pierce (1953), is given in figure 3. These results confirm those obtained by MIP. The BET specific surface areas have been also computed from the adsorption isotherms (table 1). The carbonation causes a drastic decrease of specific surface area which is at least divided by 2 . This is not surprising as finer pores have higher specific surfaces. The increase in volume of pores larger than 300 µm can be explained by their being filled with carbonation products and therefore not taken into account by MIP. According to these results as the the carbonation causes only a little drop of total porosity for mortars of C/S = 0.25 and 0.43 and as specially the volume of finer pores is decreased, the transport properties of these carbonated mortars should not be considerably modified. But, it should not be the case for the morttar of C/S = 0.67, which has a significant drop of porosity.

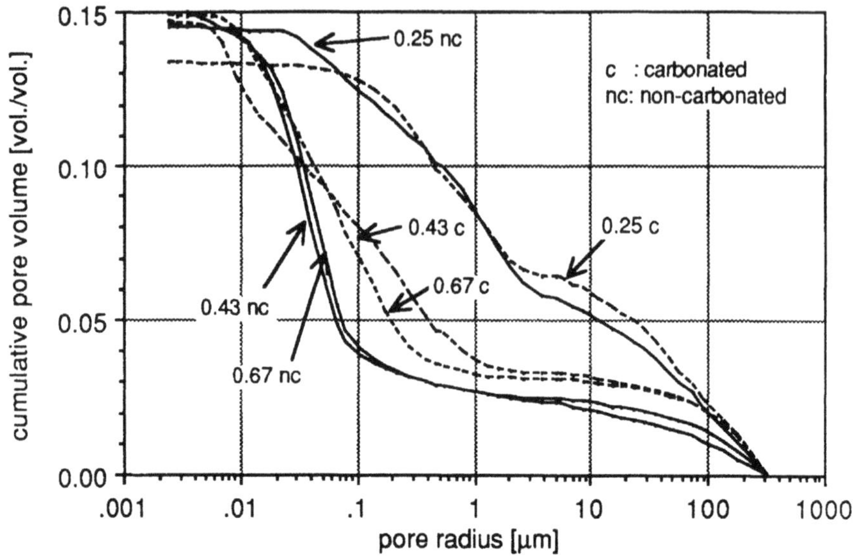

Fig. 2. Pore size distribution of the mortars measured by mercury intrusion porosimetry

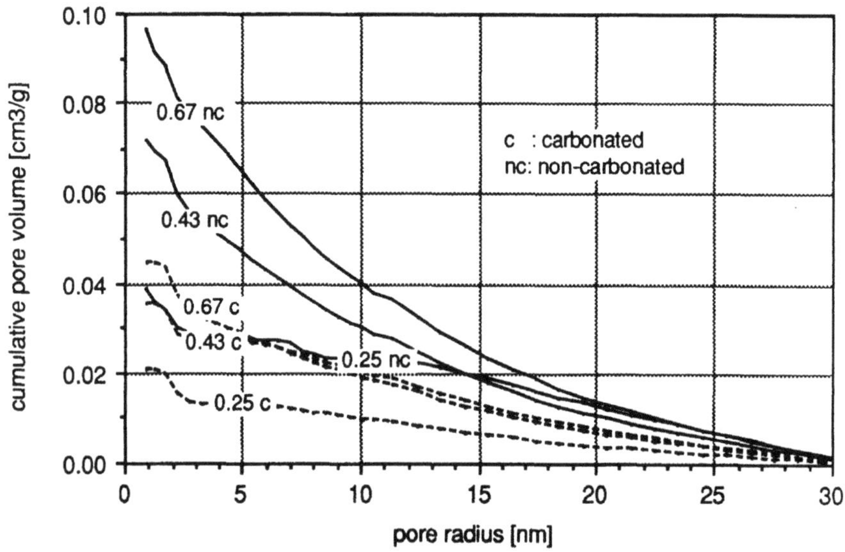

Fig. 3. Meso- and micro-pore size distribution calculated from adsorption isotherms data of the mortars (0.25, 0.43, 0.67 are C/S)

The difference between total porosity, measured by water absorption under vacuum, and porosity, measured by MIP is about 10% to 13%. This difference is essentially due to the the presence of air pores, entrained by the fine aggregate. The volume of the interfacial zones bas been computed on the basis of the particle size distribution of the sand and by assuming spherical particles, for interfacial zones from 30 μm to 10 μm thick. The corresponding volumes are 63.6% (C/S=0.25) and 11.9% (C/S=0.67). By assuming an interfacial zone of 20 μm thick, the interfacial volume and the aggregate volume are respectively 93.25, 82.2%, 65,1% for C/S=0.25, 0.43 and 0.67. With this simple model, it is shown that the interfacial zones represent the main part of cement paste. Therefore, the high air-pores content of the mortars must reduce the effect of transition volumes in gas transport.

4.2 Water permeability

The results of water permeability measurements on non-carbonated mortars are given in table 2. The one-dimentional permeation of water through a porous system can be described by equation (4):

$$J = -\lambda \frac{dH}{dx} \tag{4}$$

Where J stands for the flux of the gas and $\frac{dH}{dx}$ the gradient of moisture and λ the coefficient of permeability. If the water content in equilibrium at a given relative humidity w is used instead of the relative humidity H, the flux of moisture can be writen as follow:

$$J = -D \frac{dw}{dx} \tag{5}$$

Where D is the diffusion coefficient. By means of equations (4) and (5), one can write equation (6):

$$D = \lambda \frac{dH}{dw} \tag{6}$$

$\frac{dH}{dw}$ being the inverse of the adsorption isotherm derivative.

Table 2. H_2O permeability and diffusion coefficients of the non-carbonated mortars
C_S: sand concentration by volume

Δ R.H. [%]	C_S = 55.1%		C_S = 48.6%		C_S = 38.5%	
	λ [kg/msPa]	D_{H2O} [m²/s]	λ [kg/msPa]	D_{H2O} [m²/s]	λ [kg/msPa]	D_{H2O} [m²/s]
97-86	$1.7\cdot10^{-12}$	$7.5\cdot10^{-12}$	$2.0\cdot10^{-12}$	$5.3\cdot10^{-12}$	$2.0\cdot10^{-12}$	$4.4\cdot10^{-12}$
86-76	$1.7\cdot10^{-12}$	$1.8\cdot10^{-11}$	$1.1\cdot10^{-12}$	$6.9\cdot10^{-12}$	$9.7\cdot10^{-13}$	$4.3\cdot10^{-12}$
76-55	$1.2\cdot10^{-12}$	$3.6\cdot10^{-11}$	$7.8\cdot10^{-13}$	$1.3\cdot10^{-11}$	$6.8\cdot10^{-13}$	$8.4\cdot10^{-12}$
55-33	$2.0\cdot10^{-12}$	$7.0\cdot10^{-11}$	$9.3\cdot10^{-13}$	$2.4\cdot10^{-11}$	$6.2\cdot10^{-13}$	$1.6\cdot10^{-11}$

The diffusion coefficient of water as a function of the relative humidity, computed according to equation (6), is also given in table 2. These diffusion coefficients decrease as the relative humidity increases; this behavior being more pronounced for mortars with higher total porosity and equilibrium water content. This is because adsorbed and condensed water hinders the transport of vapor. The diffusion coefficient decreases as the aggregate content increases. This is due to more bigger pores associated with mortars having not a sufficient cement content to fill all voids between aggregates. As the adsorption isotherm of carbonated mortar are lower than those of non-carbonated mortars and as the total porosity is little lowered by carbonation, the diffusion coefficient would be only hardly influenced by carbonation, at least for R.H. < 90%.

4.3 CO_2 and O_2 diffusion coefficients

The results of the experiments to determine the diffusion coefficients are shown in table 3. It is obvious that the moisture content has a little influence on the measured diffusion coefficients in the covered range. Measurements have also been carried out on neat hardened cement paste. In figure 4, the diffusion coefficients for CO_2 and O_2 are plotted as a function of the sand concentration. As can be expected at low concentrations, the diffusion coefficients decrease with the sand content. Above 50%, however, a steep increase is observed. Similar behaviour has already been observed with moisture diffusion in concrete.

Table 3. Diffusion coefficient of CO_2 and O_2 of the carbonated mortars
C_S: sand concentration by volume

R.H. [%]	C_S = 55.1%		C_S = 48.6%		C_S = 38.5%	
	D_{O2} [m²/s]	D_{CO2} [m²/s]	D_{O2} [m²/s]	D_{CO2} [m²/s]	D_{O2} [m²/s]	D_{CO2} [m²/s]
55	$1.0 \cdot 10^{-7}$	$7.0 \cdot 10^{-8}$	$1.8 \cdot 10^{-8}$	$1.1 \cdot 10^{-8}$	$3.2 \cdot 10^{-8}$	$2.7 \cdot 10^{-8}$
76	$9.8 \cdot 10^{-8}$	$6.7 \cdot 10^{-8}$	$1.5 \cdot 10^{-8}$	$7.0 \cdot 10^{-9}$	$3.5 \cdot 10^{-8}$	$2.2 \cdot 10^{-8}$
86	$1.0 \cdot 10^{-7}$	$6.2 \cdot 10^{-8}$	$2.0 \cdot 10^{-8}$	$8.0 \cdot 10^{-9}$	$4.6 \cdot 10^{-8}$	$1.5 \cdot 10^{-8}$

This phenomenon can be explained by the role of the interfacial zones or the auréoles de transition in the diffusion process. The composite structure can be simulated in a two dimentional model (see figure 5). As long as the interfacial zones are well separated from one another, the diffusion process is governed by the porosity of the matrix. Above a certain concentration the auréoles de transition become interconnected and thus create channels for easy diffusion.

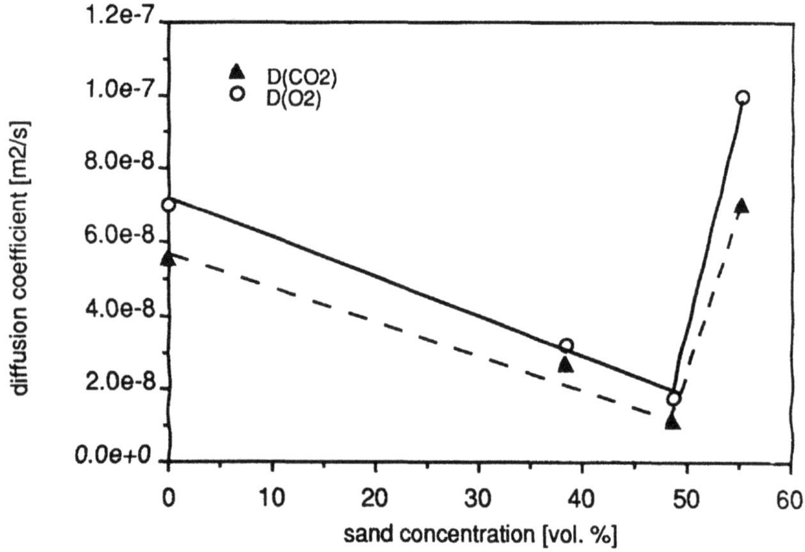

Fig. 4 Diffusion coefficients for O_2 and CO_2 of the mortars as a function of sand concentration

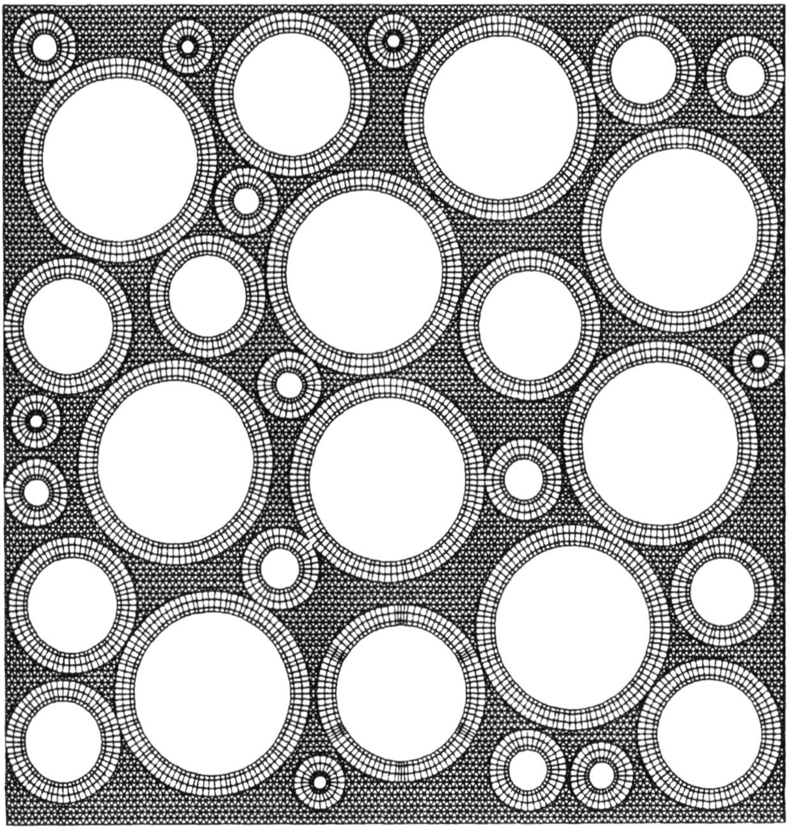

Fig. 5 Finite volume representation of a computer-generated random composite structure (3 phases are present: *matrix, auréole de transition and inclusions*)

5 References

Diamond, S. Mindess, S. and Lovell J. (1982) On the Spacing between Aggregates Grains in Concrete and the Dimentions of the Aureole de Transition in **Rilem Colloquium Liaisons Pâtes de Ciment Matériaux associés**, Toulouse, pp. C42-C46.

Farran, J. (1956) Contribution minéralogique à l'étude de l'adhérence entre les constituants *hydratés des ciments et les matériaux enrobés*, **Rev. Mat. Constr.** , *No 490-*491,155-172, and No 492, 191-209.

Houst, Y.F. Roelfstra, P.E. and Wittmann, F.H. (1983) A model to predict service life of concrete structure in **International Colloquium on Materials Science and Restoration** (ed F.H. Wittmann), Edition Lack + Chemie, Filderstadt , pp. 181-186.

Houst, Y.F. and Wittmann, F.H. (1986), The diffusion of carbon dioxide and oxygen in aerated concrete in **2nd International Colloquium: Materials Science and Restoration** (ed F.H. Wittmann), Technische Akademie Esslingen, Ostfildern, pp. 629-634.

Houst, Y.F. (1991) Influence of Microstructure and Water on the Diffusion of CO_2 and O_2 through cement paste in **Second CANMET/ACI International Conference on Durability of Concrete** (ed V.M. Malhotra), CANMET, Ottawa, Supplementary Papers pp. 141-159.

Pierce, C. (1953) Computation of Pore Sizes from Adsorption data, **J. Phys. Chem.**, 57, 149-152.

Scrivener, K.L. and Pratt, P.L. (1987) The characterization and quantification of cement and concrete microstructures in **Pore Structure and Construction Materials Properties vol. 1** (ed J.C. Maso), Chapman and Hall, London-New York, pp. 61-68.

Scrivener, K.L. and Gartner, E.M. (1988) Microstructural gradients in cement paste around aggregates particles in **Bounding in Cementitious Composites** (eds S. Mindess and S.P. Shah), Material Research Society, Pittsburgh, pp. 77-85.

Scrivener, K.L. Crumbie, A.K. and Pratt, P.L. (1988) A Study of the Interfacial Region between Cement Paste and Aggregates in Concrete in **Bounding in Cementitious Composites** (eds S. Mindess and S.P. Shah), Material Research Society, Pittsburgh, pp. 87-88.

Tuuti, K. (1982) **Corrosion of Steel in Concrete**, Swedish Cement and Concrete Institute, Stockholm.

30 INTERFACE INFLUENCE UPON THERMAL CONDUCTIVITY OF LIGHTWEIGHT CONCRETES

V. S. GRYSLOV
R.I. Laboratory for Testing Building Materials,
Cherepovets, Russia

Abstract
This article contains the results of experimental research of interface influence upon the thermal conductivity of lightweight concrete. A theoretical analysis of this influence has been carried out and a physical and mathematical model has been created. The model makes it possible to predict lightweight concrete thermal conductivity taking into account its aggregate structure and interface areas.
Keywords : Thermal, Conductivity, Interface, Physical and Mathematical Model.

1 Introduction

Modern levels of quality control of lightweight concrete manufacture require more precise prediction of its properties. Therefore, along with improving techniques of increasing lightweight concrete porosity and researching new ones, we need a scientifically grounded concept of regulation and minimization of lightweight concrete thermal conductivity through other structure forming factors, such as : friction composition of aggregates, interface, binder microstructure etc..., the pre-set strength being minimal.

At present this concept is developing based upon common laws of the general theory of materials as a structural theory of lightweight concrete conductivity.

2 Experimental research

Common principles of mathematical modelling for the evaluation of thermal conductivity of materials are based upon the analysis of two-component systems (matrix-inclusion) such as models made by Krishcher, Ribeant, Missenard, Russel, Odelevsky et al.

We have performed experimental testing of the models mentionned with slag concrete taken as a sample.

At the first stage samples of different phase composition were used for our research. Preparing the slag samples included grinding, melting at 1200-1300 ° C and cooling under various conditions. As a result, the degree of slag crystallisation ranged from 30 % to 80 %.

Phase composition was defined by the light microscopy technique as well as the X-ray technique. Thermal conductivity was defined by the technique of stationary thermal flow on flat samples cut out of slag specimens.

Interfaces in Cementitious Composites. Edited by J. C. Maso. © RILEM.
Published by E & FN Spon, 2 - 6 Boundary Row, London SE1 8HN. ISBN 0 419 18230 6.

The analysis of the results obtained (Table 1) showed that a common relationship of the following type exists and the models proposed by Missenard (1965) and Odelevsky (1951) can be transformed into this type :

$$\frac{1}{\lambda_{eff}} = \frac{1}{\lambda_m}\left[1+(\beta-1)V\right] \quad (1)$$

where λ_{eff} - the model (sample) thermal conductivity;
λ_m, λ_i - thermal conductivity of the matrix, the inclusion ;
β - relative thermal conductivity of λ_m/λ_i ;
V - relative volume of the inclusion

However, experimental values differ from calculated ones acording to Equ. (1).

Table 1. Results of experiments with slag samples

	Phase composition		λ exp.	λ_m/λ_{eff} experiment	λ_m/λ_{eff} according to [2]
	crystalline phase	glass-like phase			
1	-	100	0,704	1,000	1,000
2	30	70	0,958	0,734	0,830
3	55	45	1,080	0,651	0,730
4	80	20	1,140	0,617	0,650
5	100	-	1,170	0,601	0,600

This is evidence of statically inhomogeneous phase distribution in samples tested and of zones differing from the matrix and the inclusion.

The physical sense of this phenomenon can be explained through the existence of an interface (a contact layer) which is sure to be observed in models with physical and chemical mechanisms of component interaction.

This conclusion was proved at the second stage by experiments on the influence of the interface upon thermal resistance of concrete models.

The research was carried out with pumice cement samples of laminated structure having :

- one layer of slag pumice plate 70 mm in diameter and 15-20 mm thick;
- a second layer of cement stone made of grout of normal density, 10-15 mm thick.

Sample hardening was performed under various conditions acording to proper schedules. After 28 days of hardening the dry samples were tested for thermal conductivity by the stationary thermal flow technique. Test results and calculated values are shown in Table 2. The theoritical average thermal conductivity values of sample irrespective of hardening conditions were 15 % lower than the experimental ones, proving the interface influence upon thermal conductivity of composition materials.

The microstructural analysis of slag pumice concrete samples showed that at the interface of binder and aggregate there appears a clearly visible zone, 1-1.5 mm thick, resulted from the physical and chemical interaction of aggregate and cement. It should be noted that the thickness of this zone does not change with the increase of the hydration period (40-50 days). Newly formed structures appearing in the contact zone are in a colloidal or submicrocrystalline state which hampers thermal wave propagation.

Regarding the results of the microstructural analysis and thermal engineering test, we propose the following inequality :

$$\lambda_{c.z.} \leq \lambda_i < \lambda_m \tag{2}$$

3 Theoretical research

The experimental results were used for determining and creating conditions for physical and mathematical models of lightweight concrete thermal conductivity.

First of all, we have drawn the conclusion that there exists a limited field of application of the common model approach for solving the problem of thermal conductivity of multi-component systems in the physical sense, for it presents the substance as a mechanical mixture, without taking into account physical and chemical interaction occuring at the binder-aggregate contact zone. Therefore equation (1) needs further consideration.

Interface influence can be determined in the following way. Let us assume that in concrete there are n aggregate fractions, each consisting of cubes with edges d ; ($d_1 < d_2 < \ldots d_n$). The cubes are oriented so that the thermal flow is perpendicular to an edge. The relative portion i of the aggregate fraction is αi ; ($\Sigma\alpha i = 1$). Then, if the aggregate unit volume V is known, the specific thermal resistance of a concrete parallelepiped is :

$$\frac{1}{\lambda} = \frac{1}{\lambda_i} V + \frac{2rV}{\sum_i \alpha_i d_i} \cdot \frac{1}{\lambda_{c.z.}} + \left(1 - V\right) - \frac{2rV}{\sum_i \alpha_i d_i} \Big) \lambda_m \tag{3}$$

For equation (3) to acquire the appearance corresponding (1) to the common structure, let us introduce the following parameters :

$\bar{\mathcal{D}} = r/\bar{d}$ - relation of the contact layer thickness r to the average grain size of the aggregate ($\bar{d}$)

$\rho = \lambda_m / \lambda_{c.z.}$ - relative thermal conductivity of the contact zone.

After the transformation we have :

$$\frac{1}{\lambda} = \frac{1}{\lambda_m} \left[1 + (\beta - 1) V \left(1 + 2\bar{\mathcal{D}} \frac{\rho - 1}{\beta - 1}\right)\right] \tag{4}$$

In case (2) $\rho > \beta$, then $\frac{\rho - 1}{\beta - 1} > 1$

Let us introduce $\beta' = (\rho - 1)/(\beta - 1)$ and express $(1 + 2\bar{\mathcal{D}}\beta')$ through χ_2, which will correspond to the increase of concrete thermal conductivity at the expense of the contact zone.

Then the physical and mathematical model will be expressed as :

$$\frac{1}{\lambda} = \frac{1}{\lambda_m} \left[1 + (\beta - 1) V \chi_1 \chi_2\right] \tag{5}$$

where

χ_1 - coefficient of the inclusion position influence, reflecting the mechanical aspect of the model ;

χ_2 - coefficient of the interface influence reflecting the physical aspect of the model.

The suggested model approach interprets exactly enough lightweight concrete thermal conductivity at the stage of the composition design. When determining coefficients χ_1 and χ_2, one can take into consideration the folowing : the value $\bar{\mathcal{D}}$ increases proportionally with the increase of the portion of chalk fraction provided the contact zone thickness does not depend on the aggregate size $\left(\frac{M}{M+K}\right)$;

Table 2. Experimental and calculated values of thermal conductivity of slag pumice cement samples

Hardening condition	Performance of samples						λ_{exp}	λ_{calc}	ΔR % $\frac{R_{exp}-R_{calc}}{R_{exp}}$
	L_m	h_m	V_{rel}	λ_m	λ_i	β			
Water	0.0205	0.0140	0.68	0.464	0.288	1.61	0.286	0.329	13.1
Normal	0.0196	0.0133	0.68	0.576	0.246	2.34	0.266	0.301	11.6
Air	0.0208	0.0146	0.070	0.484	0.266	1.80	0.264	0.308	14.4

L - common sample thickness ; h - slag pumice layer thickness ; V_{rel} - relative inclusion volume ; Δ R - relative change of sample thermal resistivity

Table 3. Experimental and calculated values of thermal conductivity of slag pumice concrete, grade B 5

$\frac{M}{M+K}$	Concrete density, Kg/m	V	Cement performance Kind	Grade	λ	$\bar{D}$	β	β'	χ_1	χ_2	λ sale by (5)	λ exp., when dry,
0.3	1410	0.732	*s.p.c.	400	0.484	0.108	1.68	1.31	0.85	1.30	0.32	0.321
0.35	1380	0.743	**p.c.	400	0.504	0.118	1.75	1.29	0.84	1.34	0.39	0.305
0.40	1420	0.745	p.c.	450	0.579	0.125	2.01	1.25	0.80	1.33	0.31	0.334
0.50	1400	736	s.p.c.	300	0.464	0.145	1.61	1.33	0.86	1.40	0.31	0.314

*s.p.c. - slag portland cement ; **p.c. - portland cement

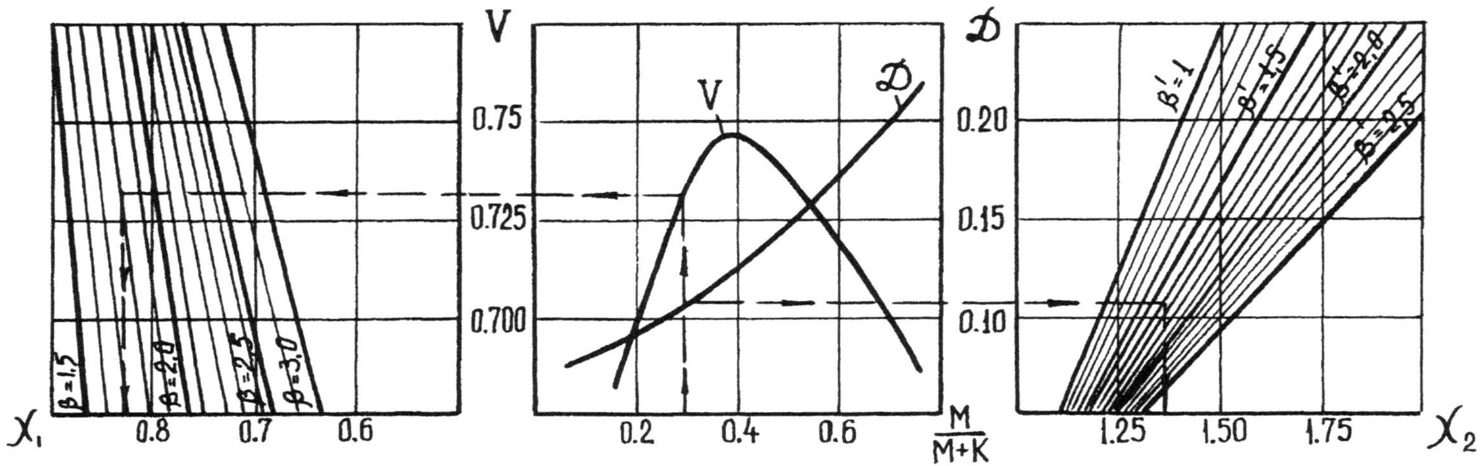

Fig. 1 Nomogram for defining coefficients

- the sum of relative aggregate volumes per fraction $\sum_i V_i$ is connected in regression with its granulometric composition through $\left(\frac{M}{M+K}\right)$.

Based upon the given statements, a monogram determining coefficients χ_1, χ_2 is plotted (in Fig. 1).

The comparison results of the calculated (5) and experimental (Table 3) research results show that the thermal conductivity model created can be applied in lightweight concrete composition design.

4 Conclusion

A universal physical and mathematical model of lightweight concrete thermal conductivity has been worked out taking into account aggregate structural, physical and chemical component interaction.

The interface influence upon lightweight concrete thermal conductivity has been proved experimentally and the technique of taking into consideration this influence in the prediction of concrete thermal proof properties has been shown.

5 References

1. Missenard A., (1965), Conductivité thermique des solides, liquides, gas et de leurs mélanges, (éditions Eyrolles).
2. Odelevsky V., (1951), Generalized conductivity calculation of heterogeneous system, Lh. T. Ph., vol. 21, issue 6.

PART 2.3

RELATIONSHIPS BETWEEN MICROSTRUCTURE OF THE INTERFACIAL ZONE AND PROPERTIES OF THE COMPOSITE: ENVIRONMENTAL CONDITIONS

(Relations entre la microstructure de l'auréole et les propriétés du composite: Influence de l'environnement)

31 INFLUENCE OF THE INTERFACIAL ZONE BETWEEN CEMENT PASTE AND AGGREGATE ON THE CORROSIVE ATTACK ON CONCRETE

K. P. GROSSKURTH, W. MALORNY and
W. RUTHMANN
Institut fur Baustoffe, Massivbau und Brandschutz,
Technical University of Braunschweig, Germany

Abstract
Investigations into the structure and chemical composition of the cement paste-aggregate interfacial layers before and after the action of sulfate solution resp. sulfuric acid were carried out on sandwich-like specimens, consisting of a flat polished aggregate substrate and a cement paste layer. They confirmed that transport processes and subsequent chemical reactions take place in a narrow, highly porous layer below the substrate covering textured portlandite film containing CSH-gel domains. Early investigations on samples from real concretes reveal that contiguous seems of sulfate containing compounds around the aggregate grains are not to be registered. Nevertheless there are obviously weak areas where penetrating ions accumulate. A novel interesting technique for the non-destructive examination of the interface between cement paste and aggregate is IR-spectroscopy utilizing multiple attenuated total reflectance.
Keywords: Cement Paste-Aggregate Interfacial Layers, Sulfate Attack, Sulfuric Acid Attack, IR-Spectroscopy

1 Introduction

The deterioration of concrete structures by aggressive media is to a large extent governed by transport processes in the pore space of the cement paste. It is well known that the paste-aggregate interfacial layers of normal concrete constitute a weak zone of lower mechanical strength and higher permeability. Yet, little is known about the influence of the cement paste-aggregate interfacial layers on concrete corrosion. A thorough understanding of concrete deteriorating processes, however, requires research into the connection between chemical reactions and transport processes in these layers. The following contribution, though it is far from comprehensive, is an attempt to present some interesting aspects with respect to this discussion.

Interfaces in Cementitious Composites. Edited by J. C. Maso. © RILEM.
Published by E & FN Spon, 2 - 6 Boundary Row, London SE1 8HN. ISBN 0 419 18230 6.

2 Specimen preparation

Investigations into the nature of the cement paste-aggregate interfacial layers were primarily carried out on small sandwich-like specimens consisting of a flat substrate onto which the cement paste was cast. This kind of preparation provides the preconditions required for the subsequent investigations by means of x-ray analytical methods and microscopy.

Polished substrates of either quartz, granite or dolomite, of about 25 mm x 25 mm, were used as model aggregates. Ordinary portland cement paste (w/c = 0.4 to 0.6) was cast from above onto the substrate below. The cement composition is given in Table 1.

Table 1. Composition of the portland cement used in the investigations (weight %)

Elemental content		Calculated phase content	
Na_2O	0.22	C_3S :	49.36
MgO	0.55	C_2S :	28.31
Al_2O_3	5.25	C_3A :	11.28
SiO_2	22.80	C_4AF:	4.75
SO_3	3.29		
K_2O	0.83		
CaO	67.40		
Fe_2O_3	1.56		

Specimens were cured at a constant climate of 23 °C and nearly 100 % relative humidity to prevent hydration from stopping. Additional investigations on concrete samples were carried out on epoxy pressure-impregnated polished sections (diameter: 20 mm).

3 Characterization of the interfacial zone

The microstructural characterization of the cement paste-aggregate interfacial zone was done both by x-ray methods, and by means of scanning electron microscopy (SEM) in combination with energy-dispersive microanalysis (EDX). For this, cement paste and substrate of the double layer samples were separated by means of two wedges so that both sides of the interface could be examined.

3.1 X-ray diffraction

Phase identification was accomplished by means of the JCPDS powder diffraction file. It was noted that the AFm phases $C_3A \cdot CaCO_3 \cdot 11H_2O$ and $C_4A \cdot 1/2CO_2 \cdot 12H_2O$ were present regardless of the kind of substrate used.

Studies on the variations in phase composition as well as determinations of the degree of preferential orientation of portlandite depending on the distance from the cement paste-aggregate interface were accomplished by successive abrasion by means of a milling machine.

According to the findings of other researchers published earlier there is a strong preferential orientation of portlandite crystals in the so-called transition zone that decreases with increasing distance from the aggregate surface. Using the well known orientation index it was found that deviations from a random statistical distribution only occurred in the first 60 μm from the aggregate's surface. By far the highest orientation index was found for the cement paste still adhering to the substrate, typically characterized by a value of 5. On the cement paste side the uppermost layer yielded a value of not more than 2.

3.2 X-ray spectrometry

In addition to the afore mentioned x-ray diffraction analysis x-ray spectrograms were taken on the same successive layers. Because of the nature of x-rays, it has to be kept in mind that the spatial resolution of x-ray fluorescence signals is limited by a wavelength dependent emission depth. Table 2 shows that significant variations in the registrated elemental composition as a function of increasing distance from the interface did not occur. One significant difference found was between the first and second layer in the calcium content. This effect is attributed to the higher portlandite content in close proximity to the substrate surface.

Table 2. Chemical composition of cement paste as a function of distance from the substrate

Layer [μm]	Na_2O [%]	MgO [%]	SO_3 [%]	Al_2O_3 [%]	SiO_2 [%]	K_2O [%]	CaO [%]	Fe_2O_3 [%]
0	0.48	0.95	3.16	3.81	16.67	1.72	71.42	1.75
5	0.92	1.40	3.63	5.16	21.41	2.36	61.73	3.27
10	0.86	1.50	3.47	5.86	23.72	2.09	59.19	3.20
20	0.88	1.81	3.67	5.94	23.02	2.10	59.11	3.35
40	0.79	1.71	3.76	5.60	22.88	2.00	58.97	4.25
80	0.78	1.72	3.61	5.51	22.70	1.52	59.67	4.47
1100	0.50	1.72	3.63	5.44	22.72	0.76	62.02	3.20

Furthermore a depletion of the ferritic phases in the vicinity of the aggregate surface is noticeable. This observation is probably due to the lower solubility of the ferritic phases. According to this finding it seems likely that under sulfate attack secondary ettringite of the expansive non-ferritic modification is preferentially formed at the cement paste-aggregate interface while the bulk cement could be characterized by a relatively higher content of the ferrite-containing ettringite.

3.3 Scanning electron microscopy

SEM investigations on the nature of the contact layers between portland cement paste and quartz substrates reveal that the outermost layer of the cement paste does not consist of a homogenous substrate-covering film but is rather built up of spotty domains consisting of both portlandite and CSH-gel (Fig.1). In accordance with the above mentioned findings from x-ray spectrometry, EDX investigations afirm the depletion of the ferritic phases.

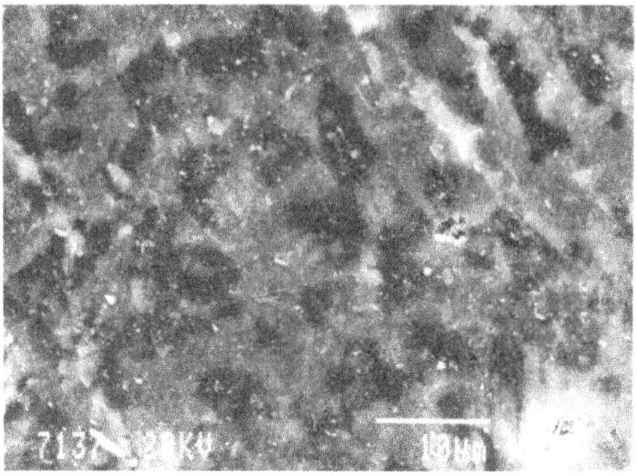

Fig.1. SEM micrograph of the contact layer (paste side)

Dolomite substrates show a somewhat greater adherence to the cement paste. When the substrate and the cement paste are separated larger parts of the substrate-covering film remain stuck to the substrate giving insights into the structure of the underlying paste. Fig.2 shows areas of loosely packed partly fibrous cement hydrates.

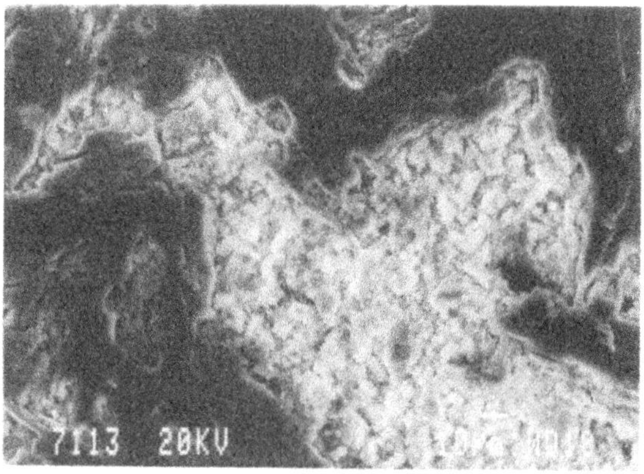

Fig.2. Porous structure underneath the contact layer

Obviously a layer of considerable pore space exists adjacent to the substrate-covering film. The latter might have a thickness of about 1 to 2 μm. Stacked plates of large portlandite crystals tend to stick to this dense film sporadically. According to the obseravations, the c-axis of these hexagonal crystals exhibits an inclination deviating from the perpendicular (Fig.3).

Fig.3. Contact film covering the dolomite substrate

3.4 IR-spectroscopy
A further way to investigate the interface between cement paste and aggregate without damaging the sandwich layers is the multiple attenuated total reflectance method (ATR-technique). The schematic assembly of the ATR-modul is shown in Fig.4. The ZnSe crystal is arranged horizontally

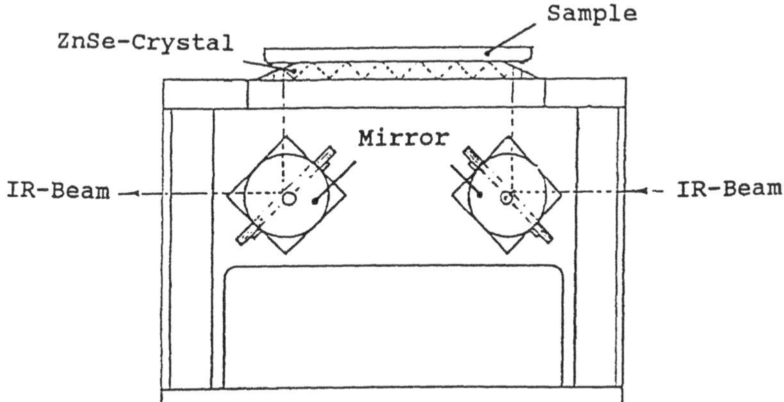

Fig.4. Schematic assembly of the ATR-Modul with infra-red beam path through ZnSe crystal.

in this assembly and allows depositing cement paste directly onto its surface and serves as an inert model aggregate. The multiple total reflectance of the infra-red beam in the ZnSe crystal analyses only the surface of the cement paste. A measuring program allows observation of multiple infra-red spectra as a function of hydration time. A qualitative phase analysis can be performed using the characteristic infra-red absorption bands of the single phases.

To verify the phase analysis the pure single phases were synthesized and pressed into tablets with KBr in a ratio of 299 mg KBr/1 mg sample. Their spectra are shown in Fig.5. The SO_4^{2-} vibration band is fundamentally different in the spectra of ettringite and monosulfate. The ettringite single band at 1115 cm^{-1} corresponds to the weak doublet at 1100 cm^{-1} and 1170 cm^{-1} in the spectrum of monosulfate.

The duration of the subsequent ATR measurement was 67 hours. During the first hour spectra were measured every minute and later every 30 minutes. Nitrogen was used as a sweep gas in the detector housing to eliminate interferences of CO_2 and H_2O.

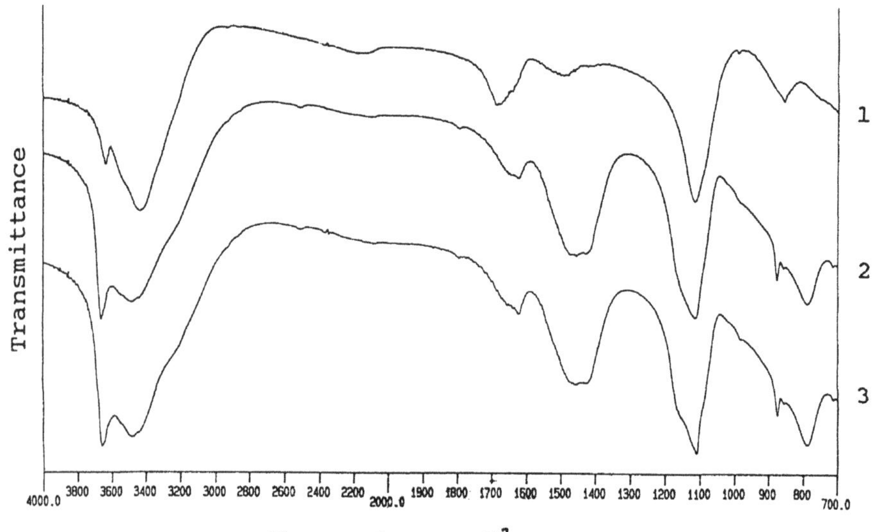

Fig.5. Infra-red spectra of synthesized ettringite(1), monosulfate(2) and a mix 1:1 of both(3)

The w/c ratio of the cement paste was 0.3 to avoid broad water bands overlapping the spectra. Fig.6 indicates that at the beginning of hydration the SO_4^{2-} band at 1115 cm^{-1} and the Al-O-H band at 870 cm^{-1} give evidence about the

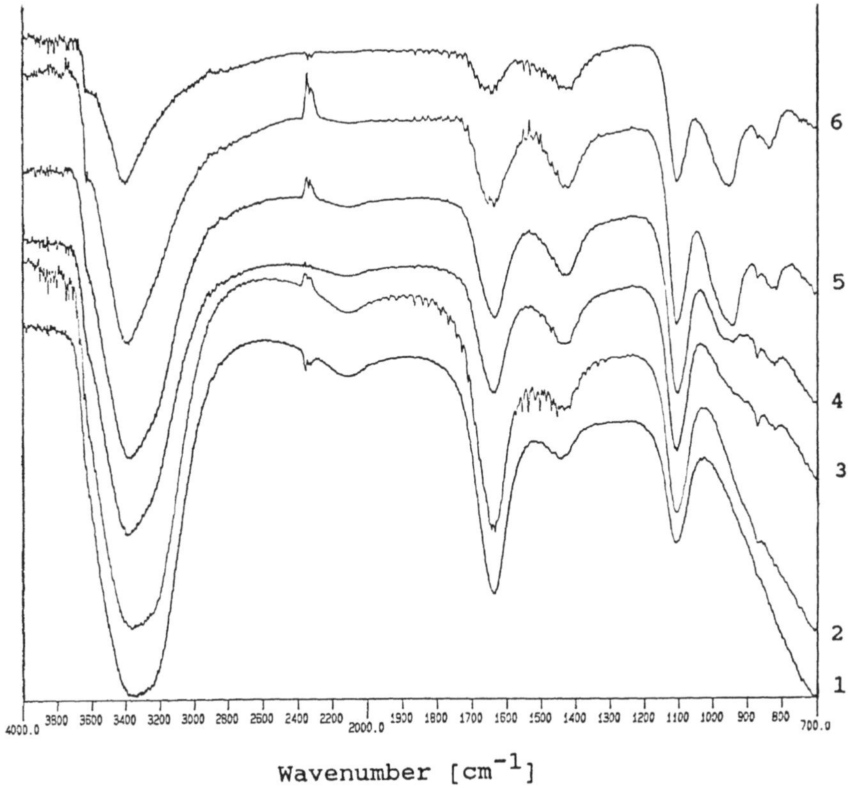

Fig.6. Infrared spectra measured during 67 hours hydration of a portland cement (1 = 2 min, 2 = 59 min, 3 = 3 h, 4 = 4 h, 5 = 10 h, 6 = 67 h)

crystal growth of ettringite. Progressive hydration is indicated by the formation of CSH (asymmetric vibration of the SiO_4 tetrahedron at ≈960 cm^{-1}).

Formation of portlandite is recognizable only by the (OH) vibration band at 3640 cm^{-1} which first overlaps the (OH) band of ettringite at 3635 cm^{-1}. Approximatly 10 hours later a weak shoulder appears at 3640 cm^{-1}, and after 67 hours this (OH) band is significant. No hints at the formation of monosulfate were detected in the infrared spectrum of cement paste after 67 hours.

Finally, the ATR-technique can be employed in situ for the study of hydration processes using the characteristic infra-red vibration bands of the cement hydrates. It still has to be proved that in situ measurements under attack of sulfate-containing solutions can be performed by means of the ATR-method, too.

4 Corrosive attack on sandwich-type specimens and concrete

Dilute sulfuric acid (1%) as well as sulfate solution ($CaSO_4$, 3700 mg/l) were used for the study of corrosive attack on the cement paste-aggregate interfacial zone. While the sulfate action provokes an expansive attack, the action of sulfuric acid causes concrete deterioration by both expansive forces and disintegration.

4.1 Sulfuric acid

Investigations into the corrosive attack of sulfuric acid were only carried out on sandwich-type specimens until now. Specimens were surface coated on all but one side and immersed into the dilute acid, thus allowing unidirectional capillary suction of the liquid into the sample. Whereas corrosive attack on the cement paste aggregate interfacial layers by concentrated sulfuric acid leads to gypsum formation, dilute sulfuric acid (1%) leads to ettringite formation. Fig.7 shows, in comparison, the diffractogram of the cement paste contact layer before and after the action of dilute sulfuric acid for a week. Obviously, not only is the ettringite peak considerably stronger, but the adjacent peak of an AFm phase is stronger as well. According to the JCPDS powder diffraction file this peak, corresponding to an approximate d-value of 0.76 nm, is attributed to the carboaluminate hydrate. The fact that the rise of this strong peak is related to the action of dilute sulfuric acid suggests that on the contrary it might be an ill-defined sulfate-bearing compound. This assumption was supported by EDX investigations of newly formed reaction products. The investigations demonstrated that the zone of greater permeability adjacent to the substrate exhibits only a small extension of about 15 to 20 μm.

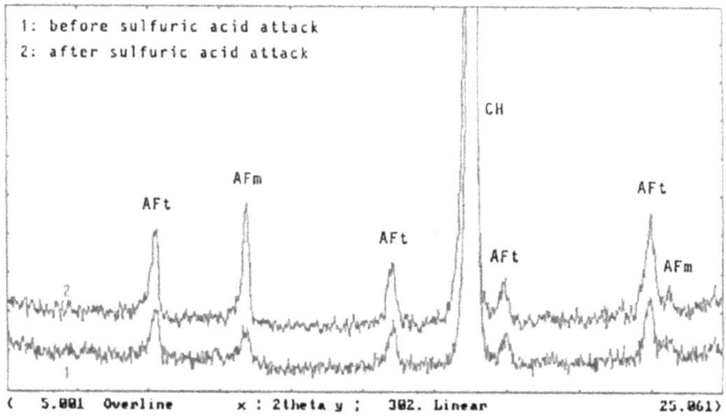

Fig.7. Diffractogram of the cement paste contact layer before / after the action of dilute sulfuric acid

4.2 Sulfate attack on concrete

Parallel to the investigations on sandwich-type specimens the action of sulfate containing solutions on concrete was simulated by means of small concrete cylinders made by drilling cores out of larger samples. As the ingress of sulfate ions from solutions low in sulfate concentration is a time consuming process, the permeation was promoted under pressures of up to 5 bar.

It is known that the mean distance between adjacent aggregate particles of properly made concretes are partly smaller than the extension of the cement paste-aggregate zone, as measured on the described sandwich structures.

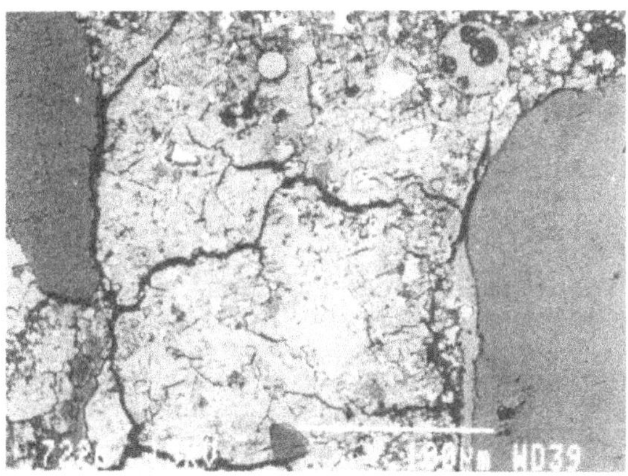

Fig.8. Micrograph of a polished section of concrete; bright grey: cement matrix, dark grey: aggregate

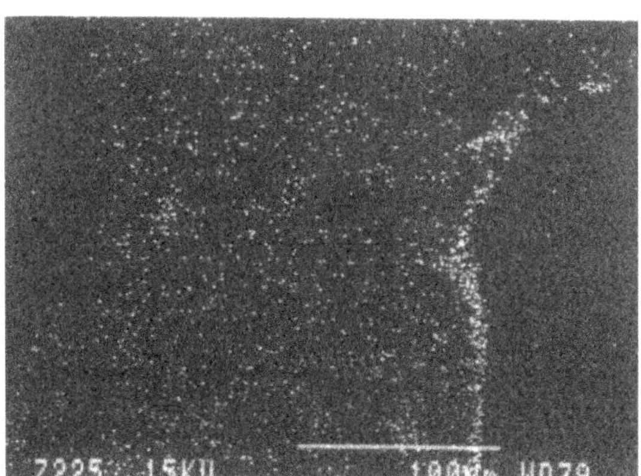

Fig.9. Corresponding sulfur distribution

First investigations into the role of the aggregate interface in the distribution of sulfate in the matrix reveal that contiguous seems of sulfate ions around aggregate particles are not to be registrated. As is demonstrated in Fig.8 and 9 by means of backscattered electron imaging and elemental mapping, segments of sulfate accumulation can be made visible, however. Apart from cracks running across the aggregate interface that are filled with precipitate products, there are obviously localized weak zones where penetrating ions accumulate.

These investigations represent the first steps in a field that requires much more attention.

5 Discussion

Investigations on double layer sandwich structures made of an aggregate substrate with cement paste cast onto it give interesting insights into the structure of the cement paste-aggregate interface zone. There is still some doubt as to the applicability of the results to real concrete structures. Early SEM-investigations on polished sections of concrete show that the build-up of an ordered series of distinctive layers does not occur. They show, however, that areas adjacent to aggregate surfaces exist that give rise to seems of accumulated elements from the ingress of concrete deteriorating agents. It appears that regions of greater permeability cannot be attributed in principal to the investigated interfacial zone.

ABSTRACTS OF FRENCH PAPERS IN ENGLISH

5 Influence of the origin of fly ashes used as partial replacement of cement in concretes on the transition zone (page 43 - 52)
B. HUSSON and A. CARLES-GIBERGUES

This paper intends to present the influence of three different origin fly ashes on interfacial microstructure : the first one, CC, is obtained in a thermal plant without desulfurization process, the two others, CD and LFC, obtained by desulfurization respectively by semi-humid way and by dry way.
Physico-chemical characteristics of these fly ashes are different : fly ash CD is distinguished by a higher fineness, fly ashes CD and LFC are enriched with sulfur (respectively calcium sulfite and calcium sulfate). Rheological behaviour is also modified : fly ashes CD and LFC incorporation in pastes increases water demand, while fly ash CC introduction has on the contrary a fluidifying effect.
But these rheological behaviour differences don't modify differentially interfacial structure : in the three cases, a portlandite orientation ratio decrease and a transition zone depth increase are observed.

9 Interfaces between synthetic fibers and cement paste (page 79 - 88)
V. HOUGET, M. CHABANNET and J. PERA

A study has been carried out on cementitious composites reinforced with various synthetic fibers, focusing on their tensile behaviour, toughness and the nature of the interfaces between fibers and matrix at different curing times : 7, 28, 90 and 180 days. The following types of fibers were examined : polypropylene, polyamide, polyester, polyacrylonitril, polyvinylalcohol, polyaramid and carbon.

21 Experimental model for mechanical, physical and chemical study of steel-concrete interface (page 197 - 206)
R. FRANCOISE, M. ARNAUD, G. ARLIGUIE and A. AYED

Steel concrete bonding is a major factor of durability of reinforced concrete. Many researchs have been performed to study steel-concrete bonding with regard to mechanical or physical and chemical characteristics, but, yet no studies interface with both kinds of investigation.

In order to make a contribution in this way, a research was started to link steel-concrete bond with physical and chemical observations. Thus, it is possible to follow bonding developpement due to, endogenous or exogenous, mechanical and physico-chemical effects.

An experimental model allows us to perform this program. The model is made with two pieces of mortar separate by a steel plate and subjected to compressive tests.

This paper deals with a theoretical study of the strain and stress fields at the steel-mortar interface of the model. Three plate orientations are studied (30°, 45°, 60°). The stress field seems complex, nevertheless for one orientation the shear stress field is almost constant. Experimental study by means of a test piece provided with strain gages, confirms the complexity of the strain field in the neighbourhood of the interface, but allows a comparison with the theoretical results.

The results show that a correlation between mechanical resistance and physico-chemical characteristics cannot be established by taking into account a global criterion based on an average stress field along the whole interface.

23 Study of the interfaces of polyester fibre reinforced cement composites (page 217 - 226)
A. JELIDI, M. CHABANNET, J. AMBROISE and J. PERA

Until now, the use of untreated polyester fibers in concrete has been delayed because these fibers are not alkali-resistant. In order to develop durable and tougher composites, the cementitious matrix was not modified by adding burnt clays (kaolin and montmorillonite).
Burnt kaolin reduced the portlandite content and burnt montmorillonite trapped potassium ions. Interfaces between fibers and matrix were modified and the polyester fiber could withstand the alkaline attack.

24 Mechanical model for the tensile behaviour of glass fiber reinforced modified cementitious composites (page 227 - 236)
J. PERA, J. DEJEAN and J. AMBROISE

In glass fiber reinforced concrete, the addition of metakaolinite reduces the growth of hydration products around the filaments and leads to a very ductile composite.
Classic methods of calculation are inadequate to describe the mechanical behaviour of such composites.
The authors present a new model to take into account the debonding of fibers which occurs during the post-cracking phase.

26 A boundary integral method taking into account the mechanical behaviour of the matrix-inhomogenity interface (page 247 - 256)
L. ULMET and S. CAPERAA

This work shows the interest of the Boundary Elements Method for the determination of stress-fields in the vicinity of inhomogeneities, taking into account various interfacial contact conditions. The method is also applied to crack-propagation problems in a composite material.

28 Diffusivity of chloride ions in the transition zone between cement paste and granite (page 269 - 278)
D. BRETTON, J.-P. OLLIVIER and G. BALLIVY

Permeation and diffusion properties of the interfacial zone between rock and cement paste are important towards durability and efficiency of cracks grouting.
Two experiments have been conducted in order to compare the transfert properties in the interfacial zone and in the bulk. Transport of chloride ions has been studied, firstly by diffusion and secondly under the action of an electric field. The results show an increase of the diffusivity in the interfacial zone. Furthermore, the effect of chemical shrinkage may be detected by these experiments.

AUTHOR INDEX

Agopyan, V. 23
Alexander, M.G. 13, 129
Ambroise, J. 217, 227
Arliguie, G. 197
Arnaud, M. 197
Ayed, A. 197
Ballivy, G. 269
Bentur, A. 53
Bentz, D.P. 107, 259
Bonora, V. 207
Breton, D. 269
Buyukozturk, O. 139
Caperaa, S. 247
Carles-Gibergues A. 43
Chabannet, M. 79, 217
Chatterji, S. 3
Chen, S.W. 99, 149
Chu Qingli 71
Courtade, R.M. 169
Dantas, F. de A.S. 23
Dejean, J. 227
Demura, K. 89
Diamond, S. 13
Folliard, K. 65
Francois, R. 197
Garboczi, E.J. 107, 259
Goldman, A. 53
Grosskurth, K.P. 99, 297
Gryzlov, V.S. 289
Hamelin, P. 169
Houget, V. 79
Houst, Y.F. 279
Husson, B. 43
Jelidi, A. 217
Jensen, A.D. 3
Kawakami, H. 179
Kawamura, M. 33
Lange, D.A. 119
Lee, K.M. 139
Li, Z. 119
Lie Qu 13, 129
Malorney, W. 297
Mindess, S. 13, 129
Mitsui, K. 119
Nilsen, U. 65
Ohama, Y. 89
Ollivier, J.P. 269
Pareek, S.N. 89
Pera, J. 79, 217, 227
Perbix, W. 99
Qi Yun 71
Qu Lie
Rosas Rosas, M. 207
Ruthmann, W. 297
Saccani, A. 207
Sadouki, H. 279
Saito, M. 33
Sandberg, P. 65
Sandrolini, F. 207
Savastano, H. 23
Schlangen, E. 237
Schneider, U. 149
Schorn, H. 159
Shah, S.P. 119
Snyder, K.A. 259
Stroeven, P. 187
Stutzman, P.E. 107
Ulmet, L. 247
Van Mier, J.G.M. 237
Winslow, D.N. 259
Wittmann, F.H. 279

SUBJECT INDEX

This index has been compiled from the keywords assigned to the papers, edited and extended as appropriate. The numbers refer to the first page number of the relevant paper.

Adhesion,
  epoxy resin 99
  polymer cement mortar 89
Admixture, mineral, modelling 107
Ageing, polyester fibre-cement composites 217
Aggregate concentration, effect on diffusion 279
Aggregate, expanded shale 71
Aggregate, modelling 107
Aggregate-paste interface 3, 119
Aggregate type, effect on mechanical behaviour 179
Alkali-silica reaction 33
Andesite 13, 129
Asbestos fibres 23
Autogenous healing 3
Autogenous shrinkage 3

Backscattered electron microscopy 13, 119
Bleeding 3
Bochum method, numerical simulation 159
Bond, damage evolution 187
Bond-slip relationship 119
Bond, steel-concrete 197
Bond strength,
  effect of aggregate type 179
  epoxy resin 99
Boundary integral method 247

Calcium hydroxide crystals, orientation 33
Calcium hydroxide, effect of mineral admixtures 65
Calcium sulphate 43
Carbonation 279

Carbon dioxide diffusion 279
Carbon black filler 53
Cement clinker aggregate 107
Cement paste-aggregate interface, corrosion 297
Cement/rock interface 13
Chloride ions, diffusion 269
Coir fibres 23
Compaction 3
Composite ductility, glass fibre 227
Composites, deformation behaviour 139
Compressive strength, effect of aggregate type 179
Computer modelling 107
Concentration intensity 65
Concrete failure, effect of aggregate type 179
Concrete fracture 237
Contact conditions, interfaces 247
Continuum model analysis 259
Corrosion, cement paste-aggregate interface 297
Crack injection 99
Crack propagation 247
Cracking 187
  glass fibre concrete 227
  mortars 139
Curing, polymer cement mortar 89

Damage evolution 187
Damage process, modelling 159
Debonding,
  glass fibres 227
  interface 187
Densification of transition zone 53
Diffusion of gases 279
Diffusivity, chloride 269
Displacement field measurements 159
Dolomite 13, 129
Drying of concrete 3

Durability 3
  polyester fibre-cement composites 217

Electric field effects on chloride diffusion 269
Electrochemical measurements 269
Energy release rate, composites 139
Ettringite formation in interface 33
Expanded shale 71
Experimental technique, steel-concrete bond 197

Fibre cement 23
Fibre/cement paste transition zone 23
Filler effect, densification 53
Finite elements 169
Flue gas desulphurization 43
Fluidized bed ash 43
Fly ash origin 43
Fracture energy 79, 119
Fracture mechanics, interfaces 139
Fracture, numerical modelling 237
Fracture properties, interfacial zone 129
Fracture surfaces, cement/rock interface 13
Fresh concrete, bleeding 3
Friedel's salt 33

Glass fibre reinforced concrete, tensile properties 227
Global features of damage evolution 187
Granite rock, chloride diffusion 269
Green mixes: see Fresh concrete 3

Hard core-soft shell 259
Hybrid finite elements 169
Hydration, computer modelling 107

Inclusions, finite element modelling 169
Infra red spectroscopy 217, 297
Interface,
  cement-organic fibres 79
  debonding 187
  expanded shale composite 149
  resin-concrete 99
  rock-cement paste 129
Interface dependant properties 159
Interfacial bond, cement-expanded shale 71
Interfacial properties, numerical modelling 237
Interfacial zone,
  aggregate-cement 119
  microstructure, modelling 107
  percolation 259

Kaolin, effect on polyester composite durability 217

Laser granularity 169
Lattice model 237
Lightweight aggregate 71, 107, 149
Lightweight concrete 289

Malva fibres 23
Matrix-inhomogeneity interface 247
Mechanical behaviour
  concrete, debonding 187
  polyester fibre-cement composites 217
Mechanical modelling, interfaces 169
Mechanical properties,
  concrete, effect of aggregate 179
  numerical simulation 159
  polymer-cement composites 207
Mechanical tests, vegetable fibre composites 23
Mercury intrusion porosimetry 259, 279
Metakaolinite, effect on hydration products 227
Microfillers, densification 53
Microhardness, expanded shale concrete 71, 149
Micromechanical properties of interface 187
Micromorphology, expanded shale 149
Microstructure,
  carbonated mortar 279
  computer modelling 107
  high strength concrete 53
  mortar 259
  organic fibre composites 79
  polymer cement composites 89, 207
Mineral admixtures,
  computer modelling 107
  effect on transition zone 65
Mineralogy, fly ash 43
Modelling,
  boundary integral method 247

computer 107
glass fibre reinforced concrete 227
interfacial zone percolation 259
material structure 159
steel-concrete interface 197
thermal conductivity 289
Montmorillonite, effect on polyester composite durability 217
Morphology, epoxy resin 99
Mortar adhesion 89
Mortar-aggregate interface 139
Mortar, polymer composites 207

Numerical modelling,
cement based composites 237
mechanical behaviour 169
Numerical simulation, mechanical properties 159

Optical tracking techniques 169
Organic fibres 79
Orientation index 43
Orientation of calcium hydroxide crystals 33
Oxygen diffusion 279

Paste-aggregate bond 129
Percolation, interfacial zones 259
Permeability, chloride 269
Petrographic techniques 3
Phase angle 139
Physico-chemical bond, steel-concrete 197
Polyester fibre-cement composites, ageing 217
Polyester fibre treatment 217
Polyester resins 207
Polymer cement, bond 89
Polymer cement mortar, mechanical properties 207
Polypropylene fibres 23
Pore solution chemistry 3
Pore size distribution 279
Porosity gradient, interface 119
Post-cracking, glass fibre concrete 227
Pozzolanas 217
densification 53
Preferred orientation, calcium hydroxide 65
Push out test 119

Resin, epoxy 99
Resin mortar 207
Rheology 43
Rice husk ash 65
Rock type, effect on mechanical behaviour 179

Sand content, effect on interfacial zone percolation 259
Sandwich specimens, mortars 139
SEM,
aggregate-cement interfaces 297
expanded shale-cement paste 149
expanded shale concrete 71
high strength concrete 53
organic fibre composites 79
polymer cement mortar 89
resin-concrete 99
rock/cement paste interface 13
Shear bond strength 119
Shrinkage 3
Silica fume 119, 129
cement paste 13
effect on transition zone 65
high strength concrete 53
Simulation,
fracture 237
interfacial zone 107, 259
numerical, mechanical properties 159
Sisal fibres 23
Slag 65
Slag pumice cement, thermal conductivity 289
Sodium chloride 33
Steel-concrete interface 197
Stereology 187
Strain, aggregate effects 179
Strain gauges 197
Strength, expanded shale concrete 71
Stress field modelling 247
Stress field, steel-concrete interface 197
Stress distribution, interface 169
Sulphate attack, cement paste-aggregate interface 297

Techniques for examining interfaces 3
Thermal conductivity, lightweight concrete 289
Thermal expansion mismatch 3

Thermogravimetry 33
Transition layer, expanded shale 149
Transition zone,
densification 53
fibre/cement paste 23
SEM 13

Vegetable fibres 23
Vibration of fresh concrete 3
Visualization, interfacial zones 3

X-ray diffraction 33, 297

RILEM, The International Union of Testing and Research Laboratories for Materials and Structures, is an international, non-governmental technical association whose vocation is to contribute to progress in the construction sciences, techniques and industries, essentially by means of the communication it fosters between research and practice. RILEM activity therefore aims at developing the knowledge of properties of materials and performance of structures, at defining the means for their assessment in laboratory and service conditions and at unifying measurement and testing methods used with this objective.

RILEM was founded in 1947, and has a membership of over 900 in some 80 countries. It forms an institutional framework for cooperation by experts to:

* optimise and harmonise test methods for measuring properties and performance of building and civil engineering materials and structures under laboratory and service environments;

* prepare technical recommendations for testing methods;

* prepare state-of-the-art reports to identify further research needs.

RILEM members include the leading building research and testing laboratories from around the world, industrial research, manufacturing and contracting interests as well as a significant number of individual members, from industry and universities. RILEM's focus is on construction materials and their use in buildings and civil engineering structures, covering all phases of the building process from manufacture to use and recycling of materials.

RILEM meets these objectives though the work of its technical committees. Symposia, workshops and seminars are organised to facilitate the exchange of information and dissemination of knowledge. RILEM's primary output are the technical recommendations. RILEM also publishes the journal *Materials and Structures* which provides a further avenue for reporting the work of its committees. Many other publications, in the form of reports, monographs, symposia and workshop proceedings, are produced.

Details of RILEM membership may be obtained from RILEM, École Normale Supérieure, Pavillon du Crous, 61, avenue du Pdt Wilson, 94235 Cachan Cedex, France.

Details of the journal and the publications available from E & F N Spon/Chapman & Hall are given below. Full details of the Reports and Proceedings can be obtained from E & F N Spon, 2-6 Boundary Row, London SE1 8HN, Tel: (0)71-865 0066, Fax: (0)71-522 9623.

Materials and Structures

RILEM's journal, *Materials and Structures*, is published by E & F N Spon on behalf of RILEM. The journal was founded in 1968, and is a leading journal of record for current research in the properties and performance of building materials and structures, standardization of test methods, and the application of research results to the structural use of materials in building and civil engineering applications.

The papers are selected by an international Editorial Committee to conform with the highest research standards. As well as submitted papers from research and industry, the Journal publishes Reports and Recommendations prepared buy RILEM Technical Committees, together with news of other RILEM activities.

Materials and Structures is published ten times a year (ISSN 0025-5432) and sample copy requests and subscription enquiries should be sent to: E & F N Spon, 2-6 Boundary Row, London SE1 8HN, Tel: (0)71-865 0066, Fax: (0)71-522 9623; or Journals Promotion Department, Chapman & Hall, 29 West 35th Street, New York, NY 10001-2291, USA, Tel: (212) 244 3336, Fax: (212) 563 2269.

RILEM Reports

1 **Soiling and Cleaning of Building Facades**
Report of Technical Committee 62-SCF. *Edited by L. G. W. Verhoef*
2 **Corrosion of Steel in Concrete**
Report of Technical Committee 60-CSC. *Edited by P. Schiessl*
3 **Fracture Mechanics of Concrete Structures - From Theory to Applications**
Report of Technical Committee 90-FMA. *Edited by L. Elfgren*
4 **Geomembranes - Identification and Performance Testing**
Report of Technical Committee 103-MGH. *Edited by A. Rollin and J. M. Rigo*
5 **Fracture Mechanics Test Methods for Concrete**
Report of Technical Committee 89-FMT. *Edited by S. P. Shah and A. Carpinteri*
6 **Recycling of Demolished Concrete and Masonry**
Report of Technical Committee 37-DRC. *Edited by T. C. Hansen*
7 **Fly Ash in Concrete - Properties and Performance**
Report of Technical Committee 67-FAB. *Edited by K. Wesche*

RILEM Proceedings

1 **Adhesion between Polymers and Concrete. ISAP 86**
Aix-en-Provence, France, 1986. *Edited by H. R. Sasse*
2 **From Materials Science to Construction Materials Engineering**
Versailles, France, 1987. *Edited by J. C. Maso*
3 **Durability of Geotextiles**
St Rémy-lès-Chevreuses, France, 1986
4 **Demolition and Reuse of Concrete and Masonry**
Tokyo, Japan, 1988. *Edited by Y. Kasai*
5 **Admixtures for Concrete - Improvement of Properties**
Barcelona, Spain, 1990. *Edited by E. Vazquez*
6 **Analysis of Concrete Structures by Fracture Mechanics**
Abisko, Sweden, 1989. *Edited by L. Elfgren and S. P. Shah*
7 **Vegetable Plants and their Fibres as Building Materials**
Salvador, Bahia, Brazil, 1990. *Edited by H. S. Sobral*
8 **Mechanical Tests for Bituminous Mixes**
Budapest, Hungary, 1990. *Edited by H. W. Fritz and E. Eustacchio*
9 **Test Quality for Construction, Materials and Structures**
St Rémy-lès-Chevreuses, France, 1990. *Edited by M. Fickelson*
10 **Properties of Fresh Concrete**
Hanover, Germany, 1990. *Edited by H.-J. Wierig*
11 **Testing during Concrete Construction**
Mainz, Germany, 1990. *Edited by H. W. Reinhardt*
12 **Testing of Metals for Structures**
Naples, Italy, 1990. *Edited by F. M. Mazzolani*
13 **Fracture Processes in Concrete, Rock and Ceramics**
Noordwijk, Netherlands, 1991. *Edited by J. G. M. van Mier, J. G. Rots and A. Bakker*
14 **Quality Control of Concrete Structures**
Ghent, Belgium, 1991. *Edited by L. Taerwe and H. Lambotte*
15 **High Performance Fiber Reinforced Cement Composites**
Mainz, Germany, 1991. *Edited by H. W. Reinhardt and A. E. Naaman*
16 **Hydration and Setting of Cements**
Dijon, France, 1991. *Edited by A. Nonat and J. C. Mutin*
17 **Fibre Reinforced Cement and Concrete**
Sheffield, UK, 1992. *Edited by R. N. Swamy*
18 **Interfaces in Cementitious Composites**
Toulouse, France, 1992. *Edited by J. C. Maso*
19 **Concrete in Hot Climates**
Torquay, UK, 1992. *Edited by M. J. Walker*

For Product Safety Concerns and Information please contact our EU
representative GPSR@taylorandfrancis.com
Taylor & Francis Verlag GmbH, Kaufingerstraße 24, 80331 München, Germany

www.ingramcontent.com/pod-product-compliance
Ingram Content Group UK Ltd.
Pitfield, Milton Keynes, MK11 3LW, UK
UKHW022010180926
475701UK00003B/295
* 9 7 8 0 3 6 7 8 6 3 4 3 2 *